PREFACE

The construction industry is one of the major employers in the nation's work force. The U. S. Department of Labor has estimated that more than 5 million people are employed in the construction industry. Several hundred thousand small contractors employ fewer than 10 people each. The greatest source of income for these small companies is the construction of small buildings and residences.

These contractors construct homes according to the working drawings and the specifications which are prepared by another designer. The contractor must have a reasonably accurate idea of the probable cost, time requirements, and profit before agreeing upon a contract price.

This book is written to give you a schedule to follow to make an accurate quantity take off after reading the specifications and studying the working drawings. So that you may understand various types of specifications, I have used a full set of specifications that would have been drawn up by an architect's office and also the Federal Housing Administration Form 2003 and the Veteran's Administration Form 26-1852 for the same set of working drawings.

Specifications are written in the sequence in which the building would be constructed, with a few exceptions. For example, the section that deals with moisture protection covers foundation damp-proofing and also roofing, although these operations are completed at different times during the construction of a house.

Every individual will use his own method of estimating. The more that you work at it, the more proficient you will become. The easiest method for the beginner is to estimate the material in the order or sequence in which the building would be constructed, starting with the site work and excavation. All materials are listed in the order in which the workmen will use them on the job. This serves two purposes. First, it gives you a list of materials that can be used directly for ordering in the sequence that will be required on the job. Second, it provides a check to make sure you have not missed anything.

The Illustrated Handbook of Home Construction walks you through the estimating process for residences. The first part is a review of prerequisite background material. After these beginning chapters, you are led through the process of doing a quantity take off. Each process presents a concise discussion of the materials and techniques involved in a certain aspect of construction. A two-story Garrison-style house has been chosen for this estimate because it offers a wide variety of estimating experiences.

Reading and understanding a set of specifications and a set of working drawings for a house will give you a background of construction techniques, details, and workmanship. With this knowledge of construction established, the book then takes you step-by-step through the process of planning, designing, sketching, and then drawing a set of working drawings for a ranch house and a two-story house.

Upon completion of the working drawings, you will then go through a quantity take off of the materials required for a ranch house. Also included is a guideline for estimating the cost of labor.

Halsey Van Orman has a solid background in the construction industry. His formal education includes a degree in Architectural Drafting from the State University of New York at Delhi, and courses in building construction and education at the State University of New York at Oswego and Russell Sage College. He has worked for more than thirty years as a mason, foreman, specifications writer, and estimator. He has been an instructor of Architectural Drafting for twelve years and is presently Vocational Supervisor.

CONTENTS

Preface iii

1. **INTRODUCTION** 1

   Geometry 1   The Architect's Scale 4   The Alphabet of Lines 5   Architectural
   Symbols and Conventions 6

2. **WORKING DRAWINGS AND SPECIFICATIONS** 10

   Scale 10   Kinds of Drawings 10   Working Drawings 12   Specifications 25

3. **SITE WORK** 57

   Plot Plans 57   Excavation 57

4. **MASONRY ESTIMATE: FOOTINGS AND CONCRETE FLOORS** 60

   Footings 60   Concrete Floors 62

5. **MASONRY ESTIMATE: FOUNDATIONS** 64

   Concrete Block 64   Concrete Walls 66   Poured Concrete 66

6. **MASONRY ESTIMATE: FIREPLACES, CHIMNEYS, AND BRICK VENEER** 68

   Fireplaces and Chimneys 68   Brick 71

7. **MASONRY ESTIMATE: DAMP-PROOFING** 74

   Foundation Damp-proofing 74   Estimating the Perimeter Drain 75   Estimating
   Damp-proofing 76

8. **CARPENTRY FRAMING: INTRODUCTION, BASEMENT, AND
   FIRST FLOOR** 77

   Lumber and Frame Construction 77   Basement Beams on Girders 78   Sill
   Seal and Sill 81   Box Sill 82   Floor Joists 82   Bridging 83   Headers and
   Trimmers 83   Subflooring 84

9. **CARPENTRY FRAMING: WALLS, CEILINGS, AND SECOND FLOOR** 86

   Plates 86   Studs 87   Headers 87   Second Floor Framing 88   Ceiling
   Joists 90

10. **CARPENTRY FRAMING: ROOF STRUCTURE 91**

Roof Trusses 91   Rafters 92   Ridge Board 93   Collar Beams 93   Gable Studs 94   Roof Sheathing 94

11. **CARPENTRY FRAMING: ROOF FINISHING 96**

Roofing 96   Estimating Roofing 98

12. **CARPENTRY FRAMING: CORNICES 100**

Cornice Construction 100   Estimating Cornices 101

13. **CARPENTRY FRAMING: OUTSIDE WALLS 103**

Sheathing 103   Siding 106   Estimating Siding 106

14. **CARPENTRY EXTERIOR FINISH 108**

Corner Boards 108   Overhang 108   Louvers 108   Entrance Roof 109

15. **CARPENTRY INTERIOR WORK: WALLS 110**

Insulation 110   Wallboard 111   Joint System 112   Plaster 112   Moldings 113   Estimating Cove Molding 113   Baseboard and Carpet Strip 113   Estimating Baseboard and Carpet Strip 113

16. **CARPENTRY INTERIOR WORK: FLOORS 115**

Underlayment 115   Resilient Flooring 116   Slate Flooring 116   Hardwood Flooring 117   Ceramic Tile 118

17. **CARPENTRY MILLWORK AND FINISHES: DOORS AND CLOSETS 120**

Exterior Doors 120   Interior Doors 120   Prehung Doors 121   Estimating Doors 121

18. **CARPENTRY MILLWORK AND FINISHES: WINDOWS 124**

Windows 124   Window Sizes 125   Window Trim 126

19. **CARPENTRY MILLWORK AND FINISHES: STAIRWORK 128**

Stairs 128   Estimating Stairs 129

20. **PAINTING AND DECORATING 132**

Paint and Other Finishes 132   Estimating Paint 133

21. **CARPENTRY MILLWORK AND FINISHES: CABINETS 136**

Types of Cabinets 136   Estimating Cabinets 137

22. **CARPENTRY MILLWORK AND FINISHES: HARDWARE AND NAILS 140**

Hardware 140   Nails 141

23. **MECHANICAL: PLUMBING 143**

Rough Plumbing 143   Finished Plumbing 145   Estimating Plumbing 146

24. **MECHANICAL: HEATING AND AIR CONDITIONING 148**

Types of Systems 148   Heating Loads 151   Air Conditioning 154

25. **ENERGY CONSERVATION 155**

Solar Hot Water Heating 156   Solar Space Heating 157

26. **ELECTRICAL WIRING 159**

Conductors 159   Service 159   Branch Circuits 160   Estimating Electrical Work 161

27. **ESTIMATING CHECKLIST 163**

Site Work 163   Masonry Estimate 164   Carpentry Estimate: Framing and Exterior Work 168   Carpentry Estimate: Interior Work and Finishes 172 Painting 176

28. **LABOR ESTIMATING GUIDELINE 178**

Site Work 178   Masonry Work 178   Carpentry Work 180   Carpentry Finish Work: Exterior 182   Carpentry: Interior Work 182   Carpentry: Millwork and Trim 183   Painting: Exterior Work 185   Painting: Interior Work 185

29. **DESIGNING A HOUSE 187**

30. **DRAWING THE FIRST FLOOR PLAN 192**

31. **DRAWING THE SECOND FLOOR PLAN 201**

32. **DRAWING THE BASEMENT AND FOUNDATION PLAN 207**

33. **DRAWING THE HALF SECTION VIEW AND FRONT ELEVATION 213**

34. **DRAWING THE SIDE AND REAR ELEVATIONS 220**

35. **DRAWING THE WALL SECTION AND SPECIAL DETAILS 225**

36. **DRAWING A RANCH HOUSE WITH BRICK VENEER FRONT 237**

37. **DRAWING KITCHEN CABINETS IN PERSPECTIVE 264**

38. **GENERAL REVIEW—ESTIMATING A RANCH HOUSE 271**

APPENDIX: Labor Estimating Tables 293

GLOSSARY 312

REFERENCES 316

INDEX 317

1
INTRODUCTION

There are a number of occupations in the building industry that rely on knowledge of building materials and quantity takeoff procedures. For the most part, buildings are designed by architects and engineers, who transmit their ideas to drafters. The drafter prepares the working drawings from which the building is constructed. The construction estimator uses the information from these drawings and specially prepared specifications to calculate the quantity of materials, labor requirements, overhead, and profit for the construction project. Many large construction companies employ purchasing agents to contact building materials salespersons and check catalogs to obtain materials and supplies for the company. Building inspectors are employed by government agencies and some large construction companies to ensure that work is properly completed and meets the requirements set forth in the building codes.

A contractor is one who agrees to complete certain work for a fixed price. Contractors in building construction fall into two categories: general contractors and subcontractors. General contractors agree to have a building constructed for a future owner. Subcontractors specialize in one phase of construction and agree to perform that part of the work for the general contractor. Subcontractors usually do the plumbing, heating, and electrical work on most construction projects. Mason work and painting are also sublet on some projects. This depends upon the general contractor. The general contractor has the final responsibility for seeing that each subcontractor's portion of the work is completed.

Before we start our actual estimating from a set of working drawings and specifications, we should refresh ourselves with a basic arithmetic review, with what working drawings really are, with the architect's scale and how to use it, and with the meaning of lines, symbols, and conventions used in construction.

GEOMETRY

To assist those who have no knowledge of geometry and mensuration (measurement of areas and volumes), definitions and explanations of new terms and propositions are as follows.

A *point* indicates position only; it has no length, thickness, or width. A *line* has one dimension only — length. It may be straight, curved, or irregular (Figure 1-1). If the direction of the line does not change, it is a straight line. If the direction changes continually, it is a curved line. An irregular line is a combination of straight and curved lines. A *perpendicular* line is a line that is drawn at right angles (90 degrees) to another line (Figure 1-2).

Plane Figures

A *surface* is the exterior part of anything that has length, width, and thickness. A *plane surface* is a flat surface with no depressions or high points. It is usually referred to as a plane, and has only two dimensions — length and width. The *perimeter* of a surface is the distance around its sides, or the sum of the length of all the sides.

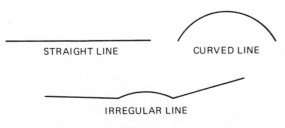

STRAIGHT LINE   CURVED LINE

IRREGULAR LINE

Fig. 1-1.

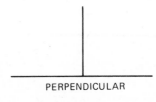

PERPENDICULAR

Fig. 1-2.

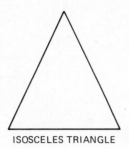

ISOSCELES TRIANGLE

Fig. 1-3.

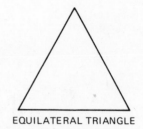

EQUILATERAL TRIANGLE

Fig. 1-4.

Triangles

All triangles have three sides and three angles. An *isoceles triangle* is a triangle having two of its sides equal (Figure 1-3). An *equilateral triangle* is a triangle having all three sides equal (Figure 1-4). A *scalene triangle* is a triangle having no two sides equal (Figure 1-5).

The *base* of any triangle is the side on which the triangle is considered to rest. The *altitude* or height of a triangle is the perpendicular distance between the base and the *vertex* of the angle opposite the base (Figure 1-6). A *right triangle* is a triangle having one right angle (90 degrees). The side opposite the right angle is called the *hypotenuse.*

One must know linear measurement rules to find the *perimeter* (distance around) of buildings, and the length of walls, baseboards, and ceiling moldings. One must also know area measurements to find the areas of floors, walls, ceilings, gables, and roofs.

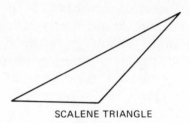

SCALENE TRIANGLE

Fig. 1-5.

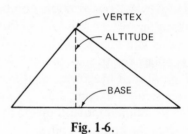

VERTEX

ALTITUDE

BASE

Fig. 1-6.

Perimeter

The distance around a square or a rectangle is found by adding the length and width and multiplying by 2, or by adding the length of all four sides. The same rule applies to buildings. The distance around the building is found by adding the length and width of all the sides. In Figure 1-7, 24'0" plus 14'0" equals 38'0"; 38 times 2 equals 76 linear feet of perimeter. If the building is as shown in Figure 1-8, then you add all the dimensions: 24'0" plus 14'0" plus 20'0" plus 7'0" plus 4'0" plus 7'0", for a total of 76 linear feet of perimeter.

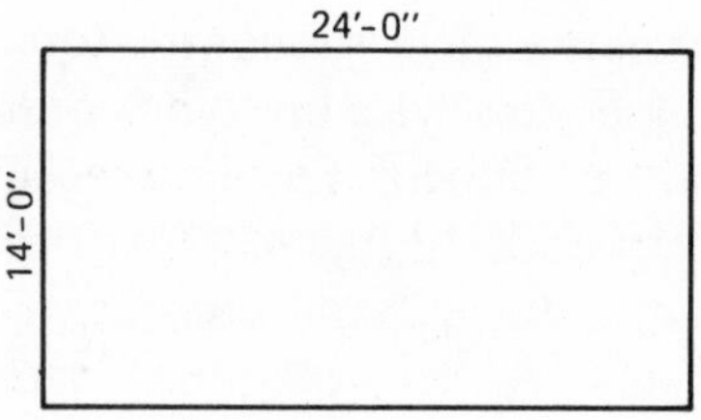

Hypotenuse

In a right triangle, the square of the hypotenuse equals the sum of the squares of the other two sides. This formula is useful in determining the length of a common rafter and in laying out a

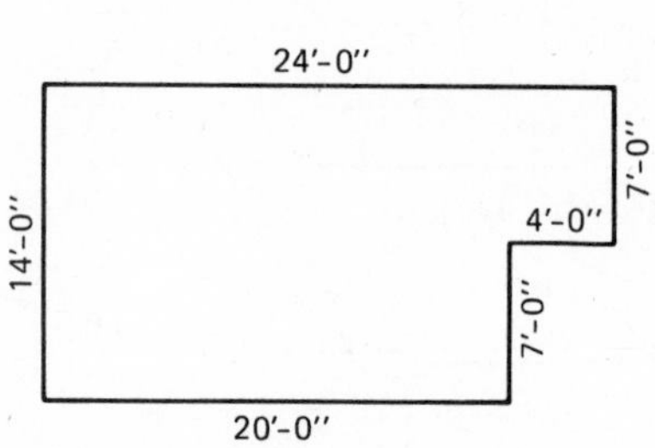

24'-0"

14'-0"

Fig. 1-7.

24'-0"

14'-0"

7'-0"

4'-0"

7'-0"

20'-0"

Fig. 1-8.

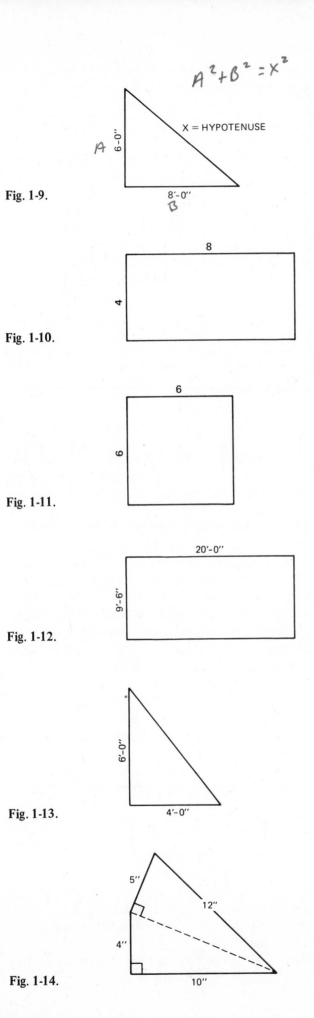

Fig. 1-9.

Fig. 1-10.

Fig. 1-11.

Fig. 1-12.

Fig. 1-13.

Fig. 1-14.

square (right angle) corner of a foundation. To find the length of X in Figure 1-9: Square 6 (6 $\times$ 6) equals 36; square 8 (8 $\times$ 8) equals 64; 36 plus 64 equals 100. Find the square root of 100, which is 10. So X equals 10.

Areas

The *area of a square or a rectangle* is found by multiplying the length by the width. In Figure 1-10, the area equals 8 multiplied by 4, or 32 square feet. In Figure 1-11, find the area by multiplying one side (6) by the other side (6) to get 36 square feet for the area.

Notice that in Figures 1-10 and 1-11 only the numbers, 8 by 4 and 6 by 6, are given. It is not indicated if they are inches, feet, or yards. One must be careful in estimating to be sure that feet are multiplied by feet, inches by inches, and yards by yards. If the measurement has feet and inches, make the inches fractional parts of a foot and then multiply. In Figure 1-12, a building is 20 feet long and 9 feet 6 inches wide. To find the area, multiply 20'0" by 9'6". 6 inches is one-half of 12 inches, or one-half of a foot. Multiply 20 by $9\frac{1}{2}$ to get an area of 190 square feet.

The *area of a triangle* is found by multiplying the length of the base by one-half the altitude. In Figure 1-13, multiply $\frac{1}{2}$ by 6'0", which equals 3, then multiply this figure by the base (4'0") to get an area of 12 square feet. Another method is to multiply the altitude by the base and then divide by 2.

The *area of a circle* may be found by using any of these three formulas:

Area = square of radius $\times$ 3.1416 — *Radius* is from the center of a circle to one side

Area = square of diameter $\times$ 0.7854 — *Diameter* is a straight line passing through the center of a circle and touching each side

Area = square of circumference $\times$ 0.07958 — *Circumference* is the distance around the outside of a circle

To find the area of an irregular polygon, divide the figure into triangles, parallelograms, or other shapes, and then find the area of each. The sum of the partial areas is the area of the irregular polygon. To find the area in Figure 1-14, divide the polygon into triangles, find the area of each and then add them together: 10 times 4 equals 40; 40

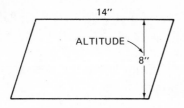

Fig. 1-15.

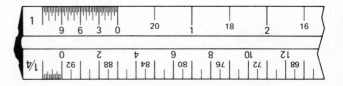

Fig. 1-16.

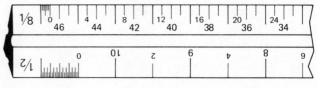

Fig. 1-17.

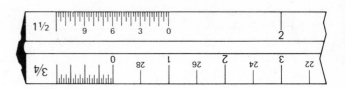

Fig. 1-18.

divided by 2 equals 20; 12 times 5 equals 60; 60 divided by 2 equals 30. Add 20 and 30 for a total of 50 square inches.

To find the area of a parallelogram, multiply the length of either side by the altitude (Figure 1-15). (The altitude equals the perpendicular distance to the opposite side.) Multiply the length (14) by the altitude (8) to get an area of 112 square inches.

THE ARCHITECT'S SCALE

Triangular scales are made of boxwood or plastic. A plastic scale works best because it has sharp edges and distinct machine markings which are necessary to read measurements accurately. The triangular scale combines eleven different scales and it is easy to handle while estimating. The architect's scale is *open divided* (Figure 1-16). This means the scales have main units undivided, and a fully subdivided extra unit placed at the zero end of the scales.

The following are the eleven scales found on the architect's triangular scale:

Full scale

$\frac{1}{8}'' = 1'0''$	$\frac{1}{4}'' = 1'0''$
$\frac{3}{8}'' = 1'0''$	$\frac{3}{4}'' = 1'0''$
$\frac{1}{2}'' = 1'0''$	$1'' = 1'0''$
$1\frac{1}{2}'' = 1'0''$	$3'' = 1'0''$
$\frac{3}{32}'' = 1'0''$	$\frac{3}{16}'' = 1'0''$

Two scales are combined on each face, except the full-size scale, which is fully divided into sixteenths. The combined scales are compatible because one is twice as large as the other and their zero points and extra-divided units are on opposite ends of the scale.

Architectural drawings use feet and inches as the major units of measurement. The architect's scale is calibrated into these units in conventionally reduced scales so that large buildings and details can be conveniently drawn on paper. This makes the drawings smaller and more easily handled.

The fraction, or number near the zero at each end of the scale, indicates the unit length in inches that is used on the building to represent one foot of the actual building. The extra unit near the zero end of the scale is subdivided into twelfths of a foot, or inches, as well as fractions of inches on the larger scales.

Most house plans and small buildings are drawn to the $\frac{1}{4}$-inch scale. This means that each quarter of an inch on the plans equals one foot of the actual size of the building. The scale of the drawing is noted on the plans and is usually given in the title box on each page of the plans. Sometimes when special details are given, the scale is placed directly under the detail.

To read the architect's triangular scale, turn it to the $\frac{1}{4}$-inch scale. The scale is divided on the left from the zero toward the $\frac{1}{4}$ mark so that each line represents one inch. Counting the marks from the zero toward the $\frac{1}{4}$ mark, there are 12 lines marked on the scale. Each one of these lines is one inch on the $\frac{1}{4}'' = 1'0''$ scale.

The fraction $\frac{1}{8}$ is on the opposite end of the same scale (Figure 1-17). This is the $\frac{1}{8}$-inch scale and is read from the right to the left. Notice that the divided unit is only half as large as the one on the $\frac{1}{4}$-inch end of the scale. There are only six divisions between the zero and the $\frac{1}{8}$ mark. This means that each line represents 2 inches at the $\frac{1}{8}$-inch scale.

Now look at the $1\frac{1}{2}$-inch scale (Figure 1-18). The divided unit is broken into twelfths of an inch and also a fractional part of an inch. Reading from the zero toward the number $1\frac{1}{2}$, notice the figures 3, 6, and 9. These represent the measurements of 3 inches, 6 inches, and 9 inches at the $1\frac{1}{2}'' = 1'0''$ scale. From the zero to the first long mark that represents one inch (which is the same length as the mark shown at 3), there are 4 divisions. This means that each gap between lines on the scale is equal to $\frac{1}{4}$ inch. Reading from the zero to the number 3, read each line as follows: $\frac{1}{4}$; $\frac{1}{2}$; $\frac{3}{4}$; 1; $1\frac{1}{4}$; $1\frac{1}{2}$; $1\frac{3}{4}$; 2; $2\frac{1}{4}$; $2\frac{1}{2}$; $2\frac{3}{4}$; and 3 inches.

THE ALPHABET OF LINES

To read a set of working drawings, it is important that the estimator know the different types of lines commonly used in making a set of plans. The mechanical drafter and the architectural drafter use the same alphabet of lines: mainly object lines, hidden lines, centerlines, extension lines, and dimension lines. The types of lines most commonly used on architectural drawings are described below.

Borderline. This is a solid line drawn medium heavy to give an even border around the drawing. The title box in the lower right-hand corner of the drawing is made the same weight and becomes a part of the border line.

Object line. The shape of the drawing is always drawn with a heavy, solid line to show the outline of the building.

Extension line. This is a light, solid line used to show the exact size or dimension. These lines extend away from the object lines at the exact points between which the dimensions are given.

Dimension line. Dimension lines are light lines, with arrowheads at their ends. They can be solid, with the dimension placed on the top side, or they can be broken, with the dimension given in the break. The arrowheads touch the extension lines or centerlines and give the exact distance referred to by the dimension line in feet and inches (Figure 1-19).

Equipment line. This is a light, solid line that is used to show the outline of equipment on a floor plan. Equipment lines are used to show such things as sinks, tubs, toilets, and cabinets.

Symbol section line. Two light, solid lines are drawn with a space between them. The symbol for the type of material to be used is then drawn between these lines (Figure 1-20). Symbols for various materials are now standardized and are easily found in most reference books, both in plans and elevations (see Figure 1-27).

Break line. This is a straight, solid line with a short up-and-down break in it (Figure 1-21). This indicates that either parts have been left out or that the full length of some part is not drawn.

Invisible line. To be complete, a drawing must include all lines that represent walls and surfaces. Sometimes these lines cannot be seen because they are covered by other sections of the object. To show that a line or surface is hidden, the drafter uses a series of short dashes (Figure 1-22). Invisible lines are used mostly on basement and foundation plans to show footings.

Centerline. A centerline is drawn as a light, broken line of long and short dashes spaced alternately (Figure 1-23). Centerlines are used to indicate the center of door and window openings and the center of partitions.

Cutting plane line. A cutting plane line is a solid line turned at right angles on each end with an arrowhead showing the direction of the section cut (Figure 1-24). The letter at the end of the arrowhead identifies the section that is referred to in the details. The cutting plane line is used to refer to a detail that shows exactly what is to be done on that part of the construction.

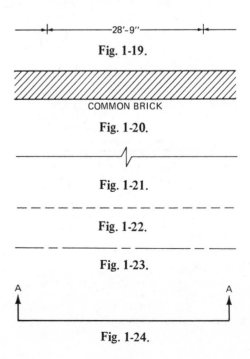

Fig. 1-19.

COMMON BRICK

Fig. 1-20.

Fig. 1-21.

Fig. 1-22.

Fig. 1-23.

Fig. 1-24.

Fig. 1-25.

Stair indicator. A stair indicator is found on any plan that has stairs going from one level to another. It is a straight line with an arrowhead on one end showing the direction of the stairs (Figure 1-25). The other end is sometimes closed to represent the tail of an arrow. Near the indicator (over the top in Figure 1-25) the words UP or DOWN are given, depending upon the direction. Sometimes the number of risers is also given.

Figure 1-26 shows where the various lines described here might be found on a set of plans.

ARCHITECTURAL SYMBOLS AND CONVENTIONS

Symbols

In order for the architect to make drawings useful, there must be a way to show materials and the type of construction procedures used so they are easily understood and drawn. This is done by the use of symbols. Figure 1-27 lists some examples of the symbols most commonly found on a set of working drawings.

When two symbols are very similar, knowledge of the material involved helps make the distinction between them. For example, the symbols for gypsum plaster and cement plaster are almost identical. From experience, it is apparent that gypsum plaster is not used on the exterior of a building where it would be exposed to the sun, rain, snow, and freezing temperatures. Cement plaster might be used on the exterior of a building where it is in contact with the elements.

On drawings it is also necessary for the architect to use abbreviations instead of spelling out each word. The estimator should be acquainted with the abbreviations most commonly used on drawings. These abbreviations are shown in Figure 1-28.

Conventions

In drawing a set of plans or working drawings, the architect must have some way of showing the sizes and location of equipment, doors, windows, rooms, kitchen cabinets, appliances, and so forth. These are

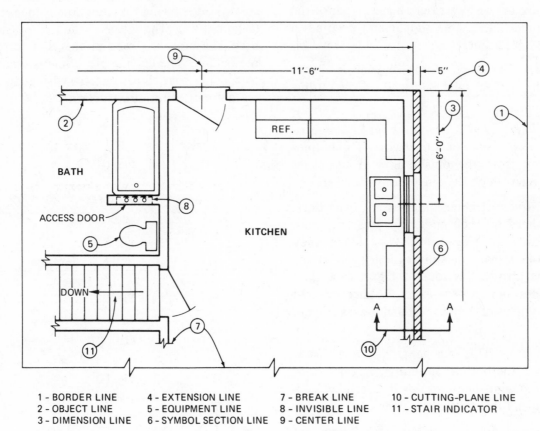

1 - BORDER LINE	4 - EXTENSION LINE	7 - BREAK LINE	10 - CUTTING-PLANE LINE
2 - OBJECT LINE	5 - EQUIPMENT LINE	8 - INVISIBLE LINE	11 - STAIR INDICATOR
3 - DIMENSION LINE	6 - SYMBOL SECTION LINE	9 - CENTER LINE	

Fig. 1-26.

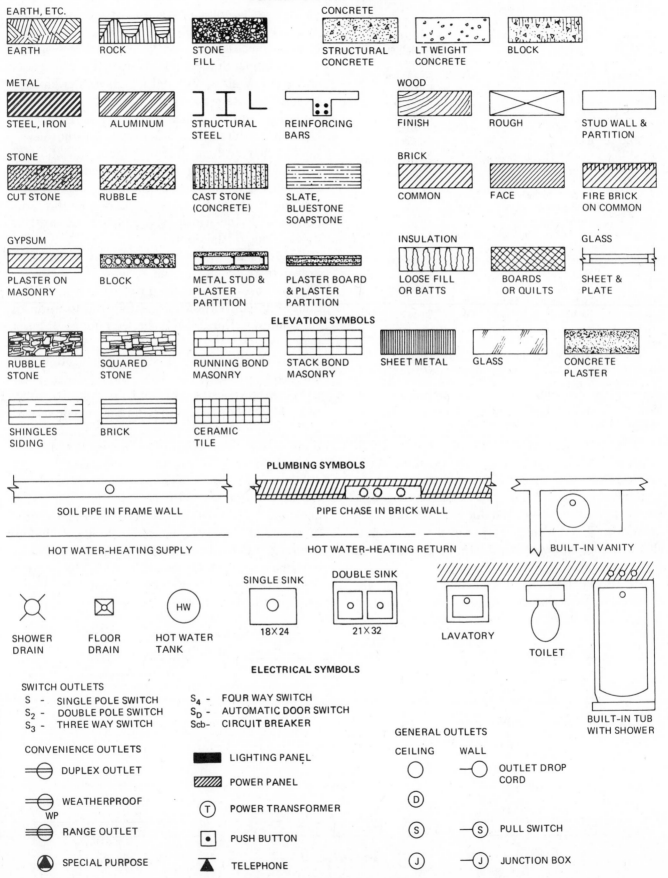

Fig. 1-27. Architectural symbols.

AWG	American Wire Gauge	GL	Glass
B	Bathroom	HB	Hose Bibb
BR	Bedroom	C	Hundred
BD	Board	INS	Insulation
BM	Board Measure	INT	Interior
BTU	British Thermal Unit	KD	Kiln Dried
BLDG	Building	K	Kitchen
CLG	Ceiling	LAV	Lavatory
C to C	Center to Center	LR	Living Room
CL or ℄	Centerline	MLDG	Molding
CLO	Closet	OC	On Center
COL	Column	REF	Refrigerator
CONC	Concrete	R	Riser
CFM	Cubic feet per minute	RM	Room
CU YD	Cubic Yard	SPEC	Specification
DR	Dining Room	STD	Standard
ENT	Entrance	M	Thousand
EXT	Exterior	T & G	Tongue and Groove
FIN	Finish	UNFIN	Unfinished
FL	Floor	WC	Water Closet
FTG	Footing	WH	Water Heater
FDN	Foundation	WP	Waterproof
GA	Gauge	WD	Wood

Fig. 1-28. Abbreviations used on drawings.

shown by using centerlines, extension lines, dimension lines, object lines, and symbols in a conventional or standard way. The architect uses conventions to show such things as the direction in which doors swing, their size, and their location. Conventions are also used by the architect to draw windows in plan, showing their exact location, size, and style. All windows and doors on exterior walls are shown. The dimensions are given to the center of all door and window openings. Doors are shown on plans as in Figure 1-29.

Notice that the swing of the door is shown. On some plans the architect designates the door size on the line showing the swing of the door. On some plans a circled number is shown at each door opening. This indicates that there is a schedule which gives the specifications for the door that corresponds to the number in the circle. Note that in giving door and window sizes, the width is always given first. Example: a door shown 3^0 x 6^8 means the door is three feet and no inches wide and it is six feet and eight inches high. If the doorway does not have a door, it may be designated with the abbreviation CO (cased opening). Also, notice that the exterior door has a line extending across the front, parallel with the line of the wall. This represents the threshold and extends the width of the outside door casing on each side of the door.

Windows are shown on the plan by a centerline drawn to the center of the opening (Figure 1-30). Window sizes are usually given on the plans just under the convention designating the location of the window. Here, also, a circled number or letter may appear. This indicates that there is a schedule located on the plans which lists the window size and style.

In veneer construction and in frame construction, dimensions are given from the outside face of the exterior sheathing to the center of the partitions or openings. All dimensions must add up to equal the overall dimensions (Figure 1-31).

Figure 1-31 shows doors and windows in a brick veneer exterior wall. Notice that the dimension lines are drawn from the face of the exterior sheathing to the centerline of the door and window openings. Figure 1-32 shows doors and windows in frame construction on exterior walls. Notice that the dimensions for the location of the picture window are given to the center of the opening. In

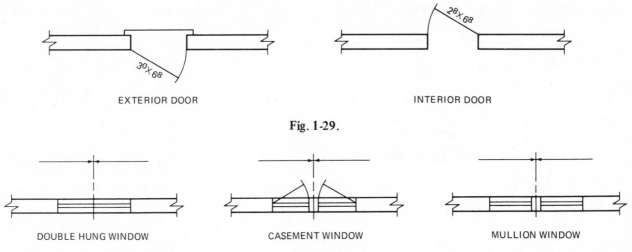

EXTERIOR DOOR   INTERIOR DOOR

Fig. 1-29.

DOUBLE HUNG WINDOW   CASEMENT WINDOW   MULLION WINDOW

Fig. 1-30.

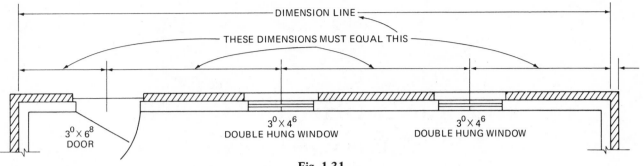

Fig. 1-31.

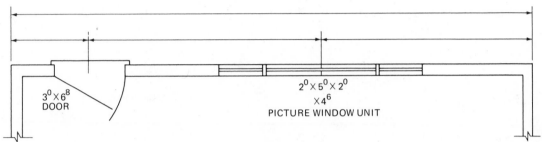

Fig. 1-32.

architectural drawing, the width is always given
first and the height second. This applies to all
windows and doors. For example, the picture
window in Figure 1-32 is a five-foot-wide center
sash with a two-foot-wide double hung window on
each side. All the windows in this unit are four feet
six inches high. The same conventions are used for
interior wall and door openings with only minor
changes. On masonry partitions and door openings, the dimensions are given from the face of the
masonry wall to the edge of the opening. The
opening is dimensioned from one side of the opening
to the other, as shown in Figure 1-33. On wood
frame partitions (Figure 1-34), the dimensions
are given from the face of the exterior sheathing to
the center of the door opening, or from the center
of one partition to the center of the door opening.

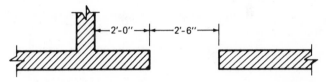

Fig. 1-33.

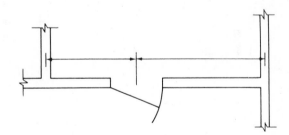

Fig. 1-34.

2
WORKING DRAWINGS AND SPECIFICATIONS

Plans are drawn by the architect or drafter to give the builder exact information about how to build a particular structure. Such information cannot easily be conveyed in words. Working drawings, together with specifications, contain all the necessary information. This information is presented in the form of drawings, notes, dimensions, and any other indications necessary to guide the builder in the construction of the building.

A set of working drawings shows the size and thickness of every part of the building. On working drawings, parts farther away are not drawn smaller to show them in perspective, as they are in a picture; every part is drawn to scale. Working drawings and specifications often indicate the material to be used and the finish to be applied. To make drawings easy to understand, architects have developed certain symbols, signs, and abbreviations to represent various materials, fixtures, and forms of construction.

SCALE

Since drawings cannot be made the full size of the building, they are scaled down. A fraction of an inch on the drawings stands for a foot on the actual building. The plans of an average-size house are usually drawn to the $\frac{1}{4}$-inch-to-the-foot scale. This is written $\frac{1}{4}'' = 1'0''$. This is the scale normally used for floor plans and elevations. Drawings made to this scale are one-forty-eighth the size of the building.

On wall sections and details, however, a larger scale, such as $\frac{3}{4}'' = 1'0''$, is used to show the drawings in greater detail. The architect or draftsman chooses a scale depending upon the space available for the drawings and the nature of the part to be described. Simple details do not have to be drawn as large as more complex ones.

KINDS OF DRAWINGS

Plans, elevations, sections, and special details are all needed to understand a set of drawings. In other words, plans of every floor and the basement are needed, as well as the elevations of the four sides of the building and a section showing the construction of the wall from the basement to the attic. In addition, details which give more precise information on particular parts or sections are needed.

The architect or drafter has two general ways of presenting three-dimensional objects on two-dimensional paper: (1) pictorial drawings, which may be perspective, isometric, or oblique; and (2) orthographic projection drawings.

An orthographic projection drawing of a block of wood would include a drawing as seen by looking straight at the front of the block, a drawing as seen by looking straight down from the top, and a drawing as seen by looking straight at one end. This method, sometimes referred to as third-angle projection, is standard practice for all forms of architectural drawing.

Elevations

A building is represented in much the same way as the block of wood. To show a building in orthographic projection, a drawing is made of the building as it is seen by looking straight at the front, the rear, and the sides. These drawings are known as elevations.

An elevation is the easiest part of a drawing to visualize. Elevations are drawn as one looks at the building. Different elevations are indicated by different names. For example, in Figure 2-1, the front of the house would be drawn as the front elevation, the right-hand side would be drawn as

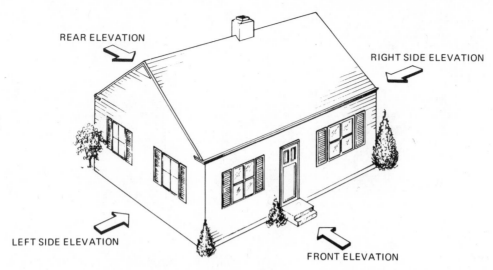

Fig. 2-1.

the right side elevation, the side on the left would be drawn as the left elevation, and the back of the building, facing front, would be drawn as the rear elevation.

Plan Views

Next, assume that a horizontal cut is made through the building about three feet from the floor and the top section is removed. Then, looking straight down from the top at the remaining part of the building is what is seen on the floor plan. The cut through the building is made at the proper height so that it passes through doors, windows, and other wall openings. This permits them to be drawn as visible edge or object lines. In Figure 2-2, the view of the cut through the building shown on the right becomes the floor plan shown on the left. This allows each floor and basement plan to be

drawn separately. Only the layout for that particular floor is shown. This provides plans that can be easily read and understood. The floor plan shows the location of the outside walls and the centers of all door and window openings and their sizes.

Sectional Views

Now looking at the same building as it was before the horizontal line was cut through it to make the floor plans (Figure 2-1), assume that a vertical cut is made through the building and one part is removed. Looking straight at one end of the building, one sees a sectional view. In Figure 2-3, notice the front elevation on the left with the cutting plane line A-A cut vertically through the building. Removing the right section of the cut and looking at the remaining end would give us the section view on the right.

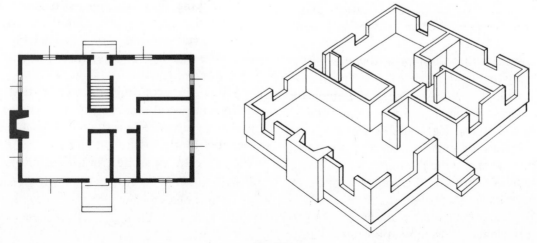

Fig. 2-2.

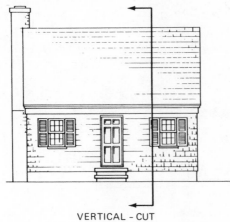

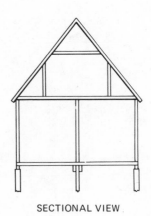

VERTICAL - CUT

SECTIONAL VIEW

Fig. 2-3.

WORKING DRAWINGS

The house we are going to be working with through-out the bulk of this book is known as a two-story Garrison. The protrusion of the second floor beyond the face of the front of the first floor front wall, allowing more space on the second floor, is the distinguishing feature of this style of house. The house is a two-story, three-bedroom home. It has a living room, dining room, kitchen, and half bath on the first floor and three bedrooms and two baths on the second floor. There is a fireplace in the living room on the outside wall. The working drawings for this house are given in Figures 2-4 through 2-15.

When studying the working drawings, start at the foundation, and see what is called for in the base-ment outside walls. Note that the basement area is divided into a laundry, a recreation room, and a workshop. Note that a partition is called for that divides the recreation room from the remainder of the basement area. In studying the basement plan, note the type of supporting beams, the direction and size of the floor joists, the location of the stairs into the basement area and any other interior finishes.

In studying the first floor plan, observe the en-trance and any hallways, room locations, partitions, any large openings, windows, doors, plumbing fix-tures, fireplaces, kitchen cabinets, the type of floor finish, and any special wall finishes or built-ins such as bookcases or china cabinets. Also notice the direction of the ceiling framing. For a two-story house, make sure that location, type, and style of the main stairs are identified.

In studying the scond floor plan, notice the lay-out of the bedrooms and their relationship to baths, closets, stairs, etc. Again notice any special cabinets in baths, vanities, mirrors, louver closet doors, linen closets, and special closets such as cedar-lined storage. Again notice the direction of the ceiling joists and any special wall finishes.

After you have studied the plans for the base-ment and floor areas, study the elevations. Here you will find the style of architecture, the cornice, siding, roofing, windows, chimneys, porches, walks, railings, entrance-door styles and door models.

What does not show on the working drawings, you will have to look up in the specifications. For example, the working drawings may call for quarry tile; the size of tile and how it is to be installed are found in the specifications. Remember, it is im-possible to put all of the information necessary to build a house on a set of working drawings. What cannot be drawn must be conveyed to the builder by the specifications.

If you find any discrepancies between the plans and the specifications, you should contact the per-son who made the working drawings and the speci-fications. If you are unable to do this, remember that the specifications *always* take precedence over the working drawings. Specifications become a part of the contract and cannot be changed once the contract is agreed upon and signed by both par-ties. The plans or working drawings can be changed by a work order.

Study the working drawings so that you fully understand what is called for on the plans. The first drawing (Figure 2-4) is the plot plan and shows the location of the house on the land of the owner. The other drawings are grouped as follows:

- Figures 2-5 through 2-7 are basement and floor plans.
- Figures 2-8 through 2-11 are elevations.
- Figures 2-12 through 2-15 are drawings of details.

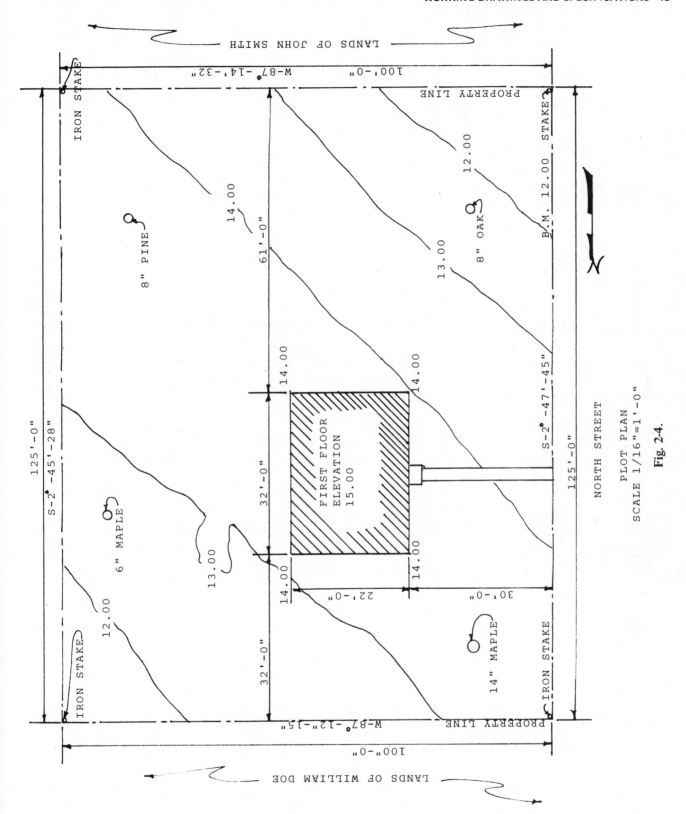

PLOT PLAN
SCALE 1/16"=1'-0"

Fig. 2-4.

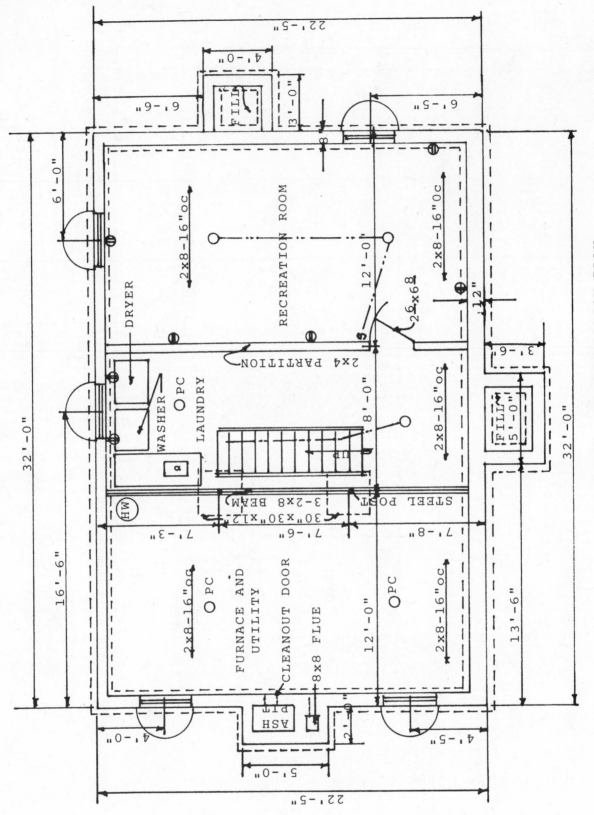

BASEMENT AND FOUNDATION PLAN
SCALE 3/16"=1'-0"

Fig. 2-5.

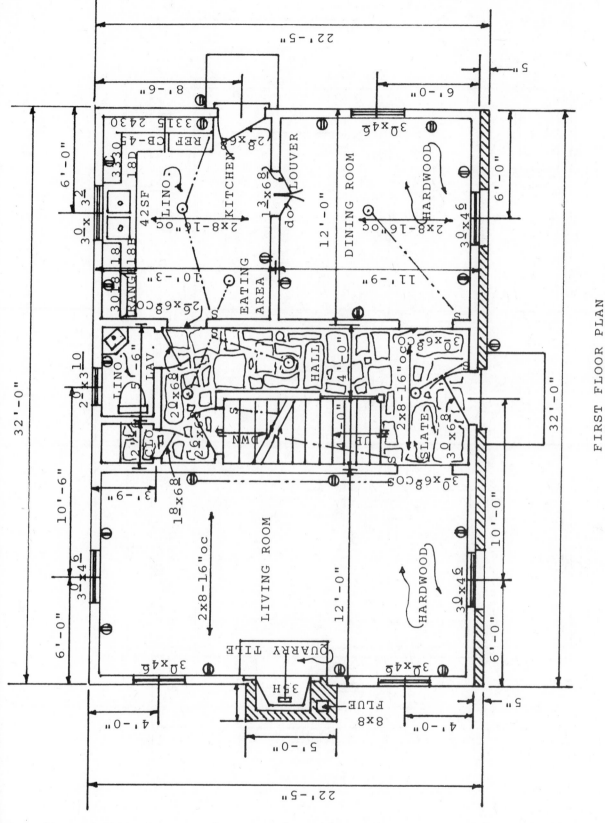

FIRST FLOOR PLAN
SCALE 3/16"=1'-0"

Fig. 2-6.

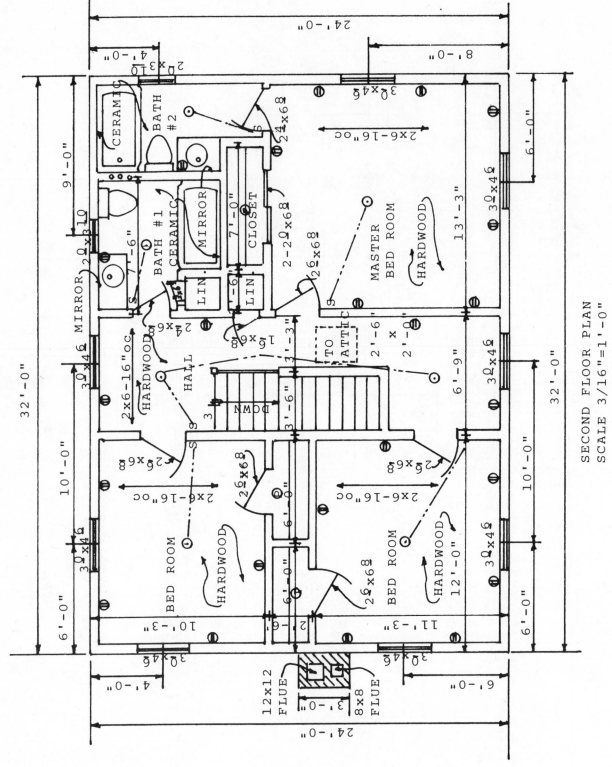

SECOND FLOOR PLAN
SCALE 3/16"=1'-0"

Fig. 2-7.

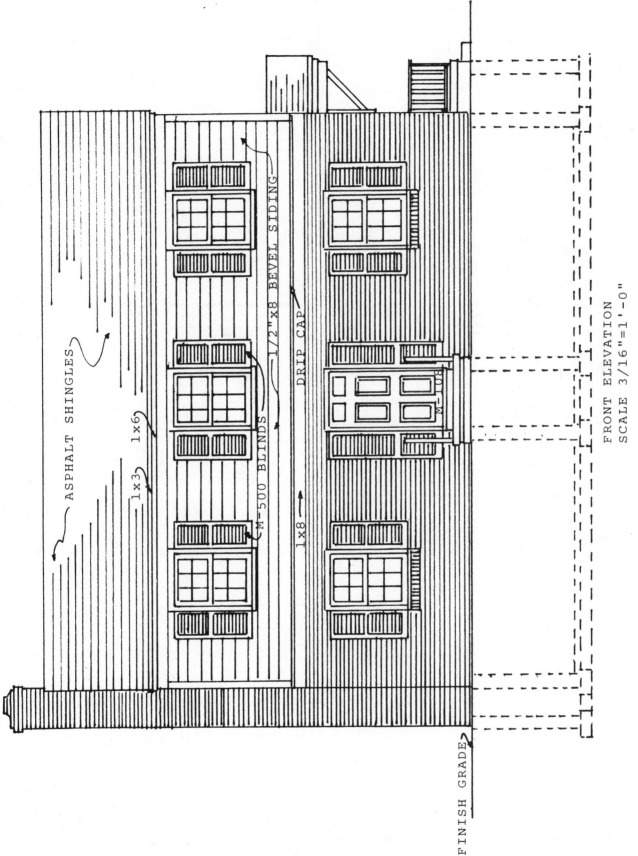

ASPHALT SHINGLES

1x3

1x6

M-500 BLINDS

1/2"x8 BEVEL SIDING

DRIP CAP

1x8

M-108

FINISH GRADE

FRONT ELEVATION
SCALE 3/16"=1'-0"

Fig. 2-8.

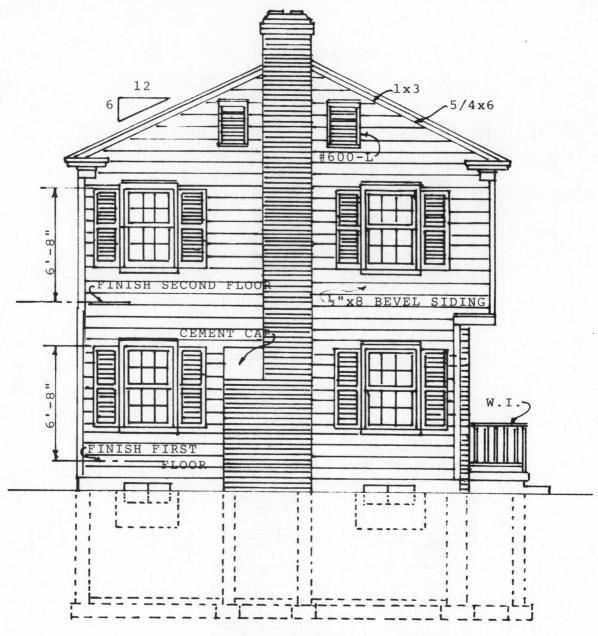

LEFT ELEVATION
SCALE 3/16"=1'-0"

Fig. 2-9.

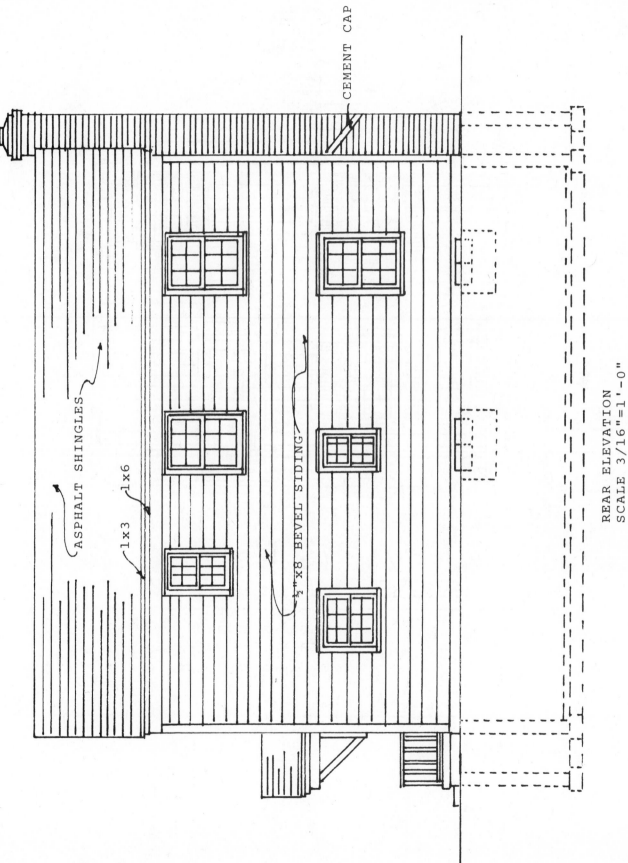

CEMENT CAP

ASPHALT SHINGLES

1x6

1x3

½"x8 BEVEL SIDING

REAR ELEVATION
SCALE 3/16"=1'-0"
Fig. 2-10.

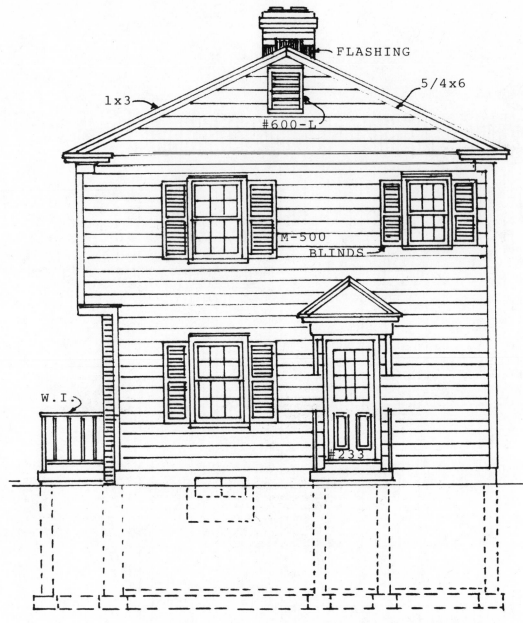

RIGHT ELEVATION
SCALE 3/16"=1'-0"

Fig. 2-11.

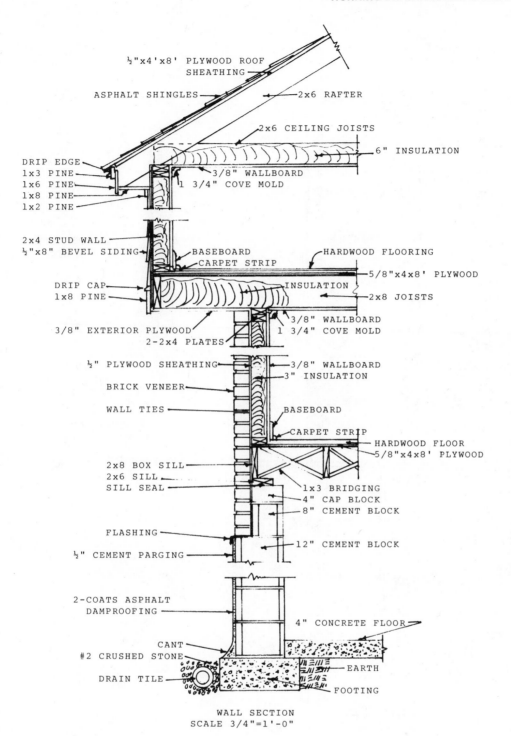

½"x4'x8' PLYWOOD ROOF
SHEATHING

ASPHALT SHINGLES ——————— 2x6 RAFTER

2x6 CEILING JOISTS

6" INSULATION

DRIP EDGE
1x3 PINE
1x6 PINE
1x8 PINE
1x2 PINE

3/8" WALLBOARD
1 3/4" COVE MOLD

2x4 STUD WALL
½"x8" BEVEL SIDING

BASEBOARD
CARPET STRIP
HARDWOOD FLOORING

DRIP CAP
1x8 PINE

INSULATION
5/8"x4x8' PLYWOOD

2x8 JOISTS

3/8" EXTERIOR PLYWOOD
2-2x4 PLATES

3/8" WALLBOARD
1 3/4" COVE MOLD

½" PLYWOOD SHEATHING

3/8" WALLBOARD
3" INSULATION

BRICK VENEER

WALL TIES

BASEBOARD
CARPET STRIP
HARDWOOD FLOOR
5/8"x4x8' PLYWOOD

2x8 BOX SILL
2x6 SILL
SILL SEAL

1x3 BRIDGING
4" CAP BLOCK
8" CEMENT BLOCK

FLASHING

12" CEMENT BLOCK

½" CEMENT PARGING

2-COATS ASPHALT
DAMPROOFING

4" CONCRETE FLOOR

CANT
#2 CRUSHED STONE

DRAIN TILE

EARTH

FOOTING

WALL SECTION
SCALE 3/4"=1'-0"

Fig. 2-12.

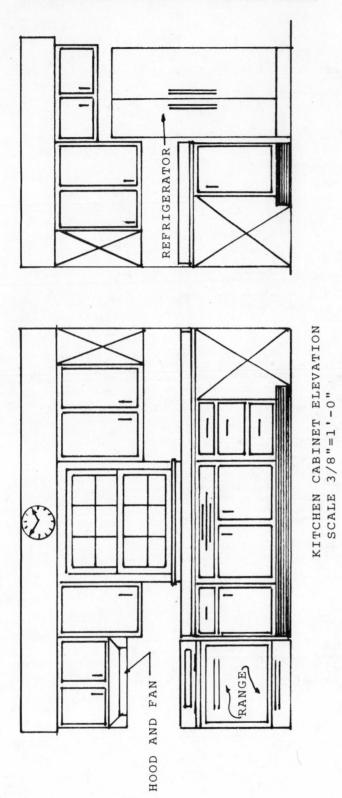

KITCHEN CABINET ELEVATION
SCALE 3/8"=1'-0"

HOOD AND FAN

REFRIGERATOR

RANGE

½"x4'x8' PLYWOOD
ASPHALT SHINGLES
2x6 RAFTER
2x6's
1x2 PINE
DRIP EDGE
1x3 PINE
1x6 PINE
1x8 PINE

CORNICE DETAIL
SCALE 3/8"=1'-0"

3-2x8's BEAM
STEEL CAP
STEEL POST
4" CONCRETE FLOOR
BASE PLATE
FOOTINGS

SUPPORT POSTS DETAIL
SCALE 3/8"=1'-0"

Fig. 2-13.

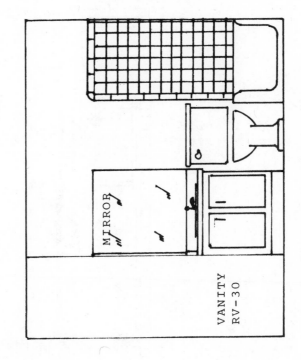

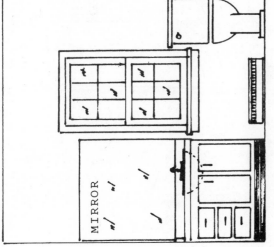

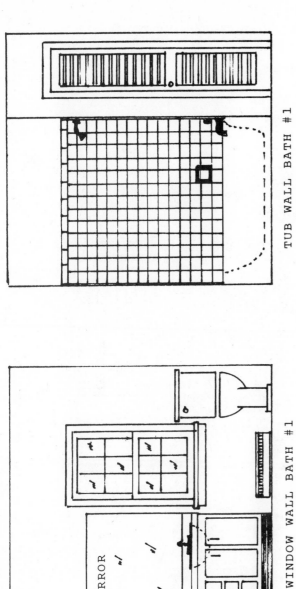

TUB WALL BATH #1

WINDOW WALL BATH #1

VANITY
RV-36

BATH #2 MASTER BEDROOM BATH

VANITY
RV-30

MIRROR

MIRROR

Fig. 2-14.

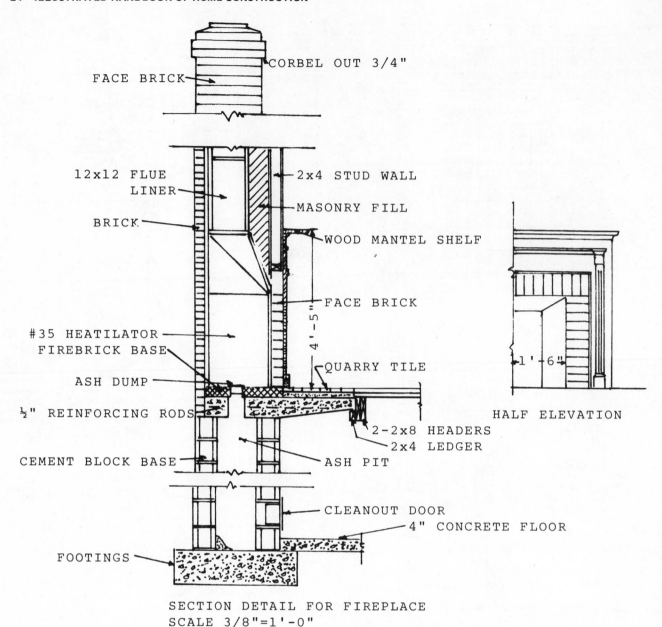

FACE BRICK

CORBEL OUT 3/4"

12x12 FLUE LINER

2x4 STUD WALL

MASONRY FILL

BRICK

WOOD MANTEL SHELF

FACE BRICK

#35 HEATILATOR FIREBRICK BASE

4'-5"

QUARRY TILE

ASH DUMP

½" REINFORCING RODS

2-2x8 HEADERS
2x4 LEDGER

CEMENT BLOCK BASE

ASH PIT

HALF ELEVATION

1'-6"

CLEANOUT DOOR
4" CONCRETE FLOOR

FOOTINGS

SECTION DETAIL FOR FIREPLACE
SCALE 3/8"=1'-0"

Fig. 2-15.

SPECIFICATIONS

When a set of working drawings for a building is made, it is impossible to include all the information necessary to build the structure. For example, if the plans show wood floors, they might be oak, maple, or vertical-grain fir. Tile flooring on the plans might be ceramic tile, asphalt tile, or vinyl tile. Where the plans show flashing, it could be galvanized steel, aluminum, or copper. Roof shingles can be asphalt, wood, or slate. The types of materials used and the quality of these materials must be described in some way.

Information that cannot be clearly shown on the drawings is conveyed to the builder by written specifications. The working drawings for a building give shape, size, and location. Specifications describe the quality and type of materials, colors, finishes, and workmanship required. Inspections, rejections, and approvals are stated so that the contractor knows what is to be expected.

Legal Aspects

The legal aspects of building construction may be confusing to the unacquainted reader. The estimator should understand the legal documents involved, because they affect the cost of building construction; these documents are described below.

A *contract* is an agreement in which one party agrees to perform certain work and the other party agrees to pay for the work and services.

A *performance bond* is a guarantee issued by a bonding company stating that the work will be done according to the plans and specifications. If the contractor fails to finish the contract, the bond ensures that money will be provided to hire another contractor to complete the job.

The *specifications* form a legal document which becomes a distinct part of the contract. Once the contract has been signed, any corrections that must be made in the specifications must be accompanied by a *change order*. In case of a discrepancy between the drawings and the specifications, the specifications are binding.

Specifications are prepared by the architect or contractor and cover the entire project. The amount of detail and the exact form of the specifications may vary. Specifications serve several purposes:

- They give instructions for the bids, owner–contractor agreements, insurance, and bond forms that are necessary
- They help prevent disputes between the builder and the owner, or between the contractor and the architect
- They eliminate conflicting opinions about the grade and quality of the material to be used
- They help the contractor estimate the material and labor
- Together with the working drawings, they are necessary to complete a contract
- Combined with the drawings, they may be used in a court of law as binding evidence.

Divisions of Specifications

In general, specifications for a residence are broken down into divisions that cover work by the different trades. It is standard practice to write the specifications in the order in which the house will be constructed. This makes it easier for the estimator to write a material list and for the contractor and subcontractors to locate specifications for their particular trade, material, or work. The following are typical divisions of specifications for a residence:

- General conditions
- General requirements
- Excavating and backfill
- Grading
- Concrete
- Masonry
- Carpentry and millwork
- Sheetmetal and roofing
- Glass
- Painting
- Hardware
- Heating and air conditioning
- Plumbing
- Electrical

The general conditions section is usually first. In the specifications for house construction, the AIA (American Institute of Architects) short-form agreement between owner and contractor is sometimes included as part of the general conditions. This form is reproduced here for reference.

The specifications for our two-story Garrison house follow. These specifications with the working drawings are necessary to complete the contract. Study the specifications the same as you did the working drawings so that you have a clear understanding of what is expected and required in the construction of this house.

THE AMERICAN INSTITUTE OF ARCHITECTS

AIA Document A107

Standard Form of Agreement Between Owner and Contractor

Short Form Agreement for **Small Construction Contracts**

Where the Basis of Payment is a

STIPULATED SUM

*THIS DOCUMENT HAS IMPORTANT LEGAL CONSEQUENCES; CONSULTATION WITH
AN ATTORNEY IS ENCOURAGED WITH RESPECT TO ITS COMPLETION OR MODIFICATION*

*For other contracts the AIA issues Standard Forms of Owner-Contractor Agreements and Standard General Conditions
of the Contract for Construction for use in connection therewith.*

This document has been approved and endorsed by The Associated General Contractors of America.

AGREEMENT

made this   day of   in the year Nineteen
Hundred and

BETWEEN the Owner:

and the Contractor:

the Project:

the Architect:

The Owner and Contractor agree as set forth below.

AIA DOCUMENT A107 • SMALL CONSTRUCTION CONTRACT • JANUARY 1974 EDITION • AIA® • ©1974
THE AMERICAN INSTITUTE OF ARCHITECTS, 1735 NEW YORK AVE., N.W., WASHINGTON, D. C. 20006

ARTICLE 1
THE WORK

The Contractor shall perform all the Work required by the Contract Documents for
(Here insert the caption descriptive of the Work as used on other Contract Documents.)

ARTICLE 2
TIME OF COMMENCEMENT AND COMPLETION

The Work to be performed under this Contract shall be commenced

and completed

ARTICLE 3
CONTRACT SUM

The Owner shall pay the Contractor for the performance of the Work, subject to additions and deductions by Change Order as provided in the General Conditions, in current funds, the Contract Sum of
(State here the lump sum amount, unit prices, or both, as desired.)

AIA DOCUMENT A107 • SMALL CONSTRUCTION CONTRACT • JANUARY 1974 EDITION • AIA® • ©1974
THE AMERICAN INSTITUTE OF ARCHITECTS, 1735 NEW YORK AVE., N.W., WASHINGTON, D. C. 20006

ARTICLE 4
PROGRESS PAYMENTS

Based upon Applications for Payment submitted to the Architect by the Contractor and Certificates for Payment issued by the Architect, the Owner shall make progress payments on account of the Contract Sum to the Contractor as follows:

ARTICLE 5
FINAL PAYMENT

The Owner shall make final payment   days after completion of the Work, provided the Contract be then fully performed, subject to the provisions of Article 16 of the General Conditions.

ARTICLE 6
ENUMERATION OF CONTRACT DOCUMENTS

The Contract Documents are as noted in Paragraph 7.1 of the General Conditions and are enumerated as follows:
(List below the Agreement, Conditions of the Contract (General, Supplementary, and other Conditions), Drawings, Specifications, Addenda and accepted Alternates, showing page or sheet numbers in all cases and dates where applicable.)

GENERAL CONDITIONS

ARTICLE 7
CONTRACT DOCUMENTS

7.1 The Contract Documents consist of this Agreement (which includes the General Conditions), Supplementary and other Conditions, the Drawings, the Specifications, all Addenda issued prior to the execution of this Agreement, all modifications, Change Orders, and written interpretations of the Contract Documents issued by the Architect. These form the Contract and what is required by any one shall be as binding as if required by all. The intention of the Contract Documents is to include all labor, materials, equipment and other items as provided in Paragraph 10.2 necessary for the proper execution and completion of the Work and the terms and conditions of payment therefor, and also to include all Work which may be reasonably inferable from the Contract Documents as being necessary to produce the intended results.

7.2 The Contract Documents shall be signed in not less than triplicate by the Owner and the Contractor. If either the Owner or the Contractor do not sign the Drawings, Specifications, or any of the other Contract Documents, the Architect shall identify them. By executing the Contract, the Contractor represents that he has visited the site and familiarized himself with the local conditions under which the Work is to be performed.

7.3 The term Work as used in the Contract Documents includes all labor necessary to produce the construction required by the Contract Documents, and all materials and equipment incorporated or to be incorporated in such construction.

ARTICLE 8
ARCHITECT

8.1 The Architect will provide general administration of the Contract and will be the Owner's representative during construction and until issuance of the final Certificate for Payment.

8.2 The Architect shall at all times have access to the Work wherever it is in preparation and progress.

8.3 The Architect will make periodic visits to the site to familiarize himself generally with the progress and quality of the Work and to determine in general if the Work is proceeding in accordance with the Contract Documents. On the basis of his on-site observations as an architect, he will keep the Owner informed of the progress of the Work, and will endeavor to guard the Owner against defects and deficiencies in the Work of the Contractor. The Architect will not be required to make exhaustive or continuous on-site inspections to check the quality or quantity of the Work. The Architect will not be responsible for construction means, methods, techniques, sequences or procedures, or for safety precautions and programs in connection with the Work, and he will not be responsible for the Contractor's

failure to carry out the Work in accordance with the Contract Documents.

8.4 Based on such observations and the Contractor's Applications for Payment, the Architect will determine the amounts owing to the Contractor and will issue Certificates for Payment in accordance with Article 16.

8.5 The Architect will be, in the first instance, the interpreter of the requirements of the Contract Documents. He will make decisions on all claims and disputes between the Owner and the Contractor. All his decisions are subject to arbitration.

8.6 The Architect will have authority to reject Work which does not conform to the Contract Documents.

ARTICLE 9
OWNER

9.1 The Owner shall furnish all surveys.

9.2 The Owner shall secure and pay for easements for permanent structures or permanent changes in existing facilities.

9.3 The Owner shall issue all instructions to the Contractor through the Architect.

ARTICLE 10
CONTRACTOR

10.1 The Contractor shall supervise and direct the work, using his best skill and attention. The Contractor shall be solely responsible for all construction means, methods, techniques, sequences and procedures and for coordinating all portions of the Work under the Contract.

10.2 Unless otherwise specifically noted, the Contractor shall provide and pay for all labor, materials, equipment, tools, construction equipment and machinery, water, heat, utilities, transportation, and other facilities and services necessary for the proper execution and completion of the Work.

10.3 The Contractor shall at all times enforce strict discipline and good order among his employees, and shall not employ on the Work any unfit person or anyone not skilled in the task assigned to him.

10.4 The Contractor warrants to the Owner and the Architect that all materials and equipment incorporated in the Work will be new unless otherwise specified, and that all Work will be of good quality, free from faults and defects and in conformance with the Contract Documents. All Work not so conforming to these standards may be considered defective.

10.5 The Contractor shall pay all sales, consumer, use and other similar taxes required by law and shall secure all permits, fees and licenses necessary for the execution of the Work.

10.6 The Contractor shall give all notices and comply with all laws, ordinances, rules, regulations, and orders of any public authority bearing on the performance of the Work, and shall notify the Architect if the Drawings and Specifications are at variance therewith.

10.7 The Contractor shall be responsible for the acts and omissions of all his employees and all Subcontractors, their agents and employees and all other persons performing any of the Work under a contract with the Contractor.

10.8 The Contractor shall review, stamp with his approval and submit all samples and shop drawings as directed for approval of the Architect for conformance with the design concept and with the information given in the Contract Documents. The Work shall be in accordance with approved samples and shop drawings.

10.9 The Contractor at all times shall keep the premises free from accumulation of waste materials or rubbish caused by his operations. At the completion of the Work he shall remove all his waste materials and rubbish from and about the Project as well as his tools, construction equipment, machinery and surplus materials, and shall clean all glass surfaces and shall leave the Work "broom clean" or its equivalent, except as otherwise specified.

10.10 The Contractor shall indemnify and hold harmless the Owner and the Architect and their agents and employees from and against all claims, damages, losses and expenses including attorneys' fees arising out of or resulting from the performance of the Work, provided that any such claim, damage, loss or expense (1) is attributable to bodily injury, sickness, disease or death, or to injury to or destruction of tangible property (other than the Work itself) including the loss of use resulting therefrom, and (2) is caused in whole or in part by any negligent act or omission of the Contractor, any Subcontractor, anyone directly or indirectly employed by any of them or anyone for whose acts any of them may be liable, regardless of whether or not it is caused in part by a party indemnified hereunder. In any and all claims against the Owner or the Architect or any of their agents or employees by any employee of the Contractor, any Subcontractor, anyone directly or indirectly employed by any of them or anyone for whose acts any of them may be liable, the indemnification obligation under this Paragraph 10.10 shall not be limited in any way by any limitation on the amount or type of damages, compensation or benefits payable by or for the Contractor or any Subcontractor under workmen's compensation acts, disability benefit acts or other employee benefit acts. The obligations of the Contractor under this Paragraph 10.10 shall not extend to the liability of the Architect, his agents or employees arising out of (1) the preparation or approval of maps, drawings, opinions, reports, surveys, Change Orders, designs or specifications, or (2) the giving of or the failure to give directions or instructions by the Architect, his agents or employees provided such giving or failure to give is the primary cause of the injury or damage.

ARTICLE 11
SUBCONTRACTS

11.1 A Subcontractor is a person who has a direct contract with the Contractor to perform any of the Work at the site.

11.2 Unless otherwise specified in the Contract Documents or in the Instructions to Bidders, the Contractor, as soon as practicable after the award of the Contract, shall furnish to the Architect in writing a list of the names of Subcontractors proposed for the principal portions of the Work. The Contractor shall not employ any Subcontractor to whom the Architect or the Owner may have a reasonable objection. The Contractor shall not be required to employ any Subcontractor to whom he has a reasonable objection. Contracts between the Contractor and the Subcontractor shall be in accordance with the terms of this Agreement and shall include the General Conditions of this Agreement insofar as applicable.

ARTICLE 12
SEPARATE CONTRACTS

12.1 The Owner reserves the right to award other contracts in connection with other portions of the Project or other work on the site under these or similar Conditions of the Contract.

12.2 The Contractor shall afford other contractors reasonable opportunity for the introduction and storage of their materials and equipment and the execution of their work, and shall properly connect and coordinate his Work with theirs.

12.3 Any costs caused by defective or ill-timed work shall be borne by the party responsible therefor.

ARTICLE 13
ROYALTIES AND PATENTS

The Contractor shall pay all royalties and license fees. The Contractor shall defend all suits or claims for infringement of any patent rights and shall save the Owner harmless from loss on account thereof.

ARTICLE 14
ARBITRATION

All claims or disputes arising out of this Contract or the breach thereof shall be decided by arbitration in accordance with the Construction Industry Arbitration Rules of the American Arbitration Association then obtaining unless the parties mutually agree otherwise. Notice of the demand for arbitration shall be filed in writing with the other party to the Contract and with the American Arbitration Association and shall be made within a reasonable time after the dispute has arisen.

ARTICLE 15
TIME

15.1 All time limits stated in the Contract Documents are of the essence of the Contract.

15.2 If the Contractor is delayed at any time in the progress of the Work by changes ordered in the Work, by labor disputes, fire, unusual delay in transportation, unavoidable casualties, causes beyond the Contractor's control, or by any cause which the Architect may determine justifies the delay, then the Contract Time shall be extended by Change Order for such reasonable time as the Architect may determine.

ARTICLE 16
PAYMENTS

16.1   Payments shall be made as provided in Article 4 of this Agreement.

16.2   Payments may be withheld on account of (1) defective Work not remedied, (2) claims filed, (3) failure of the Contractor to make payments properly to Subcontractors or for labor, materials, or equipment, (4) damage to another contractor, or (5) unsatisfactory prosecution of the Work by the Contractor.

16.3   Final payment shall not be due until the Contractor has delivered to the Owner a complete release of all liens arising out of this Contract or receipts in full covering all labor, materials and equipment for which a lien could be filed, or a bond satisfactory to the Owner indemnifying him against any lien.

16.4   The making of final payment shall constitute a waiver of all claims by the Owner except those arising from (1) unsettled liens, (2) faulty or defective Work appearing after Substantial Completion, (3) failure of the Work to comply with the requirements of the Contract Documents, or (4) terms of any special guarantees required by the Contract Documents. The acceptance of final payment shall constitute a waiver of all claims by the Contractor except those previously made in writing and still unsettled.

ARTICLE 17
PROTECTION OF PERSONS AND PROPERTY

The Contractor shall be responsible for initiating, maintaining, and supervising all safety precautions and programs in connection with the Work. He shall take all reasonable precautions for the safety of, and shall provide all reasonable protection to prevent damage, injury or loss to (1) all employees on the Work and other persons who may be affected thereby, (2) all the Work and all materials and equipment to be incorporated therein, and (3) other property at the site or adjacent thereto. He shall comply with all applicable laws, ordinances, rules, regulations and orders of any public authority having jurisdiction for the safety of persons or property or to protect them from damage, injury or loss. All damage or loss to any property caused in whole or in part by the Contractor, any Subcontractor, any Sub-subcontractor or anyone directly or indirectly employed by any of them, or by anyone for whose acts any of them may be liable, shall be remedied by the Contractor, except damage or loss attributable to faulty Drawings or Specifications or to the acts or omissions of the Owner or Architect or anyone employed by either of them or for whose acts either of them may be liable but which are not attributable to the fault or negligence of the Contractor.

ARTICLE 18
CONTRACTOR'S LIABILITY INSURANCE

The Contractor and each separate Contractor shall purchase and maintain such insurance as will protect him from claims under workmen's compensation acts and other employee benefit acts, from claims for damages because of bodily injury, including death, and from claims for damages to property which may arise out of or result from the Contractor's operations under this Contract, whether such operations be by himself or by any Subcontractor or anyone directly or indirectly employed by any of them. This insurance shall be written for not less than any limits of liability specified as part of this Contract, or required by law, whichever is the greater, and shall include contractual liability insurance as applicable to the Contractor's obligations under Paragraph 10.10. Certificates of such insurance shall be filed with the Owner and each separate Contractor.

ARTICLE 19
OWNER'S LIABILITY INSURANCE

The Owner shall be responsible for purchasing and maintaining his own liability insurance and, at his option, may maintain such insurance as will protect him against claims which may arise from operations under the Contract.

ARTICLE 20
PROPERTY INSURANCE

20.1   Unless otherwise provided, the Owner shall purchase and maintain property insurance upon the entire Work at the site to the full insurable value thereof. This insurance shall include the interests of the Owner, the Contractor, Subcontractors and Sub-subcontractors in the Work and shall insure against the perils of Fire, Extended Coverage, Vandalism and Malicious Mischief.

20.2   Any insured loss is to be adjusted with the Owner and made payable to the Owner as trustee for the insureds, as their interests may appear, subject to the requirements of any mortgagee clause.

20.3   The Owner shall file a copy of all policies with the Contractor prior to the commencement of the Work.

20.4   The Owner and Contractor waive all rights against each other for damages caused by fire or other perils to the extent covered by insurance provided under this paragraph. The Contractor shall require similar waivers by Subcontractors and Sub-subcontractors.

ARTICLE 21
CHANGES IN THE WORK

21.1   The Owner without invalidating the Contract may order Changes in the Work consisting of additions, deletions, or modifications, the Contract Sum and the Contract Time being adjusted accordingly. All such Changes in the Work shall be authorized by written Change Order signed by the Owner or the Architect as his duly authorized agent.

21.2   The Contract Sum and the Contract Time may be changed only by Change Order.

21.3   The cost or credit to the Owner from a Change in the Work shall be determined by mutual agreement.

ARTICLE 22
CORRECTION OF WORK

The Contractor shall correct any Work that fails to conform to the requirements of the Contract Documents where such failure to conform appears during the progress of the Work, and shall remedy any defects due to faulty materials, equipment or workmanship which appear within a period of one year from the Date of Substantial Completion of the Contract or within such longer period of time as may be prescribed by law or by the terms of any applicable special guarantee required by the Contract Documents. The provisions of this Article 22 apply to Work done by Subcontractors as well as to Work done by direct employees of the Contractor.

ARTICLE 23
TERMINATION BY THE CONTRACTOR

If the Architect fails to issue a Certificate of Payment for a period of thirty days through no fault of the Contractor, or if the Owner fails to make payment thereon for a period of thirty days, the Contractor may, upon seven days' written notice to the Owner and the Architect, terminate the Contract and recover from the Owner payment for all Work executed and for any proven loss sustained upon any materials, equipment, tools, and construction equipment and machinery, including reasonable profit and damages.

ARTICLE 24
TERMINATION BY THE OWNER

If the Contractor defaults or neglects to carry out the Work in accordance with the Contract Documents or fails to perform any provision of the Contract, the Owner may, after seven days' written notice to the Contractor and without prejudice to any other remedy he may have, make good such deficiencies and may deduct the cost thereof from the payment then or thereafter due the Contractor or, at his option, may terminate the Contract and take possession of the site and of all materials, equipment, tools, and construction equipment and machinery thereon owned by the Contractor and may finish the Work by whatever method he may deem expedient, and if the unpaid balance of the Contract Sum exceeds the expense of finishing the Work, such excess shall be paid to the Contractor, but if such expense exceeds such unpaid balance, the Contractor shall pay the difference to the Owner.

ARTICLE 25
MISCELLANEOUS PROVISIONS

AIA DOCUMENT A107 • SMALL CONSTRUCTION CONTRACT • JANUARY 1974 EDITION • AIA® • ©1974
THE AMERICAN INSTITUTE OF ARCHITECTS, 1735 NEW YORK AVE., N.W., WASHINGTON, D. C. 20006

This Agreement executed the day and year first written above.

OWNER _____

CONTRACTOR _____

_____

_____

General Construction, Heating and
Plumbing, Electrical and Painting

S P E C I F I C A T I O N S

for the

Construction

of a

Residence

for

Mr. & Mrs. _____

_____

_____

O W N E R

OFFICE OF

_____

ARCHITECT

_____

JOB NO. _____ SET NO. _____

Seal

Date: _____ 19_____

NOTE: Proposals shall be submitted in duplicate.

BID SHEET

for

GENERAL CONTRACT

_____
(Name of Contractor)

TO: _____

_____   Date _____

Dear Sir:

Having carefully examined the "Instruction to Bidders," the "General Conditions of the Contract" and the Specifications entitled "A residence for Mr. & Mrs. _____ of _____ prepared by the office of _____ Architect, and the drawings similarly entitled, as well as the premises and conditions affecting the work, the Undersigned proposes to furnish all materials and labor called for by them for the "General Contract" including General Construction, Heating, Plumbing, Electrical work, and Painting in accordance with said documents as issued by the architect and mailed to the Undersigned prior to the date of receipt of bids, for the sum of

_____

_____ $_____

as follows:

General Construction work as per plans and specifications governing the same for the sum of

_____

_____ $_____

Heating work as per plans and specifications governing the same for the sum of

_____

_____ $_____

Name of Heating Contractor

_____

Plumbing work as per plans and specifications governing same for the sum of

_____

_____ $_____

Name of Plumbing Subcontractor

_____

Electrical work as per plans and specifications governing same for the sum of

_____

_____ $_____

Name of Electrical Subcontractor

_____

Painting work as per plans and specifications governing same for the sum of

_____

_____ $_____

Name of Painting Subcontractor

_____

UNIT PRICES

Should the amount of excavation, concrete, etc., required be increased or decreased due to special conditions found at the site, the Undersigned agrees that the following supplemental unit prices will be the basis of his compensation for addition or deduction as the case may be, for such increases or decrease in the work.

General earth excavation—machine—including backfill or disposal	$_____	per cubic yard.
Earth excavation—hand—including backfill or disposal	$_____	per cubic yard.
Concrete in place	$_____	per cubic yard.
Rock Excavation	$_____	per cubic yard.

The above proposal is subject to acceptance within thirty (30) days after the time set for the opening of bids.

Very truly yours,

_____

_____

Specifications

DIVISION 1: GENERAL REQUIREMENTS

A. **ARCHITECT'S SUPERVISION**

The architect will have continual supervisory responsibility for this job.

B. **TEMPORARY CONVENIENCES**

The general contractor shall provide suitable temporary conveniences for the use of all workers on this job. Facilities shall be within a weather tight, painted enclosure complying with legal requirements. The general contractor shall maintain all temporary toilet facilities in a sanitary condition.

C. **PUMPING**

The general contractor shall keep the excavation and the basement free from water at all times and shall provide, maintain, and operate at his own expense such pumping equipment as shall be necessary.

D. **PROTECTION**

The general contractor shall protect all existing driveways, parking areas, sidewalks, curbs, and existing paved areas on, or adjacent to, the owner's property.

E. **GRADES, LINES, LEVELS, AND SURVEYS**

The owner shall establish the lot lines.

The general contractor shall:

1. Establish and maintain bench marks.
2. Verify all grades, lines, levels, and dimensions as shown on the drawings, and report any errors or inconsistencies before commencing work.
3. Lay out the building accurately under the supervision of the architect.

F. **FINAL CLEANING**

In addition to the general room cleaning, the general contractor shall do the following special cleaning upon completion of the work:

1. Wash and polish all glass and cabinets.
2. Clean and polish all hardware.
3. Remove all marks, stains, fingerprints, and other soil or dirt from walls, woodwork, and floors.

G. GUARANTEES

The general contractor shall guarantee all work performed under the contract against faulty materials or workmanship. The guarantee shall be in writing with duplicate copies delivered to the architect. In case of work performed by subcontractors where guarantees are required, the general contractor shall secure written guarantees from those subcontractors. Copies of these guarantees shall be delivered to the architect upon completion of the work. Guarantees shall be signed by both the subcontractor and the general contractor.

H. FOREMAN

The general contractor shall have a responsible foreman at the building site from the start to the completion of construction. The foreman shall be on duty during all working hours.

I. FIRE INSURANCE

The owner shall effect and maintain builder's risk completed-value insurance on this job.

DIVISION 2: SITE WORK

A. Work Included

This work shall include, but not be limited by the following:
1. Cleaning site.
2. Excavation, backfilling, grading, and related items.
3. Removal of excess earth.
4. Protection of existing trees to remain on the site.

All excavation and backfilling required for heating, plumbing, and electrical work will be done by the respective contractors and are not included under site work.

B. Cleaning the Site

1. Clean the area within the limits of the building of all trees, shrubs, or other obstructions as necessary.
2. Within the limits of grading work as shown on the drawings, remove such trees, shrubs, or other obstructions as are indicated on the drawings to be removed, without injury to trunks, interfering branches, and roots of trees to remain. Do cutting and trimming only as directed. Box and protect all trees and shrubs in the construction area to remain; maintain boxing until finished grading is completed.
3. Remove all debris from the site; do not use it for fill.

C. Excavation

1. Carefully remove all sod and soil throughout the area of the building and where finish grade levels are changed. Pile on site where directed. This soil is to be used later for finished grading. Do not strip below topsoil.
2. Do all excavation required for footings, piers, walls, trenches, areas, pits, and foundations. Remove all materials encountered in obtaining indicated lines and grades required. Beds for all foundations and footings must have solid, level, and undisturbed bed bottoms. No backfilling will be allowed and all footings shall rest on unexcavated earth.
3. The general contractor shall notify the architect when the excavation is completed so that he or she may inspect the soil before concrete is placed.
4. Excavate to elevations and dimensions indicated, leaving sufficient space to permit erection of walls, drain tile, waterproofing, masonry, and the inspection of foundations. Protect the bottom of the excavation from frost. Do not place the foundation, footings, or slabs on frozen ground.

D. Backfill

1. All outside walls shall be backfilled to within 6 inches of the finished grade with clean fill. Backfill shall be thoroughly puddled and tamped solid.
2. Backfill under basement floor slabs and elsewhere as required to bring the earth to proper levels and grades for subsequent work. Use only earth without rubbish. All fill shall be well tamped and puddled to prevent settling.
3. Unless otherwise directed by the architect, no backfill shall be placed until all walls have developed such strength to resist thrust due to filling operations.

E. Interior Grading
    1. Furnish and place graded gravel or bank run gravel fill as approved under all basement floor slabs.
    2. Fill under all slabs shall be sand not less than 6 inches deep. Where existing grades are lower than 6 inches below the bottom of the slabs, sand shall be used for the entire depth.
    3. All fill shall be shaped to line and grade and brought to proper elevations without voids and depressions to receive the concrete slabs.

F. Exterior Grading
See Plot Plan for area limits under contract.
    1. Do all excavating, filling and rough grading to bring entire area outside of the building to levels shown on the Plot Plan and at the building as shown on the drawings.
    2. Where existing trees are to remain, if the new grade is lower than the natural grade under the trees, a sloping mound shall be left under the base of the tree extending out as far as the branches; if the grade is higher, a dry well shall be constructed around the base of the tree to provide the roots with air and moisture. Sand and, or gravel, removed from the excavation for the building may be used for grading.
    3. After rough grading has been completed and approved, spread the topsoil evenly to the previously stripped area. Prepare the topsoil to receive grass seed by removing stones, debris, and unsuitable materials. Hand rake to remove water pockets and irregularities.
    4. Seeding will be done by the owner.

DIVISION 3: CONCRETE

A. Work Included
This work shall include but shall not be limited by the following:
    1. Concrete footing for walls, supporting posts, and fireplace.
    2. Basement concrete floors.
    3. Porch entrance floors.
    4. Sidewalks.

B. Concrete Proportions
    1. Concrete for footings shall develop an ultimate 28-day compressive strength of 3000 pounds per square inch.
    2. Concrete for floors and walks shall develop an ultimate 28-day compressive strength of 2500 pounds per square inch.

C. Concrete Footings
    1. Notify the architect for inspection of footing beds before any concrete is placed.
    2. All footings shall be of sizes indicated on the drawings.
    3. All footings shall start at the elevations indicated on the drawings and at least 3'0" below finished grade. Where excavation has been carried below required level, start footings from the bottom of the excavation.

D. Mixing Concrete
    1. All concrete shall be mixed in an approved-type power batch mixer equipped with a water measuring device.
    2. For freezing weather: sand, stone, and water for concrete shall be properly heated before using.

E. Concrete Finish
Concrete shall be finished as follows:
    1. Basement floor — smooth trowel finish
    2. Entrance porches — float finish with steel trowel
    3. Sidewalks — wood float finish
    4. Slabs on fill — After the concrete footings have been placed, remove the forms and backfill to proper level to receive the floor construction.
Under all floor areas place a 6-inch layer of graded gravel fill; puddle, tamp, and level properly to a reasonable true, even surface.

DIVISION 4: MASONRY

A. Work Included
This work shall include but shall not be limited by the following:
1. Brickwork.
2. Concrete block work.
3. Mortar for brick and block work.

B. Masonry Materials
1. Delivery and Storage
All materials shall be delivered, stored, and handled so as to prevent the inclusion of foreign materials and the damage of the materials by water or breakage. Packaged materials shall be delivered and stored in the original packages until they are ready for use.
2. Materials showing evidence of water or other damage shall be rejected.

C. Brick and Block Work
1. The brick shall be chosen by the owner from samples provided by the contractor. Brick is to be used on the fireplace face, exterior face of the fireplace chimney, and veneer work on the front as shown on drawings.
Common brick shall be hard-burned and uniform in size. No soft brick will be allowed. The brick shall be reasonably free from cracks, pebbles, particles of lime, and other substances which will affect their serviceability or strength. The brick shall be carefully protected during transportation and shall be unloaded by hand; dumping is not permitted.
2. Mortar used for laying brick and concrete block shall consist of one (1) part portland cement, one (1) part hydrated mason's lime, and six (6) parts washed sand. The mortar ingredients shall comply with the following requirements:
a. Aggregates: ASTM C-144-52T.
b. Water: Clean, fresh, free from acid, alkali, sewage, or organic material.
c. Portland Cement: ASTM C-150-55 Type I.
d. Masonry Cement: ASTM C-91-55T Type 2.
e. Hydrated Lime: ASTM C-207-49T Type S.
3. Concrete block shall be load-bearing, hollow, concrete masonry units and shall conform to the standard specifications of ASTM C-90.
4. Wall reinforcing: Reinforcing material for masonry walls shall be prefabricated welded steel.
5. Installation: Erect all block and brickwork in accordance with the following:
All work shall be laid true to dimensions, plumb, square, and in bond, and properly anchored. All courses shall be level and joints shall be uniform width; no joints shall exceed the size specified.
Joints shall be finished as follows: All brick shall be laid on a full mortar bed with a shoved joint. All joints shall be completely filled with mortar. All horizontal and vertical joints shall be raked $\frac{3}{8}$ inch deep.
All mortar joints for concrete block masonry shall have a full mortar coverage on vertical and horizontal face shells. Vertical joints shall be shoved tight. Full mortar bedding shall have ruled joints.
Horizontal reinforcing shall be placed in every other bed joint of block work. Reinforcement shall be placed in the first and second bed joints above and below all openings. Concealed work shall have cut flush joints.
Protection: Cover the wall each night and when the work is discontinued due to rain or snow. Do not work during freezing weather without permission from the architect.

D. Cleaning and Pointing
1. Concrete Block: Point up all the voids and openings in joints with mortar. Remove all excess mortar and dirty spots from the surface.
2. Brickwork: Upon completion, all brickwork shall be thoroughly cleaned with clean water and stiff fiber brushes and then rinsed with clear water. The use of acids or wire brushes is not permitted.

DIVISION 5: METALS

A. Work Included
The work shall include but shall not be limited by the following:
1. Steel basement columns shall be 4-inch diameter with $\frac{3}{8}$-inch X 4-inch X 6-inch welded cap and base plates. Location and length are shown on plans.
2. Basement windows shall be 15-inch X 12-inch, two light, glazed, with combination storm sash and screens.
3. Galvanized areaways shall be 20-gauge galvanized steel, with horizontal corrugations, as shown on drawings. All areaways shall be anchored to foundation wall. Furnish 6 inches of crushed stone below the areaway to a level 3 inches below the bottom of the basement window sill.
4. Provide wrought iron hand rails for front and side entrance porches as shown on plans.

DIVISION 6: CARPENTRY AND MILLWORK

A. Materials
1. All materials are to the best of their respective kind. Lumber shall bear the mark and grade of the association under whose rules it is produced. Framing lumber shall be thoroughly seasoned with a maximum moisture content of 19 percent. All millwork shall be kiln dried.
2. Properly protect all materials. All lumber shall be kept under cover at the job site. Material shall not be delivered unduly long before it is required for work.
3. Lumber for various uses shall be as follows:
Framing: No. 2 dimension, Douglas fir or yellow pine.
Exterior Millwork: No. 1 clear white pine.
The lumber must be sound, thoroughly seasoned, well manufactured and free from warp which cannot be corrected by bridging or nailing. All woodwork which is exposed to view shall be S4S.

B. Installation
1. All work shall be done by skilled workers.
2. All work shall be erected plumb, true, square, and in accordance with the drawings.
3. Finish work shall be blind nailed as much as possible, and surface nails shall be set.
4. All work shall be securely nailed to studs, nailing blocks, grounds, furring, and nailing strips.

C. Grades and Species of Lumber
1. Framing lumber, except studs and wall plates, shall be No. 1 Douglas fir.
2. Studs, shoes, and double wall plates shall be Douglas fir, utility grade.
3. Bridging shall be 1 X 3 spruce.
4. Joists shall be 2 x 8s spaced 16 inches o.c. except where otherwise indicated. All joists are to be doubled under partitions and around stairways and fireplace openings.
5. Subflooring shall be $\frac{5}{8}$-inch plywood APA grade C-D.
6. Ceiling joists and rafters shall be 2 x 6s spaced 16 inches o.c. Rafters are to have overhang as noted on drawings. Rafters shall have 1 x 8 collar beams 32 inches o.c.
7. Wall sheathing and roof sheathing shall be $\frac{1}{2}''$ x 4' x 8' plywood APA grade C-D exterior.
8. Exterior siding shall be $\frac{1}{2}''$ x 8'' bevel siding with 6-inch exposure.
9. Cornice material: Soffit, 1 x 8 clear pine; fascia, 1 x 6 B or better pine; rake, 5/4'' x 6''; fascia on rake, 1 x 3; bedmold, use 1 x 2 B or better pine.
10. Interior woodwork shall be kiln-dried clear pine. All interior woodwork is to be machine sanded at the mill and hand sanded on the job.
11. Hardwood flooring shall be 1 x 3 select oak strip flooring.
12. Underlayment shall be $\frac{5}{8}''$ x 4' x 8' plywood APA underlayment.
13. Backing for ceramic tile in tub areas shall be $\frac{1}{2}''$ x 4' x 8' plywood APA grade A-C exterior.
14. Cabinet and vanity countertops shall be $\frac{1}{16}$-inch standard grade, laminated plastic surfacing material conforming to NEMA standards. Color to be selected by the owner.

D. Workmanship

1. Framing: All framing members shall be substantially and accurately fitted together, well secured, braced, and nailed. Plates and sills shall be halved together at all corners and splices. Studs in walls and partitions shall be doubled at all corners and openings. Joists over 8 feet in span shall be bridged with one row of 1 x 3 spruce cross bridging, cut on bevel and nailed up tight after the subfloor has been installed.

2. Interior trim and millwork: All exposed millwork shall be machine sanded to a smooth finish, with all joints tight and formed to conceal any shrinkage. Miter exterior angles, butt and cope interior angles, and scarf all running joints in moldings.

3. Stairwork shall be as follows:

a. Risers: $\frac{3}{4}$-inch oak.

b. Stringers: Clear pine housed for risers and treads.

c. Treads: Clear oak, $1\frac{1}{16}''$ x 10" x 3'0".

d. Balusters: Morgan M-836.

e. Sub-Rail: Morgan M-834.

f. Fillet: Morgan M-835.

g. Handrail: Birch M-720.

All stairs shall be housed with oak risers and oak treads. Nosing shall have $\frac{3}{4}''$ stair cove molding under each tread overhang. All treads and risers to be wedged and glued in place. Birch handrail shall be continuous from floor to floor and shall be fastened with handrail brackets on solid walls. Open stairs and second floor handrail return shall be Morgan "Abbington" stairway M-844 style.

4. Closet rods: Furnish and install where indicated on drawings. Rods are to be adjustable chrome, fitted, and supported at least every 4 feet.

E. Cleanup

Upon completion of work, all surplus and waste materials shall be removed from the building, and the entire structure and involved portions of the site shall be left in a neat, clean, and acceptable condition.

DIVISION 7: MOISTURE PROTECTION

A. Work Included

This work shall include but shall not be limited by the following:

1. Damp-proofing of basement walls below grade.

2. Aspahlt shingle roofing, 235 pounds per square.

3. Aluminum flashing and galvanized drip edge.

4. Insulation: Ceilings of second floor 6-inch fiberglass; all side walls $3\frac{1}{2}$-inch fiberglass.

B. 1. Damp-proofing basement walls below grade: All concrete block walls below grade shall be parged with $\frac{1}{2}$-inch thick layer of portland cement plaster. Use a mix of 1 part cement, 1 part hydrated lime, and 6 parts sand. After the wall is completely cured, mop two coats of asphalt coating. Cant the plaster over the edge of the footings as shown on drawing details. All concrete block walls above grade shall be plastered with $\frac{1}{2}$-inch thick layer of masonry cement plaster. Use a mix of 1 bag masonry cement, 2 shovels of portland cement, and 18 shovels of sand. Leave with a smooth wood float finish.

2. Drain tile, crushed stone, and sump pit: The contractor shall install a perforated plastic drain pipe around the perimeter of the footings fully embedded in a trench of No. 2 crushed stone and sloping $\frac{1}{8}$ inch to the foot. The drain shall run completely around the perimeter of the building and then into a sump pit. The drain pipe and crushed stone are to be installed completely free of any loose sand or earth.

3. Insulation: Fit each batt snugly and securely between joists with vapor barrier toward the room side. Cut batts to fit irregular spaces, leaving flange on inside for stapling. In removing fiberglass to fit around pipes and other openings, do not cut vapor barrier unless necessary.

4. Roofing: Roofing shall be Johns-Manville Asphalt saturated mineral surface 235-pound strip shingles, applied in accordance with the manufacturers specifications.

5. Flashing: Flashing shall be 24-gauge, best commercial grade aluminum 28 inches wide, and shall be used for all valleys, chimneys, and general flashing. Galvanized iron drip edge shall be used on rake and eves under the asphalt shingles.

DIVISION 8: DOORS, WINDOWS, AND GLASS

A. Work Included
This work shall include but shall not be limited by the following:
1. Wood doors and wood door frames.
2. Hanging and fitting of doors.
3. All window units.
4. All window and door trim.
5. Mirrors in the bathrooms over the vanities.

B. Wood Doors
Exterior doors: Doors shall be as follows:
Front door – 3'0" x 6'8" x $1\frac{3}{4}$" Morgan Design M-108.
Kitchen Door – 2'8" x 6'8" x $1\frac{3}{4}$" Iroquois Design #233.
Interior Doors: All interior doors are to be birch flush except where shown on plans. Sizes and thicknesses as shown on drawings.

C. Wood Door Frames
All wood door frames shall be clear pine, dadoed together at the head with joints set with white glue. All frames shall be accurately set, plumb, level, and true. Hinge locations shall be solidly shimmed.

D. Hanging and Fitting of Doors
All doors shall be accurately cut, trimmed, and fitted to their frames. All doors shall operate freely without binding, and all hardware shall be adjusted properly. Exterior doors shall be fitted with weather stripping.

E. Interior Door and Window Trim
All interior door and window trim shall be clear pine, machine sanded. Casing to be $\frac{11}{16}$" x $2\frac{1}{4}$", base to be $\frac{9}{16}$" x 3". Door and window trim shall be clamshell design.

F. Wood Windows
All window units shall be double hung with removable grilles. Sizes and styles shall be as shown on the drawings. Wood combination storm and screen units will be furnished for each window.

G. Bathroom Mirrors
Mirrors shall be $\frac{1}{4}$-inch silver plated glass with copper backing, of first quality, as shown on drawings. Mirrors shall be neatly installed directly on wall with mastic and carefully fitted to the top of the vanity. Mirrors shall be the same length as the laminated vanity top and shall be 36 inches high.

DIVISION 9: FINISHES

A. Work Included
This work shall include but shall not be limited by the following:
1. Gypsum wallboard.
2. Hardwood strip flooring.
3. Ceramic and quarry tile.
4. Resilient flooring.
5. Painting.

B. Gypsum Wallboard
Gypsum wallboard material shall be $\frac{3}{8}$ inch thick. Gypsum wallboard shall be nailed to wood framing in strict accordance with manufacturer's recommendation. Space nails not more than 7 inches apart on ceilings and not more than 8 inches apart on sidewalls. Dimple the nailheads slightly below the surface of the wallboard, taking care not to break the paper surface.
Joint Finishing: After gypsum wallboard has been installed, finish all joints and nail heads using the Perf-A-Tape joint system as recommended by the US Gypsum Company. Perf-A-Tape system to be installed according to the manufacturer's recommendations.

C. Hardwood Flooring

All subfloors are to be broom cleaned and covered with deadening felt before the finished floor is laid. Wood flooring, where scheduled, is to be 1 x 3 tongue-in-groove and end-matched select oak flooring. Flooring is to be laid evenly and blind nailed or stapled every 16-inches without tool marks.

D. Ceramic Tile

All ceramic tile shall be standard grade of approved quality. The contractor shall submit samples of all ceramic and quarry tile to the owner for approval as to color and texture.
1. Wall tile shall be matte faced $4\frac{1}{4}''$ x $4\frac{1}{4}''$ in size. Setting shall be by conventional tile adhesive over waterproof plywood. All cove, bullnose, and trim pieces shall be furnished and neatly finished to drywall construction according to the drawings.
2. Floor tile shall be unglazed, 1" x 1" smooth mosaic tile, of a similar color to the wall tile.
3. Accessories shall be vitreous china to match the wall tile. The second floor baths are to have a soap dish and grab bar, and one grab bar in the tub area.
4. Marble thresholds shall be installed the same width as the bathroom door opening. Thresholds shall be of approved quality, $\frac{7}{8}$ inch thick and $3\frac{1}{2}$ inches wide.
5. Installation: Lay out ceramic tiles on floors and walls so that no tiles less than one-half size occur. Maintain full courses to produce nearest obtainable heights as shown on drawings without cutting the tile. Align joints in the wall and trim vertically and horizontally without staggered joints.

E. Quarry Tile
1. Quarry tile for the fireplace hearth shall be 4" x 4" x $\frac{1}{2}''$ standard red.
2. Set quarry tile on approved mortar bed with $\frac{1}{4}$-inch aligned joints in both directions. Mortar bed on cement base shall bring the tile surface to the level of the finished floors. Joints shall be tooled until smooth and flush with the surface of the tile.

F. Slate Flooring
1. The entrance hall shall be $\frac{3}{8}$-inch thick slate flooring and shall be laid in a random pattern.
2. Slate shall be set in conventional slate adhesive with $\frac{3}{8}$-inch joints. Slate flooring shall be placed in both the entrance hall and the closet adjacent to the half bath.
3. Slate joints shall be grouted with white cement grout. Joints shall be tooled until smooth and flush with the surface of the slate.

G. Resilient Flooring
1. The floors in the kitchen, and half bath shall be inlaid linoleum, Armstrong or equal, color and style to be selected by the owner.
2. Finished resilient flooring materials shall be installed with adhesives as recommended by the manufacturer.

H. Painting
1. All exterior paint shall be nonchalking and shall be guaranteed not to stain or otherwise discolor any adjacent work. All materials shall be used only as specified by the manufacturer. All knots and sap spots shall be treated with two coats of pure white shellac where paint or enamel is to be applied.
2. Schedule of Painting
    a. Priming: Prime all exterior work which has not been primed by the manufacturer. Prime exterior trim with an approved trim primer before installation. All primer is to be of the same manufacturer as that of the finished material which is to be applied.
    b. Exterior woodwork: Paint all exterior doors, windows, and trim with two coats of approved trim and shutter paint. Paint all other exterior woodwork with two coats of an approved nonchalking, white house paint.
    c. Interior woodwork: Paint all interior woodwork with one coat of enamel undercoating and one coat of semigloss enamel.
    d. Walls and ceilings: All walls and ceilings, except in the bathrooms and kitchen, shall be painted with two coats of flat wall paint. The walls and ceilings in the bathrooms and kitchen shall be painted with two coats of semigloss wallpaint.

DIVISION 10: SPECIALITIES

A. Work Included

This work shall include but shall not be limited by the following:

1. Fireplace.
2. Finish hardware.
3. Kitchen cabinets.
4. Bathroom vanities.
5. Kitchen appliances.

B. Fireplace

Contractor shall furnish and install a fireplace as shown on the drawings. Fireplace face shall be antique common brick. Chimney shall be faced with antique common brick. Hearth shall be quarry tile, size $4'' \times 4'' \times \frac{1}{2}''$.

1. Fireplace unit shall be #35 Heatilator form as manufactured by Vega Industries Inc. Unit to have poker control damper.
2. Fireplace mantel shall be M-1448 as manufactured by Morgan Millwork Company, Oshkosh, Wisconsin or approved equal.

C. Louvers and Vents

Contractor shall furnish and install three (3) wood louvers with aluminum wire screen backing. Wood louvers to be #600-L, $1'6'' \times 2'0''$ as manufactured by Iroquois Millwork Company, Albany, New York or approved equal.

D. Finish Hardware

The contractor shall allow in the proposal the sum of five hundred dollars ($500.00) for the purchase of all hardware. This allowance covers the net cost to the contractor and does not include any labor, overhead, or profit. Hardware shall be selected by the owner but will be purchased by the contractor where directed. The net difference in cost, if any, shall be added to or deducted from the contract as the case may be. Hardware shall be left in perfect working order, upon completion. All keys shall be delivered to the authorized representative of the owner.

E. Cabinets

1. Kitchen cabinets shall be Kingswood Oakmont, as manufactured by the B.J. Sutherland Company of Louisville, Kentucky, or equal. Sizes and styles are to be as shown on the special detail drawings.
2. Bathroom vanities shall be RV-30 and RV-36 Moonlight, as manufactured by the B.J. Sutherland Company of Louisville, Kentucky, or equal.
3. Kitchen cabinet and bathroom vanity countertops shall be $\frac{1}{16}$-inch laminated plastic bonded to $\frac{3}{4}$-inch plywood. Countertops shall be of one-piece molded construction, with 4-inch backsplash and no back seams. Color and pattern to be selected by owner.

F. Appliances

Contractor shall provide proper spaces for and do any cutting necessary to accommodate the appliances purchased by the owner.

1. Owner will purchase and dealer install one (1) refrigerator and one (1) kitchen range. Style, model and color to be selected by owner.

DIVISION 11: MECHANICAL

PLUMBING

A. Work Included

Plumbing work includes, but shall not be limited by the following:

1. Complete systems of waste drainage and vent piping, including all connections.
2. Complete systems of hot and cold water piping and all connections.
3. Furnishing, delivering, installing, and setting of all plumbing fixtures.
4. All excavation and backfilling for the plumbing.

B. General Requirements

All work shall be in accordance with local, state or federal requirements, and shall comply with all applicable codes. The contractor shall obtain and pay for all permits and certificates of inspection if any are required for the plumbing.

C. Sanitary Drainage System

1. Sanitary drainage piping, buried, shall be extra-heavy asphalt-coated hub and spigot soil pipe, of uniform weight and thickness, with corresponding cast iron soil pipe fittings. Above-ground piping shall be copper pipe with corresponding fittings.

2. Waste piping shall be uniformly pitched in the direction of the flow. Main drains shall be pitched approximately $\frac{1}{8}$ inch per foot, with branch laterals pitched $\frac{1}{4}$ inch per foot where possible.

3. Cleanouts shall be provided and installed as indicated.

D. Vent Piping

1. Vent piping shall be installed, connected to fixtures and drainage piping indicated.

2. Vents shall be extended through the roof to a height of approximately 12 inches above the roof deck. Vents shall be provided with approved flashing hub or cap.

E. Interior Water Piping

1. All inside water piping shall be of standard drawn Type L copper except that exposed branches and exposed piping to fixtures shall be chrome brass.

2. All joints in copper tubing shall be sweat type. Soldered joints shall be thoroughly cleaned with steel wool and approved, oil-free, soldering paste.

3. All water piping shall be properly and adequately supported on hangers spaced not over eight (8) feet apart.

F. Fixtures

Kohler plumbing fixtures have been used as a guide. American Standard, Crane, or approved equal will be accepted. Colors will be selected by the owner.

1. Bath tubs shall be Guardian K-786-SA recessed tub 5'0'' long, 1'4'' high. K-7004-T Dalton shower and bath faucet with valet units. K-7370 shower head with volume regulator. K-7150-R $1\frac{1}{2}$-inch pop-up drain.

2. Water closets shall be Wellington K-3512-PBA siphon jet with K-4662 Lustra seat and cover. K-7638 supply with stop.

3. Lavatories shall be Brookline K-2208-F round lavatory with K-7436-T Bancroft supply faucet with Valet units, aerator, and pop-up drain. K-7606 $\frac{3}{8}$-inch supplies with stops. K-9000 $1\frac{1}{4}''$ cast brass P trap.

4. Kitchen sink shall be Harvest, double bowl 21 x 32, Model SA-410; stainless steel with conventional trim. K-7665-T Edgewater sink supply faucet.

5. Water heater shall be Wagoner brand, glass-lined standard electric hot water heater, forty gallon, Model WG-40 as manufactured by Wagoner Corporation, Nashville, Tenn.

HEATING

A. Work Included

This work shall include but shall not be limited by the following:

1. Complete gas-fired hot air heating system.

2. Automatic controls.

3. All cutting and patching for installation.

B. General Requirements

All work shall be in accordance with local, state or federal requirements, and shall comply with all applicable codes. The contractor shall obtain and pay for all permits and certificates of inspection if any are required for the heating work.

C. Heating Unit

Furnish a Carrier upflow gas-fired furnace complete with fan and limit control. Furnace and burner shall be Carrier 58SE Model 125, Series 304, or equal. Furnace shall have an AGA input rating of 125,000 Btuh, and an output rating of 100,000 Btuh. Air delivery shall be 1550 cfm minimum with 0.5 in. wg (water glass) External Static Pressure. Blower and blower motor shall be centrifugal type, statically and dynamically balanced. Motor shall have factory-lubricated bearings, and shall be $\frac{1}{2}$ HP. Furnace shall have filter rack with permanent-type filter.

Controls: Heating contractor shall provide a complete system of automatic controls as recommended by the equipment manufacturer. Heating system control shall be Minneapolis-Honeywell T-861A clock thermostat.

DIVISION 12: ELECTRICAL

A. Work Included

In general, the electrical work includes, but shall not be limited by the following:

1. Service entrance.
2. Service panel.
3. Wiring.
4. Fixture allowance.
5. Special outlets.
6. All cutting and patching for installation of the work.

B. General Requirements

The complete installation shall be made in a neat, workmanlike manner in conformance with best modern trade practices, by competent, experienced mechanics and to the full satisfaction and approval of the architect. All work shall be in accordance with local, state, or federal requirements, and shall comply with all applicable codes.

C. Guarantee

The contractor guarantees that:

1. All work executed under this contract will be free from defects of material and workmanship for a period of one year from the date of acceptance by the owner.
2. The contractor, at his or her expense, will repair and replace all such electrical work and all other work or damage thereby, which becomes defective during the period of the guarantee.

D. Guide Specifications

1. Service supplied to the structure shall be a wire, 116/230 volts, 60-cycle, single phase.
2. Service panel shall have a 200-ampere capacity, with automatic circuit breakers. The service panel shall be flush mounting with a flush door and shall accommodate 30 circuits.
3. Wiring: All circuit wiring is to be 12-gauge or larger, Type TW for general use.
4. Boxes: Outlet boxes and junction boxes are to be galvanized steel approved for proposed use and of a suitable size to accommodate the requirements of the fixture, wiring device, or equipment, and the wiring connections.
5. Switches for general lighting shall be of the quiet AC-rated toggle type. Switches shall be installed 48 inches to center above finished floor unless otherwise specified.
6. Receptacles shall be standard duplex grounding type.
7. Location of outlets: Outlets, as shown, are in approximate locations. These must be checked on the job for possible conflicts with other trades or built-ins. Convenience outlets shall be 16 inches to center from the finished floor, unless otherwise noted.
8. Grounding: The complete electrical system will maintain a solid ground, in accordance with the National Electrical Code.

E. Special Outlets

The contractor shall furnish and install receptacles, and switches as required, for the following special outlets:

1. Exhaust fan over kitchen range.
2. Electric range in kitchen.
3. Electric hot water heater in basement.
4. Electric dryer and washer in basement.
5. Furnace control switch at head of basement stairs.
6. Waterproof outlet near front entrance porch.

F. Telephone Outlet
Furnish and install telephone wiring with one outlet in the kitchen–hall area and one in the master bedroom.

G. Exhaust Fan
Furnish and install exhaust hood and fan in the kitchen as indicated on drawings. The fan shall have at least 300 cfm free delivery.

H. Signal Chimes
Furnish and install chimes with electric connection to approved transformer and light circuit. Provide an outside push button at each door.

I. Lighting Fixtures
The contractor shall furnish and install all electrical fixtures and shall allow the sum of four hundred dollars ($400.00) for the purchase of these fixtures. This allowance covers the net cost to the contractor and does not include any labor, overhead, or profit. Fixtures shall be selected by the owner, but will be purchased by the contractor. The net difference in cost, if any, shall be added to or subtracted from the contract.

This completes the specifications for our two-story Garrison house. These working drawings and specifications would be all you would receive if you were a contractor bidding for the job of constructing his house. With the information shown on the drawings and in the specifications, you should be able to determine the cost of construction.

In many cases where financing is required in the construction of a house, the lending institution will require that the combined FHA 2005 and VA 26-1852 form be filled out and submitted with the working drawings for approval by either the Federal Housing Administration (FHA) or the Veterans Administration, if the prospective owner is a veteran. When approved, this form is used along with the working drawings and guarantee the lending institution that the United States Government will back the loan in case of default on the part of the owner. A filled-in sample of this form is reproduced on the following pages.

At first glance, these forms seem much easier to read and understand than a full set of specifications we just went over. This is not true, because there are hidden complications. Note, first, that the top of form FHA 2005 and VA 26-1852 speci-fies whether the house is proposed construction or under construction. Then read the instructions. You will find that work not specifically described or shown will not be considered unless required, and then only the minimum acceptable will be assumed. You will also find that the construction must be completed in compliance with related drawings and specifications as amended during processing. The specifications include both the Description of Materials form and the applicable minimum property standards.

Minimum Property Standards is a document of approximately 300 pages which describes fully what must be done to meet the minimum in the construction of a house; it performs the same function as a set of specifications drawn up by the architect. These standards are intended to provide a sound technical basis for FHA mortgage insurance by providing minimum standards which will assure a sound, well planned, safely constructed house. You can purchase this document from the Superintendent of Documents, Washington, D.C. 20402. The table of contents and the statement of purpose of the *Minimum Property Standards* are reproduced following the FHA/VA Description of Materials form.

TABLE OF CONTENTS

	Page
Foreword	iii
FHA Field Office Jurisdictions	iv
Purpose	vii
Scope	viii
Alternates – Design, Equipment and Construction Methods	ix
Technical Publications of the FHA	x

CHAPTER I. – Required Exhibits	Section	Page
General	100	1
Drawings for Individual Applications	101	1
Drawings for Group Applications	102	3
Description of Materials	103	4
Individual Water-Supply and Sewage–Disposal Systems	104	4

CHAPTER II. – Compliance Inspections	Section	Page
General	200	7
First Compliance Inspection	201	8
Intermediate Compliance Inspection	202	8
Second Compliance Inspection	203	9
Third Compliance Inspection	204	9
Delayed Completion – On-Site Improvements	205	9
Delayed Completion – Off-Site Improvements	206	10
Repair Compliance Inspection	207	10
Reinspections	208	10
Optional Inspection	209	11
Correction of Construction	210	11

CHAPTER III. – Definitions and Abbreviations	Section	Page
General	300	13
Definitions	301	13
Abbreviations and Symbols	302	18

CHAPTER IV. – General Acceptability Criteria	Section	Page
General	400	19
Local Codes and Regulations	401	19
Site Conditions	402	19
The Plot	403	20
Service and Facilities	404	20
Access	405	20
Types of Eligible Dwellings	406	20
Non-Residential Use	407	20
Special or Unusual Conditions	408	21

CHAPTER IV. – General Acceptability Criteria – Continued	Section	Page
Defective Conditions	409	21
Change After Issuance of Commitment	410	21
Final Acceptance	411	21
Appeal Procedure	412	21

CHAPTER V. – Plot Planning	Section	Page
General Objective	500	23
General	501	23
Lot Coverage	502	23
Yard Dimensions	503	24
Distance Between Buildings on Same Plot	504	24
Courts	505	24

CHAPTER VI. – Building Planning	Section	Page
General Objective	600	32
General	601	32
Space Standards	602	32
Light and Ventilation	603	40
Ventilation of Structural Spaces	604	41
Access	605	43
Privacy	606	43
Stairway Planning	607	44
Fire Protection and Safety	608	46
Areaways for Windows and Exterior Doors	609	46
Porches, Terraces and Entrance Platforms	610	48
Garages and Carports	611	48

CHAPTER VII. – Materials and Products	Section	Page
General Objective	700	52
General	701	52
Suitability of Alternate or Special Materials or Products	702	52
Masonry Materials	703	53
Concrete	704	55
Lumber and Wood Products	705	57
Gypsum Products	706	62
Asbestos-Cement Products	707	62
Structural Steel and Steel Products	708	63
Aluminum Products	709	64
Miscellaneous Metal Products	710	65
Glass	711	66
Bituminous Products	712	67
Vapor Barriers	713	68
Thermal Insulation	714	68
Resilient Flooring	715	69

	Section	Page
CHAPTER VII. — Materials and Products — Continued		
Ceramic Tile	716	70
Plastic Wall Tile	717	70
Screening	718	70
Special Flashing Materials	719	70
Building Paper	720	70
Adhesives	721	70
CHAPTER VIII. — Construction		
General Objective	800	72
General	801	72
Suitability of Alternate or Special Methods of Construction	802	73
Excavation and Backfill	803	75
Footings	804	76
Foundation Walls	805	78
Masonry and Concrete Piers	806	83
Dampproofing and Waterproofing	807	90
Concrete Slabs	808	92
Structural Steel	809	99
Exterior Masonry Walls	810	101
Masonry Veneer	811	103
Interior Masonry Walls	812	104
Chimneys and Vents	813	107
Fireplaces	814	110
Protection Against Termites and Decay	815	113
Wood Floor Framing	816	120
Subflooring	817	130
Exterior Wall Framing	818	131
Wall Sheathing	819	140
Sheathing Paper	820	141
Partition Framing	821	141
Ceiling Framing	822	145
Roof Framing	823	146
Roof Sheathing	824	151
Stair Construction	825	152
Garages and Carports	826	154
CHAPTER IX. — Exterior and Interior Finishes		
General Objective	900	156
General	901	156
Flashing	902	156
Exterior Wall Finish	903	166
Roof Covering	904	172
Gutters and Downspouts	905	180
Interior Wall and Ceiling Finish	906	181
Finish Floors	907	185
Glazing	908	190

	Section	Page
CHAPTER IX. — Exterior and Interior Finishes — Continued		
Millwork, Trim and Cabinets	909	191
Painting and Decorating	910	192
CHAPTER X. — Mechanical Equipment		
General Objective	1000	198
General	1001	198
Mechanical Ventilation	1002	198
Heating	1003	199
Summer Air Conditioning	1004	206
Domestic Water Heating and Storage	1005	209
Plumbing	1006	212
Electrical	1007	214
CHAPTER XI. — Individual Water Supply and Sewage-Disposal Systems		
General Objective	1100	217
General	1101	217
Individual Water-Supply Systems	1102	217
Individual Sewage-Disposal Systems	1103	219
CHAPTER XII. — Lot Improvements		
General Objective	1200	233
General	1201	233
Grading	1202	234
Drainage Structures	1203	236
Driveway and Parking Space	1204	237
Walks	1205	238
Exterior Steps Not Attached to Dwelling	1206	238
Lawns and Ground Cover	1207	239
Trees and Shrubs	1208	240
Other Lot Improvements	1209	240
APPENDIX A. — Structural Design Data		
General	A	247
Design Dead Loads	B	247
Design Live Loads	C	247
Design Deflections	D	249
Allowable Design Stresses	E	249
APPENDIX B. — Tables of Maximum Allowable Spans for Wood Joists and Rafters		
Foreword		255
Design Loads		256
Tables of Maximum Allowable Spans		258
APPENDIX C. — Reference Standards		295
Index		301

PURPOSE

The purpose of the National Housing Act, as stated in the preamble, is "to encourage improvement in housing standards and conditions, to provide a system of mutual mortgage insurance, and for other purposes."

In pursuance of this purpose, the Federal Housing Administration has established these Minimum Property Standards. They are intended to obtain those characteristics in a property which will assure present and continuing utility, durability and desirability as well as compliance with basic safety and health requirements. To provide this assurance, these standards set forth the minimum qualities considered necessary in the planning, construction and development of the property which is to serve as security for an insured mortgage.

As these standards define the minimum level of quality acceptable to FHA, a property complying with them is considered technically eligible in all FHA insuring office jurisdictions. Other factors however, such as the appropriateness of the dwelling to the site and to the neighborhood and the anticipated market acceptance of the property as a whole must also be considered in FHA underwriting analysis.

Planning and construction which exceed the minimums set forth herein and which will result in increased marketability of the property or which will reduce the expense of maintenance or early replacement of equipment, will be reflected in the FHA estimate of value.

The standards are based upon extensive study by the technical staff of FHA headquarters and field offices, and upon recommendations of builders, architects, engineers and material producers. While they represent good current practice in residential technology, they may be modified in the future as additional data and experience are gained.

The standards are not intended to serve as a building code. Such codes are primarily concerned with factors of health and safety and not the many other aspects of design and use which are included herein as essential for mortgage insurance determinations.

FHA Form 2005
VA Form 26-1852
Rev. 2/75

U. S. DEPARTMENT OF HOUSING AND URBAN DEVELOPMENT
FEDERAL HOUSING ADMINISTRATION

Form Approved
OMB No. 63–R0055

DESCRIPTION OF MATERIALS

No. _____ (To be inserted by FHA or VA)

☐ Proposed Construction

☐ Under Construction

Property address ___ REVIEW HOUSE ___ City ___ Hometown ___ State ___ USA

Mortgagor or Sponsor ___ First Main Bank _____
                              (Name)

Contractor or Builder ___ John Doe _____
                              (Name)

101 Main Street
(Address)

225 Main Street
(Address)

INSTRUCTIONS

For accurate register of carbon copies, form may be separated along above fold. Staple completed sheets together in original order.

1. For additional information on how this form is to be submitted, number of copies, etc., see the instructions applicable to the FHA Application for Mortgage Insurance or VA Request for Determination of Reasonable Value, as the case may be.

2. Describe all materials and equipment to be used, whether or not shown on the drawings, by marking an X in each appropriate check-box and entering the information called for in each space. If space is inadequate, enter "See misc." and describe under item 27 or on an attached sheet. THE USE OF PAINT CONTAINING MORE THAN ONE HALF OF ONE PERCENT LEAD BY WEIGHT IS PROHIBITED.

3. Work not specifically described or shown will not be considered unless

required, then the minimum acceptable will be assumed. Work exceeding minimum requirements cannot be considered unless specifically described.

4. Include no alternates, "or equal" phrases, or contradictory items. (Consideration of a request for acceptance of substitute materials or equipment is not thereby precluded.)

5. Include signatures required at the end of this form.

6. The construction shall be completed in compliance with the related drawings and specifications, as amended during processing. The specifications include this Description of Materials and the applicable Minimum Property Standards.

1. EXCAVATION:

Bearing soil, type ___ Sand and Loam

2. FOUNDATIONS:

Footings: concrete mix ___ Class "A" ___ ; strength psi ___ 3000 ___ Reinforcing ___ none

Foundation wall: material ___ Cement block ___ Reinforcing ___ Dur-O-Wall

Interior foundation wall: material _____ Party foundation wall _____

Columns: material and sizes ___ 4" Steel w/ base & Cap ___ Piers: material and reinforcing _____

Girders: material and sizes ___ 3–2x8's spiked together ___ Sills: material ___ 6" Sill seal and 2x6

Basement entrance areaway _____ Window areaways ___ 20 gauge galvanized steel

Waterproofing ___ 1/2" cement plaster & 2 coats asphalt ___ Footing drains ___ 10' Perforated Orangeburg

Termite protection ___ 26 gauge galvanized iron, Lock and solder seams

Basementless space: ground cover _____ ; insulation _____ ; foundation vents _____

Special foundations _____

Additional information: _____

3. CHIMNEYS:

Material ___ brick ___ Prefabricated (make and size) ___ #5 Heatilator

Flue lining: material ___ viterous tile ___ Heater flue size ___ 8x8 ___ Fireplace flue size ___ 8x13

Vents (material and size): gas or oil heater ___ oil heater ___ ; water heater ___ 40 gallon electric

Additional information: _____

4. FIREPLACES:

Type: ☒ solid fuel; ☐ gas-burning; ☐ circulator (*make and size*) _____ Ash dump and clean-out __12x12 & 8x12__

Fireplace: facing __brick__ ; lining __prefabricated__ ; hearth __Quarry tile__ ; mantel __Morgan M-1448__

Additional information: _____

5. EXTERIOR WALLS:

Wood frame: wood grade, and species __#1 Douglas Fir or Spruce__ ☐ Corner bracing. Building paper or felt __doublecraft__

Sheathing __plyscore__ ; thickness __½"__ ; width __4'__ ☒ solid; ☐ spaced ____ " o. c.; ☐ diagonal;

Siding __½" x 8"__ ; grade __clear__ ; type __cedar__ ; size __½" x 8"__ ; exposure __6¾"__ ; fastening __nailed__

Shingles ____ ; grade ____ ; type ____ ; size ____ ; exposure ____ ; fastening ____

Stucco ____ ; thickness ____ " __Lath__ ____ ; weight ____ lb.

Masonry veneer __Brick in front__ Sills __Brick__ Lintels __4 x 4 angle iron__ Base flashing __28 ga. galv.__

Masonry: ☐ solid ☒ faced ☐ stuccoed; total wall thickness ____ "; facing thickness ____ "; facing material ____

Backup material ____ ; thickness ____ "; bonding ____

Door sills ____ Window sills ____ Lintels ____ Base flashing ____

Interior surfaces: dampproofing, ____ coats of ____ ; furring ____

Additional information: __Non-chalking tinting white__

Exterior painting: material ____ ; number of coats __2__

Gable wall construction: ☒ same as main walls; ☐ other construction ____

6. FLOOR FRAMING:

Joists: wood, grade, and species __2x8 fir__ ; other ____ ; bridging __1 x 3 spruce__ ; anchors ____

Concrete slab: ☐ basement floor; ☐ first floor; ☐ ground supported; ☐ self-supporting; mix ____ ; thickness ____ ";

reinforcing ____ ; insulation ____ ; membrane ____

Fill under slab: material ____ ; thickness ____ ". Additional information: ____

7. SUBFLOORING: (*Describe underflooring for special floors under item 21.*)

Material: grade and species __plyscord 4'x8' sheets__ ____ ; size __5/8"__ ; type __CD Exterior__

Laid: ☒ first floor; ☐ second floor; ☐ attic ____ sq. ft.; ☐ diagonal; ☐ right angles. Additional information: ____

8. FINISH FLOORING: (*Wood only. Describe other finish flooring under item 21.*)

LOCATION	ROOMS	GRADE	SPECIES	THICKNESS	WIDTH	BLDG. PAPER	FINISH
First floor	Living-Dining	select	oak	25/32	3"	deadening	wood filler & varnish
Second floor	Hall & Bedrooms	select	oak	25/32	3"	deadening	wood filler & varnish
Attic floor	____ sq. ft.						

Additional information: ____

9. PARTITION FRAMING:

Studs: wood, grade, and species __Fir Utility grade__ size and spacing __2x4-16" OC__ Other ____

Additional information: ____

10. CEILING FRAMING:

Joists: wood, grade, and species __2 x 6 Fir__ Other ____ Bridging ____

Additional information: ____

11. ROOF FRAMING:

Rafters: wood, grade, and species __2 x 6 Fir__ Roof trusses (see detail): grade and species __See details__

Additional information:

12. ROOFING:

Sheathing: wood, grade, and species __½" x 4' x 8' plyscord__ ; ☒ solid; □ spaced _____ " o.c.

Roofing __315 # Asphalt shingles__ ; grade _____ ; size _____ ; type __Strip asphalt shingles__

Underlay __15 # asphalt felt__ ; weight or thickness __15 lb__ ; size __36"__ ; fastening __nailed__

Built-up roofing _____ ; number of plies _____ ; surfacing material _____

Flashing: material __Aluminum__ ; gage or weight __24 gauge__ ; □ gravel stops; □ snow guards

Additional information:

13. GUTTERS AND DOWNSPOUTS: none

Gutters: material _____ ; gage or weight _____ ; size _____ ; shape _____

Downspouts: material _____ ; gage or weight _____ ; size _____ ; shape _____ ; number _____

Downspouts connected to: □ Storm sewer; □ sanitary sewer; □ dry-well. □ Splash blocks: material and size _____

Additional information:

14. LATH AND PLASTER:

Lath □ walls, □ ceilings: material _____ ; weight or thickness _____ Plaster: coats _____ ; finish _____

Dry-wall ☒ walls, ☒ ceilings: material __gypsium wallboard__ ; thickness __3/8"__ ; finish __Painted__

Joint treatment __Dry wall joint system according to manufacturers instructions__

15. DECORATING: (Paint, wallpaper, etc.)

Rooms	Wall Finish Material and Application	Ceiling Finish Material and Application
Kitchen	gypsum wallboard painted	gypsum wallboard – painted
Bath	4¼" x 4¼" Matt finish ceramic tile	gypsum wallboard – painted
Other		

Additional information:

16. INTERIOR DOORS AND TRIM:

Doors: type __Flush__ ; material __Birch veneer__ ; thickness __1 3/8"__

Door trim: type __#37 11/16"x2¼"__ ; material __clear pine__ Base: type __#98__ ; material __clear pine__ ; size __9/16"x3"__

Finish: doors __Stain & Varnish – natural__ ; trim __primed & painted__

Other trim (*item, type and location*)

Additional information:

17. WINDOWS:

Windows: type __Double hung__ ; make __Anderson__ ; material __pine__ ; sash thickness __1 3/8"__

Glass: grade __clear__ ; □ sash weights; ☒ balances, type _____ ; head flashing __aluminum__

Trim: type __#37 clamshell__ ; material __pine__ Paint __Enamel__ ; number coats __3__

Weatherstripping: type __compression__ ; material __aluminum__ Storm sash, number __18__

Screens: ☒ full; ☒ half; type __wood combination__ ; number __18__ ; screen cloth material __aluminum__

Basement windows: type __two light hopper__ ; material __steel__ ; screens, number __5__ ; Storm sash, number __5__

Special windows

Additional information:

18. ENTRANCES AND EXTERIOR DETAIL:

Main entrance door: material __pine__ ; width __3'-0"__ ; thickness __1 3/4"__ ; Frame: material __clear pine__ , thickness __5/4"__

Other entrance doors: material __pine__ ; width __2'-8"__ ; thickness __1 3/4"__ ; Frame: material __clear pine__ ; thickness __5/4"__

Head flashing __Aluminum__ Weatherstripping: type __compression__ ; saddles __oak__

Screen doors: thickness _____ , number _____ ; screen cloth material _____ ; Storm doors: thickness _____"; number _____

Combination storm and screen doors: thickness __1 1/8"__; number __2__ ; screen cloth material __aluminum__

Shutters: ☐ hinged ☒ fixed. Railings __Wrought Iron__ , Attic louvers __#600-L wood w/ aluminum wire__

Exterior millwork: grade and species __#1 clear pine__ Paint __Trim & Shutter__ ; number coats __2__

Additional information:

19. CABINETS AND INTERIOR DETAIL:

Kitchen cabinets, wall units: material __Kingswood Oakmont__ ; lineal feet of shelves __51__ ; shelf width __12"__

Base units: material __Oak veneer__ ; counter top __Plastic laminate__ ; edging __Plastic Laminate__

Back and end splash __laminated plastic__ Finish of cabinets __Factory finished__ ; number coats _____

Medicine cabinets: make __none__ ; model _____

Other cabinets and built-in furniture __Two vanities in second floor bathrooms__

Additional information:

20. STAIRS:

STAIR	TREADS		RISERS		STRINGS		HANDRAIL		BALUSTERS	
	Material	Thickness	Material	Thickness	Material	Size	Material	Size	Material	Size
Basement	fir	1 5/8"	pine	3/4"	fir	2 x 10	pine	2"	–	–
Main	oak	1 1/16"	oak	3/4"	pine	1 x 10	birch	M-720	pine	M-836
Attic										

Disappearing: make and model number

Additional information:

21. SPECIAL FLOORS AND WAINSCOT:

FLOORS	Location	Material, Color, Border, Sizes, Gage, Etc.	Threshold Material	Wall Base Material	Underfloor Material
	Kitchen	Inlaid lineoleum		Pine	5/8" plyscord
	Bath	Ceramic Tile	marble	Pine	5/8" plyscord

WAINSCOT	Location	Material, Color, Border, Cap. Sizes, Gage, Etc.	Height	Height Over Tub	Height in Showers (From Floor)
	Bath	4¼" x 4¼" Ceramic Tile		6'	

Bathroom accessories: ☒ Recessed; material __China__ ; number __4__ ; ☐ Attached; material _____ ; number _____

Additional information:

22. PLUMBING:

Fixture	Number	Location	Make	Mfr's Fixture Identification No.	Size	Color
Sink	1	Kitchen	Kohler ss steel	Model SA-410	21x32	stainless
Lavatory	3	baths	Kohler	Brookline K-2208-F	16" round	by owner
Water closet	3	baths	Kohler	Wellworth K-3512 PB4		by owner
Bathtub	2	baths	Kohler	Guardian K-786-SA	5'-0"	by owner
Shower over tub △						
Stall shower △						
Laundry trays						

△ ☒ Curtain rod   △ □ Door   □ Shower pan: material _____

Water supply: ☒ public; □ community system; □ individual (private) system. ★

Sewage disposal: ☒ public; □ community system; □ individual (private) system. ★

★ Show and describe individual system in complete detail in separate drawings and specifications according to requirements.

House drain (inside): □ cast iron; □ tile; □ other _____   House sewer (outside): ☒ cast iron; □ tile; □ other _____

Water piping: □ galvanized steel; ☒ copper tubing; □ other _____   Sill cocks, number __2__

Domestic water heater: type __electric__ ; make and model __Wagoner WC-40__ ; heating capacity __4000 BTU__

_____ gph. 100° rise.   Storage tank: material __glass lined__ ; capacity __40__ gallons.

Gas service: □ utility company; □ liq. pet. gas; □ other _____   Gas piping: □ cooking; □ house heating.

Footing drains connected to: □ storm sewer; □ sanitary sewer; □ dry well.   Sump pump; make and model _____

_____ ; capacity _____ ; discharges into _____

23. HEATING:

□ Hot water.   □ Steam.   □ Vapor.   □ One-pipe system.   □ Two-pipe system.

□ Radiators.   □ Convectors.   □ Baseboard radiation.   Make and model _____

Radiant panel: □ floor; □ wall; □ ceiling.   Panel coil: material _____

□ Circulator.   □ Return pump.   Make and model _____ ; capacity _____ gpm.

Boiler: make and model _____   Output _____ Btuh.; net rating _____ Btuh.

Additional information: _____

Warm air: □ Gravity.   ☒ Forced.   Type of system __Gas & Fired Hot Air__

Duct material: supply __galv.iron__ ; return __galv. iron__   Insulation _____ , thickness _____   □ Outside air intake.

Furnace: make and model __Carrier__   Input __125,000__ Btuh.; output __100,000__ Btuh.

Additional information: __58SE Model 125 - Series 304__

□ Space heater; □ floor furnace; □ wall heater.   Input _____ Btuh.; output _____ Btuh.; number units _____

Make, model _____   Additional information: _____

Controls: make and types __Minneapolis-Honeywell T-861A clock thermostat__

Additional information: _____

Fuel: □ Coal; □ oil; ☒ gas; □ liq. pet. gas; □ electric; □ other _____ ; storage capacity _____

Additional information: _____

Firing equipment furnished separately: ☐ Gas burner, conversion type. ☐ Stoker: hopper feed ☐; bin feed ☐

Oil burner: ☐ pressure atomizing; ☐ vaporizing

Control _____

Make and model _____

Additional information: _____

Electric heating system: type _____ Input _____ watts; @ _____ volts; output _____ Btuh.

Additional information: _____

Ventilating equipment: attic fan, make and model _____

kitchen exhaust fan, make and model _____

Other heating, ventilating, or cooling equipment _____ ; capacity _____ cfm.

24. ELECTRIC WIRING:

Service: ☒ overhead; ☒ underground.   Panel: ☐ fuse box; ☒ circuit-breaker; make __Cutter-Hammer__ AMP's __200__ No. circuits __10__

Wiring: ☐ conduit; ☐ armored cable; ☒ nonmetallic cable; ☐ knob and tube; ☐ other _____

Special outlets: ☒ range; ☒ water heater; ☐ other __electric dryer, hot water heater, exhaust fan, furnace control__

☐ Doorbell. ☒ Chimes.   Push-button locations _____   Additional information: _____

25. LIGHTING FIXTURES:

Total number of fixtures __17__   Total allowance for fixtures, typical installation, $ __400.00__

Nontypical installation _____

Additional information: _____

26. INSULATION:

Location	Thickness	Material, Type, and Method of Installation	Vapor Barrier
Roof	6"	Fiberglas by Owen-Corning	Kraft-paper
Ceiling	6"	Fiberglass by Owen-Corning	Kraft-paper
Wall	3"	Fiberglas by Owen-Corning	Kraft-paper
Floor			

27. MISCELLANEOUS: (Describe any main dwelling materials, equipment, or construction items not shown elsewhere; or use to provide additional information where the space provided was inadequate.   Always reference by item number to correspond to numbering used on this form.) _____

_____

_____

_____

_____

_____

_____

_____

_____

_____

HARDWARE: (make, material, and finish.) __Allow $500.00 for purchase. Hardware to be Corbin, Russwin, or equal.__

SPECIAL EQUIPMENT: *(State material or make, model and quantity. Include only equipment and appliances which are acceptable by local law, custom and applicable FHA standards. Do not include items which, by established custom, are supplied by occupant and removed when he vacates premises or chattels prohibited by law from becoming realty.)*

_____
_____
_____
_____

PORCHES: Entrance porch with no roof, cement floor, concrete steps and wrought iron rail at both front and side entrances.

_____
_____

TERRACES: none

_____
_____

GARAGES: none

_____
_____

WALKS AND DRIVEWAYS:

Driveway: width _____; base material _____; thickness _____"; surfacing material _____; thickness _____"

Front walk: width 3' ; material CONCRETE ; thickness 3 ". Service walk: width _____; material _____; thickness _____"

Steps: material concrete ; treads concrete "; risers 6½ ". Cheek walls _____

OTHER ONSITE IMPROVEMENTS:

(Specify all exterior onsite improvements not described elsewhere, including items such as unusual grading, drainage structures, retaining walls, fence, railings, and accessory structures.)

_____
_____

LANDSCAPING, PLANTING, AND FINISH GRADING:

Topsoil 6 " thick: ☒ front yard; ☒ side yards; ☒ rear yard to 30 feet behind main building.

Lawns *(seeded, sodded, or sprigged)*: ☐ front yard _____; ☐ side yards _____; ☐ rear yard _____

Planting: ☐ as specified and shown on drawings; ☐ as follows:

_____ Shade trees, deciduous, _____ " caliper. _____ Evergreen trees. _____ ' to _____ ', B & B.

_____ Low flowering trees, deciduous, _____ ' to _____ '. _____ Evergreen shrubs. _____ ' to _____ ', B & B.

_____ High-growing shrubs, deciduous, _____ ' to _____ '. _____ Vines, 2-year _____

_____ Medium-growing shrubs, deciduous, _____ ' to _____ '.

_____ Low-growing shrubs, deciduous, _____ ' to _____ '.

IDENTIFICATION.—This exhibit shall be identified by the signature of the builder, or sponsor, and/or the proposed mortgagor if the latter is known at the time of application.

Date _____ Signature _____

Signature _____

FHA Form 2005
VA Form 26-1852

3
SITE WORK

Our specifications state that the contractor shall visit the site and ascertain the existing conditions. This is very important in understanding what preparatory work must be done before the excavation can be started. If there is a large amount of brush or any large trees to be removed, this must be figured in the cost. The removal of large trees can be expensive to cut down and haul away.

The contractor, after removing any trees and brush, will then stake out the location of the building on the lot and establish the grade lines for the excavation. He will establish his grades from the Bench Mark (BM) on the plot plan.

PLOT PLANS

The plot plan for a residence accompanies the specifications and is usually found on the first page of the working drawings. It may be drawn with an engineer's scale in feet and decimal part of a foot or with an architect's scale at $\frac{1}{16}' = 1'0'$.

In most areas, zoning laws require a survey and a plot plan to be submitted before a building permit is issued. Most banks and lending institutions also require a survey and plot plan before any loans can be approved on the property. Zoning laws define or restrict the type of building that may be built in specific areas of a community. They may, for example, order commercial buildings to be separated from residential buildings. Zoning laws vary from one community to another.

The following is a list of features shown on most plot plans:

- Scale and title of the drawing.
- Compass bearings.
- Location and dimension of all walks, driveways, fences, etc.
- Dimensions of the plot.
- Location and dimension of all buildings.
- Finish grade levels of main corners of the building.
- Location of a bench mark and its elevation.
- Location of utilities, such as electric, water, gas, and sewage.

In reference to plot plans, the terms location, elevation, and size have special meaning. The location of the lot is determined by a surveyor, who may be either government employed or privately employed. Surveyors get information from municipal land records, deeds, and markers established by other surveyors. The size of the lot is determined by its boundary lines. The survey indicates the length and compass bearing of each boundary line. The elevation is determined from a known coastal sea level. Almost every city or town has one or more permanent points established from this sea level. Such a point is known as a city or county datum; it is marked by a metal rod in a concrete base which is set in the gound. All such points, from which elevations are determined, are referred to as *bench marks.*

Contour Lines

On some plot plans, where the ground is irregular, the surveyor may show contour lines. These are continuous lines running through all points of the same elevation of a plot of land (Figure 3-1). The vertical distance between the contour lines is called the *vertical contour interval.* The vertical contour interval can be any measurement, but normally, it is 1 foot. For example, if two contour lines are marked 100' and 99', the vertical contour interval is 1 foot. This means that the ground is 1 foot higher along the 100' line than it is at any point along the 99' line. This 1-foot difference in elevation is the same regardless of the spacing between the contour lines. Therefore, the closer the contour lines are together, the steeper the slope is at that point.

The plan view of Figure 3-1 shows how the contour lines might appear on a plot plan. The elevation view provides a better understanding of how the slope appears. Notice that the slope is steeper in section AA than it is in section BB; this corresponds to the much closer spacing of the contour lines on the plan view at the point where section AA is taken.

On most plot plans the surveyor establishes the reference elevation mark in feet and inches for the convenience of the contractor. Frequently, however, the surveyor measures in feet and decimal parts of a foot with an engineer's scale. An engineer's scale has each foot divided into ten parts and each of these tenths divided again into ten parts. An example of a measurement taken from an engineer's scale might be 112.129 feet. To convert such a dimension into feet and inches use the following procedure.

1. Multiply the decimal part by 12 to find the number of inches and decimal parts of an inch. (.129 × 12 = 1.548 inches.)
2. Multiply the remaining decimal by 16 to find sixteenths of an inch. (.548 × 16 = 8.768/16 or $\frac{9}{16}$-inch.)
3. Add the fractional part of an inch to the whole feet and inches. (112 feet + 1 inch + $\frac{9}{16}$ inch = 112 feet, $1\frac{9}{16}$ inches.)
4. To convert to sixty-fourths of an inch instead of sixteenths of an inch, multiply the decimal fraction of an inch by 64 instead of 16. (.548 × 64 = 35.072/64 or $\frac{35}{64}$ inch.)

EXCAVATION

Referring to our plot plan (Figure 2-4) we find that the plan is drawn $\frac{1}{16}'' = 1'0''$ scale. Reading the plot plan we find that the house is to be built on North Street. It has a setback of 30'0'' from the front property line and is located 32'0'' from the property line of the lands of William Doe. Our contour lines show us that there is an approximately two-foot slope difference in the lot from the front on North Street to the back of the lot.

The specifications call for the removal of all sod and soil throughout the area that the structure occupies and where grade lines are changed. The

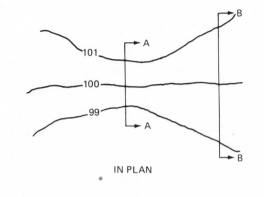

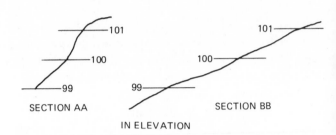

Fig. 3-1. Contour lines.

working drawings of the basement plan show that the house is 32'0'' across the front and 22'5'' from the front to the back. We need a minimum of two feet around the outside of the walls for the foundation work so our excavation should be 36'0'' across the front and 27'0'' from the front to the back.

The working drawings and specifications tell us that the foundation wall is to be cement block construction and the block is to be ten and one-half courses high. Ten and one half courses give us a total height of 84 inches or 7 feet. The top of the foundation wall is approximately 12 inches above the finish grade. The plot shows a slope of approximately 1 foot where the location of the structure is to be built. This means that our excavation will be approximately 6 feet deep.

Estimating Excavation

To determine the excavation, take the area to be excavated which in our case is 36'0'' x 27'0'' and we multiply this to get a total of 972 square feet. We figured the depth to be 6 feet. We multiply 972 square feet by 6 feet and we get a total of 5,832 cubic feet. There are 27 cubic feet to a cubic yard so dividing 5,832 by 27 we get a total of 216 cubic yards of earth to be removed. This will be used to backfill after the foundation and framing is in place.

Fig. 3-2. Excavating for a foundation.

Normally, an excavating contractor is hired for the excavation and backfill (Figure 3-2). Both operations are paid for in one payment. The operator strips the topsoil and excavates the cellar to the depth specified. After the foundation wall is either waterproofed or damp-proofed the earth is pushed back and graded so that the land slopes away from the foundation. The topsoil is then spread over the area to complete the backfill operation.

We will start our estimate for materials at the site. A contractor will sometimes drive two stakes and give the depth of the excavation from these points, or he will set up his level and set his batter boards prior to the excavation. Batter boards are stakes driven in the ground with pieces of shiplap or board nailed across them. Saw cuts are usually made in each cross board and a string is drawn which crosses at each corner of the top of the foundation

wall at the correct height for the top of the foundation wall. These cross strings show the exact location of each corner of the foundation and its correct height. Corners and tops of footings are taken from the strings that run from batter board to batter board.

Depending upon the number of corners and the layout of the house, you should figure three 2 x 4s approximately 6 feet long, two pieces of shiplap 1 x 8 about 5 feet long and one stake for each corner of the house. You will also need a few nails and a roll of twine for lines. Figure the number of corners of the building. Multiply this by three for the number of 2 x 4s for the batter boards. Take the number of corners and multiply by two and this will give you the number of pieces of 1 x 8 shiplap for the batter boards. Next count the number of corners and this will give the number of stakes to lay out the house.

Our house has just four corners. You do not include fireplace or entrance porches in laying out the main foundation batter boards. We now start our material list as shown in the table.

ESTIMATE FOR _____   Page _____

DATE _____

Job _____

Pieces	Feet	Description	Cost	Total
12		2 x 4 x 6' batter boards (would order 6–2 x 4 x 12')		
8		1 x 8 x 5' shiplap batter boards (would order 4–1 x 8 x 10')		
4		wood stakes for corner of house		
1		roll twine (500 feet) for lines on batter boards		

4
MASONRY ESTIMATE:
FOOTINGS AND CONCRETE FLOORS

FOOTINGS

Every building rests upon a footing or foundation of some sort. The footing distributes the weight of the building evenly on the earth. They vary in width and height according to the weight they have to support. For practical purposes, in house construction and small building construction, the footing is usually twice as wide as the thickness of the wall that is to be built on top of it. The height of the footing is usually the same as the thickness of the wall (Figure 4-1).

For example, a footing to support an 8-inch thick wall of either concrete or concrete block is 16 inches wide and 8 inches high. If the wall is 12 inches thick, then the footings are 24 inches wide and 12 inches high. These dimensions may be changed if unforeseen circumstances are found when the foundation and footings are dug. Footings must be

placed on solid ground or rock. Never place footings on fresh earthen fill.

The footing on the basement and foundation plan is shown as dotted lines (Figure 4-2). The dotted line is approximately four inches on each side of the solid object line that marks the foundation wall.

Footings for posts and fireplaces are also shown with a hidden line at the exact location where they are to be placed. They are usually accompanied by notes explaining the size and depth (Figure 4-3).

The size of the footing may be on the wall section view of the plans. This drawing would include a

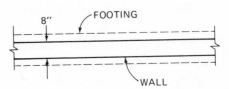

Fig. 4-2. Foundation wall footing.

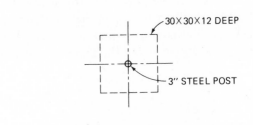

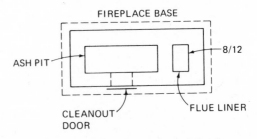

Fig. 4-3. Footings for posts and fireplace.

Fig. 4-1. The footing should be twice as wide as the thickness of the foundation wall and the same height as the thickness of the wall.

side view of the footings. Figure 4-4 is an example of the footing, concrete floor, and a wall section view on the plans.

It is standard practice in building construction that any time two dimensions are given the first is the width. For example if the opening for a cellar window is 32" x 16", the width is 32 inches and the height is 16 inches.

With the site cleared and the excavation made to the correct location as indicated on the plot plan, work is ready to begin on the footings. The basement and foundation plan is used to find the exact size and location of the footings and the foundation walls. Studying the foundation plan (Figure 2-5) and after reading the specifications we find that the foundation wall is to be cement block. The working drawings show that the walls are 8 inches wide except the front wall which is 12 inches wide. This tells us that the footings will be 16 inches wide under the eight inch wall and will be eight inches deep. The footings under the 12-inch wall will be 24 inches wide and 12 inches deep. Remember, it is standard practice to have the footings twice the width of the wall and the same depth as the wall is wide, but this may not be true on every drawing, depending upon conditions of the site. Therefore it is very important that you study the working drawings and the specifications to make double sure that you are estimating correctly according to the plans and specifications.

Once the excavation has been completed, we are ready to start our estimate for the footings. We have determined that our footings are 16 inches wide and 8 inches deep except under the front wall, where the footing is 24 inches wide and 12 inches deep. Our main foundation is 108 linear feet (perimeter). The side entrance foundation is 10 linear feet, the front entrance foundation is 12 linear feet, and the fireplace base is a total of 9

linear feet. The footings for the support posts is 30 inches square and we have two for a total of 20 linear feet.

Adding our foundation wall perimeter together we get a total of 139 linear feet. We double 139 for a total of 278 linear feet (both sides of our footing) we add 20 (our post footings) for a total of 298 linear feet of 1 x 8 shiplap needed for forms. We will need stakes to hold the sides of the footings form until the concrete has been poured. (See Figure 4-5.) If we figure one stake for every 4 linear feet of footing, we will need 7 bundles of 12 stakes to a bundle We would now add this to our material list, as shown in the table.

Pieces	Feet	Description	Cost	Total
	298	linear feet 1 x 8 shiplap – forms for footings		
7		bundles wood stakes (12 each) for forms		

Estimating Footings

To figure the amount of concrete necessary for the footings, find the number of linear feet around the outside walls of the house. Include the length of any other walls, such as porches. Multiply the width of the footings in feet by the thickness in feet, then multiply this product by the length of the footings in feet. This is the number of cubic feet in the footings.

Divide the number of cubic feet by 27 to find the number of cubic yards. The amount of con-

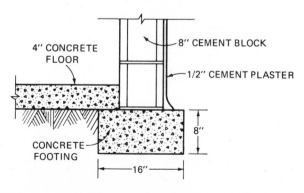

WALL SECTION DETAIL

Fig. 4-4.

4" CONCRETE FLOOR

8" CEMENT BLOCK

1/2" CEMENT PLASTER

CONCRETE FOOTING

8"

16"

Fig. 4-5. Although it is sometimes possible to place concrete directly in trenches, it is generally necessary to use side forms.

crete to be used for supporting posts and fireplace footings must be added to this amount. To figure these, multiply the length in feet times the width in feet times the thickness in feet and divide by 27.

Our footings are 16"x 8" on each side of the house, across the back of the house, and under the entrances at both the front and side. The front wall is 24" x 12". Add the length across the back of the house (32'0"), length of each side (22'0" each, for a total of 44'0"), the side entrance (10'0"), and the front entrance porch (12'0") for a total of 98 linear feet. The footings under these walls are 16 inches, which is $1\frac{1}{3}$ foot, wide, and 8 inches, which is $\frac{2}{3}$ foot, deep. You multiply $1\frac{1}{3}$ by $\frac{2}{3}$ and you get $\frac{8}{9}$ or for all practical estimating purposes 1 cubic foot per linear foot for a footing 16" x 8". There are 27 cubic feet to a cubic yard. We divide our total length of our footings where the 16" x 8" footings are by 27. Our length is 98; 98 divided by 27 gives us $3\frac{17}{27}$ or, rounded off, $3\frac{2}{3}$ yards. The footing across the front of the house is 24" x 12" x 32' or a total of 64 cubic feet. Divide 27 into 64 and we get a total of $2\frac{10}{27}$ or, rounded off, $2\frac{1}{3}$ yards. Adding $3\frac{2}{3}$ plus $2\frac{1}{3}$ we get a total of 6 cubic yards required for the foundation footings. We must figure the supporting posts and the fireplace base. The supporting posts are 30" x 30" x 12", so we multiply $2\frac{1}{2}$ × $2\frac{1}{2}$ for a total of $6\frac{1}{4}$ cubic feet or, rounded off $\frac{6}{27}$ of a yard. The fireplace base is 5' x 2' x 1' or $\frac{10}{27}$ of a yard. Adding $\frac{6}{27}$ and $\frac{10}{27}$ we get $\frac{16}{27}$ or, rounding off, $\frac{1}{2}$ yard. This means that our concrete footings will require $6\frac{1}{2}$ cubic yards of concrete. We should add this to our material list.

Pieces	Feet	Description	Cost	Total
$6\frac{1}{2}$		cubic yards concrete – footings, 1-3-5 mix		

CONCRETE FLOORS

The specifications on concrete work indicate that the cement must be a domestic portland cement type I. The basement floor is to be a smooth, metal trowel finish. The front and side entrance porch floors are to be metal-float finished with a steel trowel. The sidewalks are to be wood-float finish.

Cement

Portland cement is a type of cement, not a brand name. Every manufacturer of cement makes portland cement. Portland cements are made to meet the American Society for Testing and Materials' standard specifications ASTM Designation. The most common portland cements are divided into five types. The most commonly used in residential construction are C150 portland cement, types I and III and C175 air-entraining portland cement.

Type I, normal portland cement, is a general purpose cement suitable for all uses when the special properties of the other types are not required. It is used in pavement and sidewalk construction, reinforced concrete buildings and bridges, railway structures, tanks and reservoirs, culverts, water pipe, masonry units, and for all uses of cement or concrete not subject to sulfate attack from soil or water. Since cement always gives off a certain amount of heat as it reacts to outside chemicals, it must only be used where the heat generated while the cement cures will not cause an objectionable rise in temperature.

Type III, high-early-strength porland cement is used when high strengths are desired at very early periods — from one to three days. It is used when it is desired to remove forms as soon as possible or to put concrete into service quickly, or in cold weather construction to reduce the period of protection against low temperatures.

Air-entraining portland cements are covered by ASTM C175 and are designated as Types IA, IIA, and IIIA. These correspond to Types I, II and III, respectively, in ASTM C150. In these cements, small quantities of air-entraining materials are incorporated during manufacture. These cements have been developed to produce concrete that will resist severe frost action as well as the effects of chemicals applied for snow and ice removal. Concrete made with these cements contains tiny, well distributed, completely separated air bubbles. The bubbles are so minute that there are billions of them in a cubic foot of air-entrained concrete.

When water is added to cement, it makes a paste which binds fine and coarse materials, called *aggregates*, into a solid, rocklike mass known as *concrete*. Cement hardens through the chemical action between cement and water. This water should be clean enough to drink.

Aggregates

Aggregates are classified as *fine* or *coarse*, depending upon their size. Sand which passes through a $\frac{1}{4}$ wire screen is called a fine aggregate. Coarse aggregates, which range in size from $\frac{1}{4}$ inch to 3 inches, are usually crushed stone or cinders.

Both fine and coarse aggregates are used in concrete so that the spaces between the coarse aggregates (stone) are filled by the smaller ones (sand). In addition to the aggregates already mentioned, there are lightweight aggregates which are manufactured from vermiculite, pumice, and other minerals. These have been developed to meet the need for reduced weight in certain types of construction, such as roofs or walls.

Placing Concrete

A thickness of 4 inches if recommended for most floors, sidewalks, steps, porch floors, and most other slabs made of concrete.

Concrete is placed in forms as quickly as possible after it has been thoroughly mixed. It must be consolidated (worked to remove voids) when placed in the forms. Concrete floors are leveled off evenly with a straight board called a *screed*. The concrete is then *floated* (smoothed with a tool called a float) to bring a film of cement and sand to the top (Figure 4-6). After the concrete has been floated, it may be troweled to a smooth, flat finish.

Fig. 4-6. Floating concrete.

Estimating Concrete

The amount of concrete needed is determined by multiplying the length by the width by the height, all in feet, and then dividing the total by 27. This gives the number of cubic yards of concrete needed to do the job.

Looking at our foundation plan of the working drawings, we find the area to be covered with concrete is 32′ x 22′, for a total of 704 square feet. Our porches are 29 square feet for a total of 733 square feet. The only place we can find the sidewalk shown is on the plot plan, which shows a 30-foot setback from the property line, so our sidewalk will be 30 feet long and 4 feet wide (standard walk width). 30′ x 4′ = 120 square feet for the walk. We add this to our total of 733 square feet for a total of 853 square feet.

Remember our specifications called for a 6-inch thick layer of graded gravel fill under all concrete slabs. We have an area of 853 square feet so we divide this by 2 (6-inch gravel fill − 6 inches = $\frac{1}{2}$ foot) and we get 426. We divide 27 (cubic feet in a cubic yard) into 426 and we find we will need 16 cubic yards of gravel fill. We should add this to our estimate sheet.

Pieces	Feet	Description	Cost	Total
16		cubic yards graded gravel fill — under concrete floors, sidewalk		

Estimating our concrete for this job, we again use 853 square feet, which is our total square-foot area of the porches, basement floor, and sidewalk. Our concrete is 4 inches thick, so we would take one-third of 853 square feet for a total of $284\frac{1}{3}$ yards. We divide 27 into this and get a total of $10\frac{1}{2}$ cubic yards of concrete, rounded off, required. We would add this to our material list.

Pieces	Feet	Description	Cost	Total
$10\frac{1}{2}$		cubic yards concrete for floor, porches and walk, 1-2-4 mix		

5
MASONRY ESTIMATE: FOUNDATIONS

CONCRETE BLOCK

Concrete blocks are made by mixing portland cement and water with a dense aggregate such as sand, gravel, crushed stone, or cinders. These blocks are formed under pressure, or a combination of vibration and pressure.

Concrete blocks are made 4, 6, 8, 10, and 12 inches thick, 8 inches high, and 16 inches long. These are nominal dimensions. Actually, each dimension is $\frac{3}{8}$ inch smaller, to allow for the mortar joint.

Special sizes and forms of concrete block are made for particular uses such as half blocks, corner blocks, chimney blocks, pier blocks, jamb blocks, and decorative block. Concrete blocks are widely used for foundations of residences and for foundations where brick veneer work is fastened to frame construction. Also, blocks are used extensively in small commercial construction in combination with brick.

The specifications for block and brick work are combined, as they normally are in residential specifications. All masonry is specified in one section of the specifications.

Figure 5-1 shows several of the types and sizes of block. Corner blocks are used to make a square, straight corner. Pier blocks are closed on each end and are used to make finished support. Half and whole sash blocks are used on each side of the basement windows. Solid cap blocks are used to cap the foundation wall. These cover the holes in the block wall and make a solid top for the foundation. Chimney blocks come in two styles; whole and half blocks. Whole chimney blocks are used where an 8" x 8" flue liner is installed in a single chimney. If a larger flue, such as 8" x 13", is to be used, then the use of half chimney block is recommended,

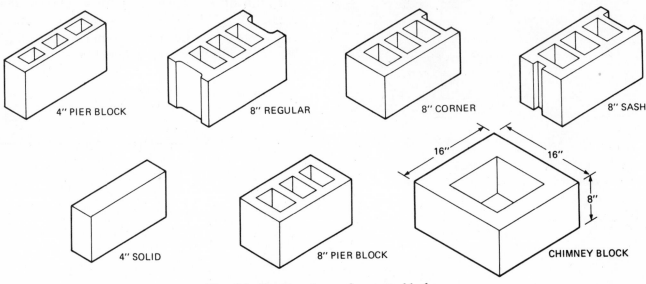

4" PIER BLOCK　　8" REGULAR　　8" CORNER　　8" SASH

4" SOLID　　8" PIER BLOCK　　CHIMNEY BLOCK

16"　　16"　　8"

Fig. 5-1. Common types of concrete blocks.

and the direction of the block is reversed on every other course.

Estimating Concrete Block

There are two methods of figuring concrete block. The *first method* is as follows. Find the area of the wall. Each 8 x 8 x 16-inch block covers 0.888 square foot, so figure $112\frac{1}{2}$ blocks for every 100 square feet of wall. Add 5–10 percent for waste. The *second method* is to figure three (3) blocks for every four (4) linear feet of wall, and three (3) courses for every two (2) feet in height. I prefer the second method.

Our specifications tell us that the cement block shall be load-bearing hollow concrete masonry units and shall conform to ASTM C-90 specifications. To find out where the blocks are to be used and their size, we must refer to the foundation plan of the working drawings. After studying the working drawings, we find that the foundation is to be constructed with 8-inch block on the sides and the back wall, and the entrance porch foundation. The front wall is to be 12-inch block. Our wall section shows that the 12-inch block is used for nine courses and the last course and cap block are 8 inch.

Estimating the number of blocks required, figure three (3) blocks for every four (4) linear feet of wall. Add the perimeter of the building. Our working drawings show we have a foundation 32′ x 22′ so we add 32 + 32 + 22 + 22 for a total of 108 linear feet. We add our porches and fireplace base. In excavation, enough space must be left around the outside of the foundation walls to allow workers to move easily; therefore, when the excavation is finished the foundation is usually made large enough to include room for the footings for the porches. This means that although there is no basement under them, the foundations for the porches have the same number of block courses as the rest of the foundation. The front entrance porch is 5 feet wide and 3′6″ deep, for a total of 12 linear feet. The side entrance porch is 4 feet wide and 3 feet deep, for a total of 10 feet. Our fireplace base is 5 feet wide and 2 feet deep, for a total of 9 feet. Our porches and fireplace base add up to 31 feet, which we add to our 108 linear feet for the main foundation wall, for a total of 139 linear feet.

Multiply 139 by 3 and divide by 4. 139 × 3 = 417 divided by 4 = 106 blocks are required for one course. Our foundation is $10\frac{1}{2}$ courses high; 10 times 106 gives us a total of 1060 cement blocks, plus 106 solid cap blocks.

Our working drawings show us that the front wall is to have brick veneer and calls for 12-inch block for nine courses. The distance across the front is 32 feet. 32 × 3 divided by 4 gives us 24 blocks for one course of 12-inch block. Nine courses high means that 24 × 9 = 216 12-inch block are required.

Our working drawings on the foundation plan show that we have five basement windows and our specifications (under Division 5: Metals) tell us that basement windows shall be steel 15″ x 12″, two light, glazed with combination storm sash and screens. We will need two half sash blocks and two whole sash blocks for every window. We multiply the number of cellar sashes by two and list these on our material list.

Pieces	Feet	Description	Cost	Total
10		half sash block — cellar windows		
10		whole sash block — cellar windows		

Next count the number of corners in the foundation and multiply this by 10 to find how many corner blocks will be needed. We have 10 corners; 10 × 10 = 100 cement corner blocks will be required. We subtract the 100 corner block from the 1060 blocks and we get a total of 960. We also have to subtract the number of 12-inch blocks: 216 from 960 gives us a total of 744 cement blocks. We would now list these on our estimate.

Pieces	Feet	Description	Cost	Total
744		8 x 8 x 16 regular cement block — foundation		
216		12 x 8 x 16 regular cement block — veneer foundation		
100		8 x 8 x 16 regular cement corner block		
106		4 x 8 x 16 regular solid cap block		

When brick veneer is used, it is common practice to lay 12-inch concrete blocks for nine courses. Then one course each of 8-inch and 4-inch blocks are laid flush with the inside face of the foundation wall. The holes in the top of the 12-inch wall are filled with cement. The bricks are laid on top of the top course of the 12-inch block (Figure 5-2). This places the brick below the finished grade and improves the appearance of the brickwork.

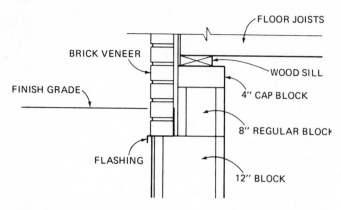

Fig. 5-2. Brick veneer the foundation is laid with larger block.

CONCRETE WALLS

Basement Sash

We already know the number and size of the cellar sashes, so we would now list them on our material list.

Pieces	Feet	Description	Cost	Total
5		15 x 12 steel sash, glazed, with storm and screen		

Our basement and foundation plan of the working drawings also show an areaway around each basement window. Again looking at the specifications (Division: 5 Metals), we find that the galvanized areaways shall be 20-gauge galvanized steel, with horizontal corrugations. We have five (5) basement sashes so we would list five (5) areaways.

Pieces	Feet	Description	Cost	Total
5		galvanized 20-gauge with horizontal corrugation – areaways		

We need a good mortar to lay the block, and our specifications (Division 4: Masonry) tell us that the mortar for laying brick and block shall be 1 part portland cement, 1 part mason's lime, and 6 parts clean sand. Estimating mortar for block, we figure 1 bag of cement for every 28 blocks. We have a total of 1186 blocks; divide 28 into 1186 and to get 42+. We would add this to our material list.

Pieces	Feet	Description	Cost	Total
42		bags regular cement – mortar		
42		bags mason's lime – mortar		
	10	cubic yards washed sand – mortar		

We need a good grade washed sand to make a good mortar. Most house construction will require approximately 10 cubic yards of sand for the foundation block mortar, chimney, and parging.

Horizontal Reinforcement

The specifications call for horizontal reinforcement in every second bed joint. The foundation is 84 inches high, which means there are 10 full courses and 1 course solid cap block. The total distance around the foundation wall is 108 linear feet. Divide 10 (courses) by 2 (every second course) and we get a total of 5. Multiply 108 × 5 = 540 linear feet. Our 12-inch wall is 9 courses high, so we will need a total of 4 times the length, which is 32 feet; our 12-inch wall will need 32 × 4 = 128 linear feet. The specifications also indicate that reinforcing shall be placed in the first and second bed joints above and below the openings. There is no cement block above the window openings. There are five (5) basement windows, which are 32 inches long. To be effective, the reinforcing should extend at least $1\frac{1}{2}$ block on each side of the opening. If there are three blocks for every four feet and the window openings are added, then the reinforcing under the openings is 80 inches long. Round this off by adding 4 inches and figure a total of 7 feet for each course under each window. Since there are 5 basement windows, multiply 14 by 5 to get 70 linear feet, which is added to our 540 linear feet for a total of 610. This is our total length of reinforcing. Our 12-inch block will require 12-inch reinforcing, so we subtract 128 from 610 to get 482 linear feet of 8-inch reinforcing and add this to our material list.

Piece	Feet	Description	Cost	Total
	482	linear feet Dur-O-Wall reinforcing – block 8″		
	128	linear feet Dur-O-Wall reinforcing – block 12″		

Figure 5-3 shows how horizontal reinforcement is used in a cement block wall.

This completes our foundation for cement block, in regards to the materials that is required.

POURED CONCRETE

If our foundation was to be poured concrete instead of cement block, we would have to estimate

Fig. 5-3. Reinforcing.

the number of yards of concrete required. To do this, multiply the length of the foundation times its thickness in feet. Multiply this times its height in feet and divide by 27 for the cubic yards required. Our foundation wall that is 8 inches, including the porches, is 98 linear feet. Take three-quarters of this, for a total of 74 feet, and multiply by the height (8 feet), for a total of 592 cubic feet. Our front wall is 32 feet long and 12 inches thick. Multiply 32 by $\frac{3}{4}$ (for the 8-inch height), for a total of 256 cubic feet. Add 256 and 592, for a total of 848 cubic feet in our foundation walls. Divide 848 by 27, for a total of $31\frac{11}{27}$, or a total of $31\frac{1}{2}$ cubic yards of concrete required in the foundation. Reinforcing rods, formwork, and bracing for the poured concrete would be listed separately.

6
MASONRY ESTIMATE: FIREPLACES, CHIMNEYS, AND BRICK VENEER

FIREPLACES AND CHIMNEYS

Outside appearances of fireplaces have varied throughout history. They range from the early colonial massive brick and hearth style to the more modern two-way corner fireplace (Figure 6-1).

The design of the interior portion of fireplaces has remained fairly constant. The proportions necessary to allow a fireplace to draw (pull smoke up the chimney) and burn properly have been established by experience. In order for a fireplace to burn properly, it must have a good draft (draw well) to prevent smoke from coming into the room. The size of the fireplace opening is determined by the damper used, the size of the flue liner, and the height of the chimney. All of these factors influence the draft.

Many fireplace forms and fireboxes are prefabricated metal units (Figure 6-2). These come in various styles and sizes. The flue size is specified by the manufacturer.

There are standard rules to follow in the construction of a fireplace. The base of the firebox is always made with firebrick. The firebricks are set in a bed of mortar without any side or end joints. These small cracks fill with fine sand and dirt. If mortar joints are used between the firebricks, the joints soon burn out and frequently need repairing. Usually a metal ash dump is placed in the floor of the firebox to clean the ashes from the firebox. The ash dump opens into an ash pit, which is built into the base of the fireplace. After a period of time, the ashes are removed through the cleanout door, which is built into the base of the fireplace.

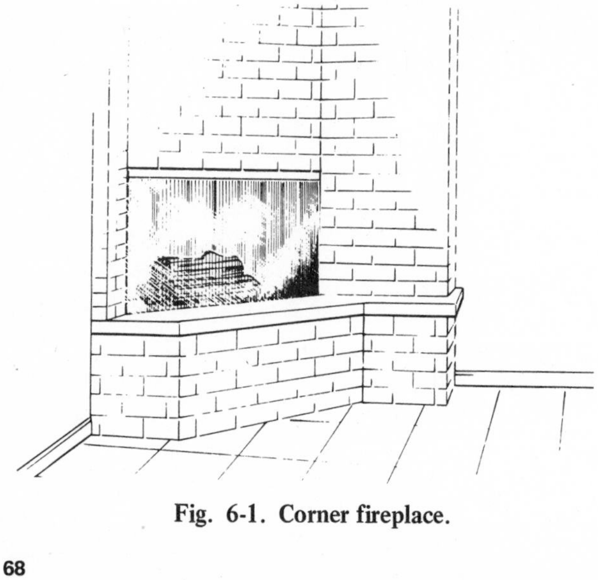

Fig. 6-1. Corner fireplace.

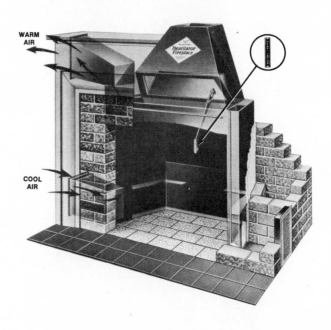

Fig. 6-2. Prefabricated metal fireplace.

If the firebox is to be built of firebrick, the total thickness of the firebox wall should be at least 8 inches.

The hearth is made of masonry fireproof construction. This is usually brick, slate, stone, marble, or quarry tile. The hearth should extend at least 16 inches in front of the fireplace opening and at least 8 inches on each side. Combustible materials should not be used within 8 inches of each side of the fireplace opening.

Chimney Construction

The chimney should have a fireclay flue lining. Standard flue liners are 24 inches long. The most common sizes are $8\frac{1}{2}''$ x $8\frac{1}{2}''$, $8\frac{1}{2}''$ x 13", 13" x 13", and 13" x 18". When more than two flues are used in the same chimney, a 4-inch masonry division is necessary. The chimney wall should be at least 4 inches thick and should be separated from the wood framing of the building by a 1-inch air space, and from the flooring and sheathing by a 1-inch air space.

The top of the chimney should be at least 2 feet above the ridge of the roof or any other nearby obstruction. The width and depth of the fireplace depends upon its style and its location. A chimney that is exposed along the outside wall of a house takes longer to warm up in cold weather, so it takes longer to get a good draft.

The plan for the chimney is found on the basement and foundation plan. If a fireplace is included, the chimney may be located either on an inside wall near the center of the house or on an outside wall. If it is a single-flue chimney, it is usually located near the center of the building. The purpose of the chimney is to carry the fumes of the heating system out of the house. The heating system is usually placed near the center of the building to save on installation costs and to provide more efficient heat distribution.

The flue for the furnace is usually an 8" x 8" flue. On the plans, the architect will specify either 8" x 8", 8" x 12", or 12" x 12". The actual flue liner is $8\frac{1}{2}''$ x $8\frac{1}{2}''$, $8\frac{1}{2}''$ x 13", or 13" x 13".

Estimating a Chimney

For a concrete block chimney, find the total height of the chimney in feet from the basement to the top of the chimney cap. To find the number

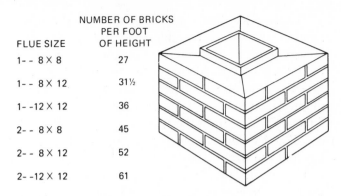

FLUE SIZE	NUMBER OF BRICKS PER FOOT OF HEIGHT
1 - - 8 × 8	27
1 - - 8 × 12	31½
1 - -12 × 12	36
2 - - 8 × 8	45
2 - - 8 × 12	52
2 - -12 × 12	61

Fig. 6-3. Brick chimney.

of flue liners needed, divide this height by 2 because standard flue liners are 2 feet long. A chimney is usually built of flue liners and whole chimney blocks up to the roof line. Brick is used where the chimney comes through the roof (Figure 6-3).

To build an 8" x 8" furnace chimney, three whole chimney blocks are needed for each flue liner. Find the height of the chimney to where it comes through the roof. From the roof line to the top of the chimney (which should be a minimum of two feet above the ridge) estimate 27 bricks for every foot of height.

For example, consider a 24-foot high chimney which extends 6 feet through the roof, with a single 8 x 8 flue. The total height of 24 feet indicates that 12 flue liners are needed. Subtract the 6 feet above the roof from the 24 feet (total height). This leaves 18 linear ft. of whole chimney block. Three blocks are required for every two feet of height; 2 divided into 18 equals 9, and 9 multiplied by 3 equals 27: 27 whole chimney blocks are required. From the roof line to the top of the chimney is 6 feet; 6 times 27 (number of bricks needed for each foot of height for a single 8 x 8 flue) equals 162 bricks.

To estimate a fireplace, it is necessary to become more involved with the construction.

Estimating a Fireplace

The floor plans, elevations, and detail drawings must be studied to determine how the fireplace is to be constructed and what materials are required. The base of the fireplace is usually built up with regular concrete block. If the foundation is concrete block, the base of the fireplace is usually figured with the foundation block. The ash pit requires a cleanout door, and this is usually 8" x 8" or 12" x 15" and the furnace flue requires an 8" x 8" inch cleanout door.

The furnace flue liner extends from the basement to the top of the chimney. The firebox flue liners extend from about one foot above the damper to the top of the chimney. The fireplace is on the first floor, so approximately three fewer flue liners of the larger size are required than are smaller flue liners. The flue from the firebox is usually 8" x 13" or 13" x 13", depending upon the size of the chimney and the size of the damper.

The top of the fireplace hearth is usually even with the finished floor. In some cases the hearth may be built up several courses above the finished floor. This is known as a raised hearth. The top of the fireplace base is framed so that the concrete can be poured over the top and extend into the room under the hearth. This is reinforced with steel rods. A hole is cut into the ash pit so that a metal ash dump can be placed in the base of the firebox. The ash dump is usually the same size as a firebrick. The cement is finished at a depth which leaves room for the firebrick and hearth material to be set in a bed of mortar.

Most fireplace bases use approximately 30 regular firebricks for the floor area. Firebricks are usually $2\frac{1}{2}$" x $4\frac{1}{2}$" x 9". If the mason is to construct the fireplace using firebrick for the sides of the firebox, study the fireplace details to make sure how the firebricks are to be laid. The firebox walls may be $4\frac{1}{2}$ inches thick, with the firebrick laid flat, (Figure 6-4) or they may be $2\frac{1}{2}$ inches thick, with the firebrick laid on edge (Figure 6-5).

To estimate the number of firebricks required, first find the area of the walls in square feet. Multiply this area by 6.4 if the walls are to be $4\frac{1}{2}$ inches thick; if they are to be $2\frac{1}{2}$ inches thick, multiply the area by 3.55. When firebricks are used in the firebox wall, they are usually set in fireclay. About 40 pounds of fireclay are required to set 30 firebrick.

When firebrick is used for the sides of the firebox, a damper must be ordered. Dampers vary in size and design; therefore, the manufacturer, catalog number, width of the fireplace opening, and type of control (poker or rotary) must be listed.

A steel angle should be included as a lintel to support the masonry work over the fireplace openings. This may not be necessary in all cases; some dampers have a built-in lintel, so a separate piece of angle iron need not be included. Unless it is specified differently on the plans a $3\frac{1}{2}$" x $3\frac{1}{2}$" steel angle iron should be used. The length of the angle iron should be equal to the fireplace opening plus at least 4 inches of bearing on each end. For example, a 30-inch fireplace opening should have at least a 38-inch angle iron.

Preformed metal fireplace units come with the firebox, damper, and controls all in one unit. These preformed units should be listed by size and manufacturer. The manufacturer furnishes details such as the necessary flue size for the specific chimney height. Firebrick for the floor of the unit must be ordered separately.

The location of the fireplace is important, as is the number of fireplaces on the same chimney. The fireplace we will estimate is in the living room and on the outside wall, which means we will have to figure face brick on the exposed face. Our working drawings and the specifications tell us that the fireplace unit is a prefabricated unit. Under Division 10: Specialities, it tells us that the fireplace unit shall be a #35 Heatilator manufactured by Vega Industries.

Our fireplace base was figured with the foundation, so we need only estimate from the top of the foundation wall, except for the 8" x 8" flue liners and the cleanout doors. The fireplace prefabricated unit sets on a firebrick base, so we should estimate 30 firebricks for the base. We will need an ash dump that goes in the firebox base. We need one 12" x 15" cleanout door for the ash pit and one 8" x 8" cleanout door for the furnace flue. We would list the prefabricated unit by size and name. Our chimney is 32 feet high, so we will need 16-

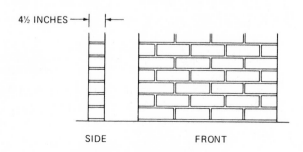

4½ INCHES

SIDE   FRONT

Fig. 6-4. Firebrick laid flat.

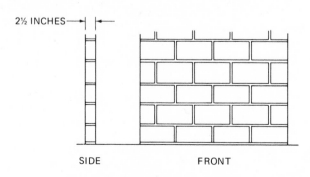

2½ INCHES

SIDE   FRONT

Fig. 6-5. Firebrick laid on edge.

8″ x 8″ flue liners for the furnace flue. The fireplace is on the first floor, so we will need fewer of the larger size flue liners. The fireplace flue starts approximately four (4) feet above the first floor. From this point to the top of the chimney is 22 feet, so we will need 11–8″ x 12″ flue liners for the fireplace. We will need a 6-inch thimble for the furnace flue. To carry the face brick across the fireplace opening we will need a $3\frac{1}{2}$″ x $3\frac{1}{2}$″ angle iron 40 inches long.

Our chimney is exposed on the outside wall. Measuring the chimney on the elevation drawings we find that the wide part of the chimney where the fireplace heatilator goes is 9 feet around (5 feet wide and 2 feet returns) by 8 feet high, for a total of 72 square feet. From the corbel to the top is 20 feet high by 7 feet, for a total of 140 square feet. The chimney above the roof is approximately 3 feet high, so we add 9 more square feet for a total of 221 square feet. Estimate $6\frac{3}{4}$ brick for every square foot; $221 \times 6\frac{3}{4} = 1,492$ bricks. We need fillers between the flue liners and the face brick, so we add 20 percent or 298 more bricks. We need brick around the face of the fireplace opening in the living room, and this will take approximately 50 bricks. This gives us a total of 1840 face bricks or, rounded off, 1,900 bricks.

You may also figure 55 bricks for every foot in height with an 8 x 8 and an 8 x 13 flue liner. Our height is 20 feet, for a total of 1,100 bricks, plus 72 square feet at $6\frac{3}{4}$ where the heatilator is located, for a total of 486, for a grand total of 1586. We still need fills around the heatilator. Our 55 bricks per foot in height includes fills, so we only need add 15 percent for fills which means we add 218 for a total of 1804 brick. We still must add our face brick around the opening on the fireplace, so we add 50 to our 1804, for a total of 1,854 or, rounded off, 1,900 bricks required.

We need regular cement to finish the fireplace cap and to set the quarry tile hearth. The average fireplace will require 4 bags of portland cement for this work. Our hearth is to be quarry tile and this totals 12 square feet.

We need mortar to build a fireplace. One bag of portland cement and one bag of mason's lime with 48 shovels of sand will lay approximately 400 bricks. We should allow one-third for filling between brick, fills, and flue liners. For mortar, you should figure 1 cubic yard of sand for every 10 bags of cement. For brick veneer with $\frac{1}{2}$-inch mortar joint we need 8 bags of cement for every 1,000 bricks. Our

chimney has 1,900 bricks, so we will need 16 bags of cement. Figuring 25% for waste and fills, we would need 4 more bags, so we should figure 20 bags of cement and 20 bags mason's lime for our chimney. We would now add our fireplace material to our material list.

Pieces	Feet	Description	Cost	Total
30		regular firebrick – firebox base		
1		ash dump – fireplace base		
1		12″ x 15″ cleanout door – ash pit		
1		8″ x 8″ cleanout door – furnace flue		
1		6″ thimble – furnace flue		
1		No. 35 heatilator fireplace unit – Vega Industries		
16		8 x 8 flue liners – furnace flue		
11		8 x 12 flue liners – fireplace flue		
1		angle iron, $3\frac{1}{2}$″ x $3\frac{1}{2}$″ x 40″ lintel		
1,900		antique common brick – fireplace and chimney		
4		bags regular portland cement – hearth and cap		
	12	square feet quarry tile, 4″ x 4″ x $\frac{1}{2}$″ – hearth		
20		bags regular portland cement – mortar for chimney		
20		bags mason's lime – mortar for chimney		

BRICK

Bricks are small, solid units made in rectangular form from clay or shale, and hardened by heat. Bricks used for structural purposes fall into three general classifications.

1. *Common brick* is a clay brick which is formed and then baked to a hard material. Common brick is porous and is used primarily as a structural material where the appearance is not of primary importance.

2. *Face brick* is a shale brick which is formed and then baked into a hard material. Face brick is

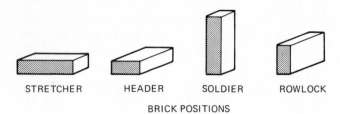

STRETCHER   HEADER   SOLDIER   ROWLOCK

BRICK POSITIONS

Fig. 6-6. Brick positions.

Fig. 6-7. Running bond.

Fig. 6-8. English cross bond.

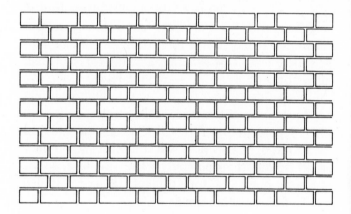

Fig. 6-9. Flemish bond.

dense shale and will not absorb water. The surface can have a smooth or matte finish or it can be finished with a wire brush to give a textured finish. Face brick is used for exterior facing and comes in various colors.

3. *Special types* are bricks made for a particular purpose, such as glazed brick, acid-resistant bricks, paving bricks, and firebricks.

The advantages of using brick is its durability, beauty, low maintenance cost, and the higher resale value of houses on which it is used.

The nominal size of common brick and face brick is $2\frac{1}{4}$ inches thick, $3\frac{3}{4}$ inches wide, and 8 inches long. Firebrick is $2\frac{1}{2}$ inches thick, $4\frac{1}{2}$ inches wide, and 9 inches long. Roman brick is $1\frac{5}{8}$ inches thick, $3\frac{3}{4}$ inches wide, and 12 inches long. These are the most common types of brick used in residential construction. Bricks are also named according to their position in the wall (Figure 6-6). By using bricks in various combinations of these positions, a variety of bond patterns can be achieved (Figures 6-7–6-9).

Estimating Brick

In estimating the quantity of common or standard face brick in veneer work, 675 bricks are assumed for every 100 square feet of wall. An allowance of 5–10 percent should be included for waste. Brick veneer refers to one pier of brick as a face pier, which is backed by other masonry units or attached to a frame wall. If bricks are laid on edge (rowlock) or standing on end (soldier) allow $4\frac{1}{2}$ brick for each linear foot. To estimate the number of bricks in a solid pier, with $\frac{3}{8}$-inch joints, see Figure 6-10. To estimate bricks for brick steps with treads 10 inches wide and $7\frac{1}{2}$-inch rise, allow $8\frac{1}{2}$ bricks on edge for each linear foot of step, Figure 6-11.

Our plans and specifications call for a brick veneer wall across the front of the building as high as the second floor projection. The length of the front is 32 feet and by measuring the elevation, we determine the height to be 10 feet. We multiply the length by the height and we get a total of 320 square feet. Our plans show that we have two windows and one door where the veneer work is to be placed. Each window is 3'0" x 4'6" and the front door is 3'0" x 6'8". If we multiply 3 feet by $4\frac{1}{2}$ feet we get a total of $13\frac{1}{2}$ square feet. We have two windows, so the total window area is 27 square feet. Our front door is 3' x 6'8", so we multiply 3 feet by 7 feet for a total of 21 square feet. We add 27 square feet to 21 square feet for a total of 48 square feet that don't need brick. We subtract 48 from 320 for a total of 272 square feet for the brick-

PIER SIZE	NUMBER OF BRICKS PER FOOT OF HEIGHT
8 × 8	13½
8 × 12	18
12 × 12	20½
16 × 16	36

SOLID PIER

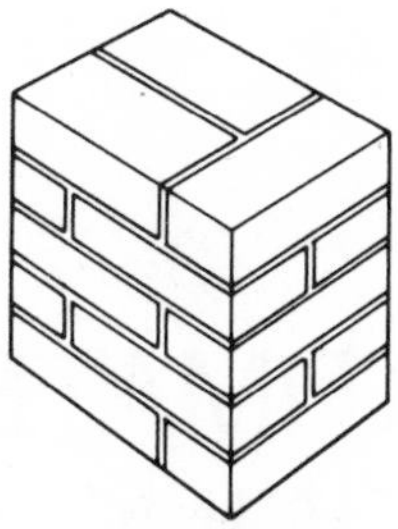

Fig. 6-10. Solid pier.

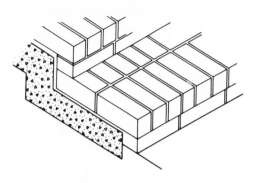

Fig. 6-11. Brick steps.

work. It takes 6.75 bricks for every square foot of face brick. We multiply 272 by 6.75, for a total of 1,836 bricks.

We should figure 70 wall ties for every 100 square feet of brick veneer. We have 272 square feet of wall where the veneer work goes; 272 divided by 100 rounds off to $2\frac{3}{4}$; $2\frac{3}{4}$ times 70 gives a total of $192\frac{1}{2}$, rounded off to 192 wall ties required.

In order to place the brick in the wall, we need mortar for the joints. We figure eight (8) bags of cement for every 1,000 bricks. We had a total of 1,836 bricks; we multiply 8 times $1\frac{3}{4}$ for a total of 14 bags of masonry cement for our veneer work.

To carry brick over an opening, we use a $3\frac{1}{2}$ x $3\frac{1}{2}$ steel angle iron. The angle iron should be long enough to rest on each side of the opening at least the length of one brick. All three openings are 3 feet wide, so we would figure three (3) $3\frac{1}{2}$ x $3\frac{1}{2}$ irons, 4′6″ long.

Our wall section detail shows that flashing is required where the brickwork starts at the top of the foundation wall. We also need flashing where the chimney meets at the cornice of the house. We would estimate 32 linear feet of 24-gauge aluminum, 28 inches wide, for this work. The gauge and width and material for flashing is specified in the specifications under Division 7: Moisture Protection, B-5. We would now add this to our material list.

Pieces	Feet	Description	Cost	Total
1,836		colonial antique common brick — veneer work		
192		galvanized wall ties		
14		bags masonry cement — brick veneer		
3		$3\frac{1}{2}$ x $3\frac{1}{2}$ x 4′6″ steel angle irons — lintels, veneer work		
	32	linear feet 24-gauge aluminum, 28″ wide — flashing		

7
MASONRY ESTIMATE: DAMP-PROOFING

We now come to the last part of our masonry estimate, the foundation damp-proofing. We find the information covering this aspect of construction in our specifications under Division 7: Moisture Protection.

FOUNDATION DAMP-PROOFING

Because of the level of the water table (the natural level of water in the ground) it is not possible to have a usable cellar in many parts of the country. In these areas most homes are built on concrete slabs or have a crawl space underneath the first-floor joists. In other sections of the country it is possible to have a usable basement, even where subsurface water exists. The building must be designed for these conditions.

Building contractors are usually aware of the water conditions in their area. However, unfavorable water conditions may not be discovered until the excavation is completed. If there is a possibility of a future water problem, it is less expensive to take care of it at the beginning of construction. Once the foundation has been backfilled it is impossible to waterproof the walls from the inside.

The specifications and working drawings for most houses require perimeter drains and foundation damp-proofing when the house is built. The most practical way to prevent a wet cellar is to use a perimeter drain (an underground drain pipe around the footings) and crushed stone to carry the water away from the foundation. If the ground has a natural slope, the pipe can be run around the foundation wall and away with the slope of the ground. In areas where there is no natural drainage, the drain is run to a sump on the inside of the foundation. A sump pump is installed to take the

water out of the area, either to a storm sewer or a dry well away from the building. The perimeter drain pipe is normally either clay tile (Figure 7-1) or plastic pipe (Figure 7-2).

Clay drain tiles are sections 12 inches long with a 3 inch inside diameter. These sections are laid end to end and the joints are wrapped with building paper. Plastic drain pipe is manufactured in 10-foot lengths and 250-foot rolls and may be either perforated (with a series of holes to allow water to pass through) or solid. Rigid plastic pipe is joined with

Fig. 7-1. Clay tile.

Fig. 7-2. Perforated plastic.

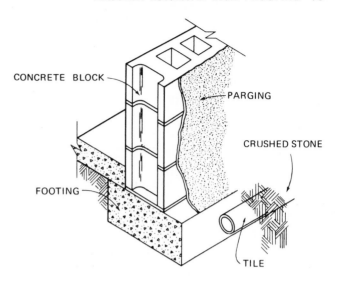

Fig. 7-3 Damp-proofing.

special couplings. An assortment of plastic fittings is available, such as tees, elbows, and a variety of angles. The advantage of plastic pipe is that the longer lengths and tighter fitting joints eliminate the possibility of soil filtering into the joints and clogging the pipe.

Both clay tile and plastic pipe are laid in the same manner. A trench the size of the footing is dug just outside the footing. The drain is embedded in crushed stone in this trench. The drain pipe or tile is pitched $\frac{1}{8}$-inch per foot so that the water will run toward the sump or drain opening (Figure 7-3).

ESTIMATING THE PERIMETER DRAIN

To figure the amount of plastic pipe needed, find the distance around the house and divide by 10. This is the number of 10-foot pieces of plastic pipe needed. A 90-degree elbow is required at every corner; and where the drain goes under the footing to a sump, a tee is required. The perimeter drain requires approximately one ton of stone for every 18 feet of pipe.

Our specifications call for No. 2 crushed stone and perforated plastic drain tile. Drain tile are 10-foot lengths. Our foundation is 108 linear feet around it so we will need 11 pieces of 6-inch perforated plastic pipe. We have four outside corners so we will need four 90-degree elbows and a tee where the drain goes under the footing into the

sump pit. In estimating stone, we need one ton for every 18 linear feet of drain tile. We divide 108 (perimeter) by 18 for a total of 6 tons of No. 2 crushed stone. We would add this to our material list.

Pieces	Feet	Description	Cost	Total
11		perforated plastic drain pipe, 6″ x 10′		
4		90-degree elbow–plastic drain pipe		
1		tee – plastic drain pipe		
6		tons No. 2 crushed stone – drain tile		

PARGING

The foundation wall is to be plastered with a $\frac{1}{2}$-inch coat of cement and lime below grade. When the plaster is dry, it is to be coated with two coats of asphalt. The mason trowels on the plaster, smoothing it as much as possible. Where the foundation rests on top of the footings, it is plastered from the top of the footing up. The plaster is canted so that it forms a curved surface from the wall out to the edge of the footing (Figure 7-4). This plastering is called parging.

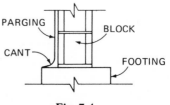

Fig. 7-4.

Portland cement mixed with sand does not work well and does not hold together to make a good plaster. This is the reason that mason's lime is added. The lime acts as a binding and fills any voids that might exist between the particles of sand and cement.

ESTIMATING DAMP-PROOFING

To estimate the quantity of materials for the foundation damp-proofing, first find the area to be covered. The perimeter of the building is 108 feet and the foundation is approximately 6 feet below grade. This means there is an area of 648 square feet to be parged. One bag of portland cement and lime will make enough plaster to cover about 40 square feet, so divide the area to be covered by 40 (648 divided by 40 equals 16 bags of cement). Our specifications also call for all concrete block above finished grade to be plastered with $\frac{1}{2}$-inch thick masonry plaster using 1 bag of masonry cement, 2 shovels portland cement, and 20 shovels of sand, the finish to be a smooth wood-float finish. The coverage for this plaster is approximately the same as below grade. Our foundation is approximately two feet above grade so 108 times 2 equals 216, which, divided by 40, tells us we should figure on 6 bags of masonry cement. Figuring 8 shovels to the bag, we still need 2 bags of portland cement to go with the 6 bags of masonry cement for the top of the foundation.

The specifications also call for two coats of asphalt damp-proofing material to be applied over the parging. The rate of coverage depends upon the manufacturer and the type of coating used. Asphalt emulsion is available in brush or trowel consistency. The brush type is the most widely used. It will cover about 30–35 square feet of wall per gallon. Divide 648 (the area to be covered) by 30 to get approximately 22 gallons. The second coat will only require about half as much as the first; add 11 more gallons, for a total of 33 gallons. Asphalt is sold in 5-gallon cans; therefore, approximately seven 5-gallon cans of coating will complete the job. We would now add this to our material list.

Pieces	Feet	Description	Cost	Total
16		bags portland cement — parging foundation wall		
16		bags mason's lime — parging foundation wall		
6		bags masonry cement — plastering top of foundation wall		
2		bags portland cement — plastering top of foundation wall		
7		5-gallon cans asphalt coating — foundation damp-proofing		
2		long-handle coating brushes		
2		yards washed sand		

8
CARPENTRY FRAMING: INTRODUCTION, BASEMENT, AND FIRST FLOOR

LUMBER AND FRAME CONSTRUCTION

Lumber is broadly grouped as either hardwood or softwood. Hardwood is not always harder than softwood, but this is generally true. Hardwood comes from deciduous (leaf-bearing) trees and softwood comes from coniferous (cone-bearing) trees. Approximately three-fourths of the total annual lumber production of the United States is softwood, which is used for all types of construction. The remainder is hardwood. Some of the most common softwoods are red cedar, cyprus, Douglas fir, eastern hemlock, western hemlock, Sitka spruce, eastern spruce, and the pines.

Lumber can be further grouped according to the amount of processing it goes through at the sawmill or the use for which it is intended. The following are some of the classifications of lumber:

1. *Rough lumber* is lumber that has not been planed or milled in any way after it is sawed.
2. *Surfaced lumber* has been surfaced on one or more sides in a planer.
3. *Worked lumber* has been shaped in a matching machine or molder.
4. *Factory* or *shop lumber* is intended to be cut up for future use. For example, clear pine, which is to be made into window or door parts.
5. *Yard lumber* is less than 5 inches thick and is graded for the future use of the entire piece.
6. *Structural lumber* is over 2 inches thick and wide and is graded for strength and the use of the entire piece.

Lumber Measurements

There are three ways to estimate quantities of lumber. These are:

1. The number of pieces required.
2. The number of linear feet required.
3. The number of board feet required.

Estimate by the number of pieces required if (1) the ends rest on bearings that are a specific distance apart; (2) a number of pieces of the same length are required; or (3) a specific length must be cut, without waste. Estimate by linear feet if the stock can be joined anywhere; for example, box sill or the sill itself, or the boards for the sides of the footing forms. Fractions of a linear foot are always counted as a full foot. Odd lengths are increased to the next even number of feet.

Lumber is commonly sold and listed by the *board foot*. A board foot is the quantity of lumber equal to 1 foot by 1 foot by 1 inch thick, or 144 cubic inches of wood. To find the board feet contained in a piece of lumber, convert all dimensions to inches, multiply the width times the length times the thickness, then divide by 144. When calculating board feet, it is customary to consider a fraction of an inch in thickness to be a full inch. For example, consider a board which measures $\frac{3}{4}''$ x 4" x 6'. After converting all dimensions to whole inches, the board measures 1" x 4" x 72" (288 cubic inches). Divide by 144 to find that the board contains 2 board feet.

Large areas such as subfloors, roof decks, and wall sheathing are usually covered by 4' x 8' panels. These panels are listed by the piece. However, if for some reason the areas are to be covered with boards, the quantity should be listed by board feet.

Nominal size is used to measure board feet. The nominal size of a piece of lumber is the size that it was sawed to at the sawmill. When the lumber is surfaced, it is further reduced in size (Figure 8-1). For example, a piece which has a nominal width of six inches is actually only $5\frac{1}{2}$ inches wide after planing.

One hundred board feet of 1 x 6 dressed (planed) lumber will not cover 100 square feet. A certain percentage must be added to make up for the loss in dressing, shiplapping, or matching. The amount of waste varies with the kind of material, where it is used, and how it is used.

Frame Construction

In order to read a set of plans and be able to make a list of materials, the estimator must know what materials are used for each specific purpose, and how to use those materials properly. There are two methods of framing a house: *balloon framing* and *platform framing*, (Figure 8-2 and 8-3).

Balloon Framing

In this type of framing, the studs are in one continuous piece from the sill or plates on the top of the foundation wall to the top plates under the roof rafters. The second floor joists are supported by a 1 x 6 strip or ribbon which is cut into the studs as shown in Figure 8-3. Notice that the floor joists and the studs rest on the sill, eliminating the box sill as the sheathing covers the ends of the joists.

Platform Framing

This type of framing is considered the fastest and the safest form of construction and it is the most widely used. Each floor is framed separately with the subfloor in place before the wall and partition studs of that floor are erected (Figure 8-2). Interior and exterior walls are framed in the same manner, assuring balanced shrinkage or settling, if any occurs.

BASEMENT BEAMS OR GIRDERS

These beams support the inner ends of the first-floor joists. The outer ends of the floor joists rest on the sill, on the top of the foundation wall. The girder and its supporting posts support the main bearing partitions, as well as part of the weight of the floors and their contents. Girders may be made of either wood or steel. When a built-up wood girder is used, it is made from three pieces of wood nailed together. Another type of wood beam which has superior strength and requires a minimum of material is known as a *box beam*. A box beam is made up of

---

The following abbreviations are in general use throughout the lumber industry. An estimator must be acquainted with them in order to read the plans and specifications.

B1S - Beaded one side
Bd. ft. - Board foot
Bev. - Beveled
B.M. or b.m. - Board (foot) measure
Btr. - Better
Clg. - Ceiling
Clr. - Clear
CM - Center matched
Com. - Common
Csg. - Casing
Dim.- Dimension
S2S - Surfaced two sides

Flg. - Flooring
Hdwd. - Hardwood
Lin. ft. - Linear foot or 12 inches
M - 1,000
Mldg. - Moulding
Qtd. - Quartered
Rfrs.- Roofers
Sdg. - Siding
Sel. - Select
T&G- Tongue and groove
S4S - Surfaced four sides

Thickness		Width	
Nominal	Dressed	Nominal	Dressed
1	3/4	2	1 1/2
1 1/4	1	3	2 1/2
1 1/2	1 1/4	4	3 1/2
2	1 1/2	5	4 1/2
2 1/2	2	6	5 1/2
3	2 1/2	7	6 1/2
3 1/2	3	8	7 1/4
4	3 1/2	9	8 1/4
		10	9 1/4
		11	10 1/4
		12	11 1/4

Fig. 8-1. Normal and minimum dressed sizes of lumber.

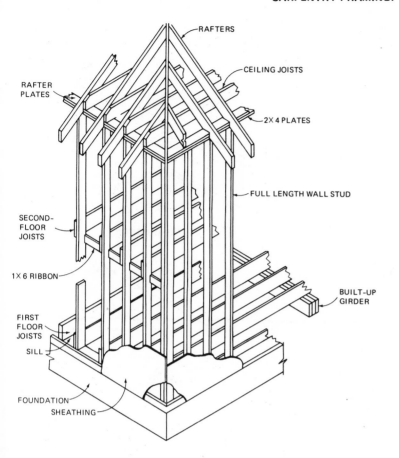

RAFTERS

RAFTER
PLATES

CEILING JOISTS

2 X 4 PLATES

FULL LENGTH WALL STUD

SECOND-
FLOOR
JOISTS

1 X 6 RIBBON

BUILT-UP
GIRDER

FIRST
FLOOR
JOISTS

SILL

FOUNDATION
SHEATHING

Fig. 8-2. Balloon framing.

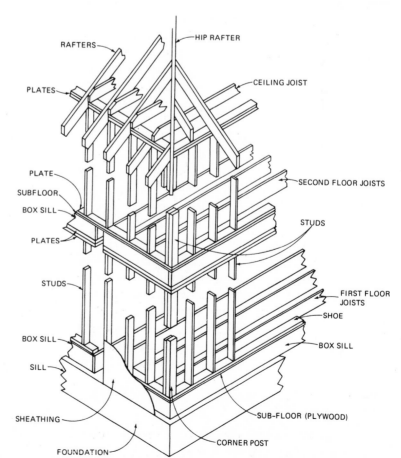

RAFTERS

HIP RAFTER

PLATES

CEILING JOIST

PLATE

SUBFLOOR

BOX SILL

SECOND FLOOR JOISTS

PLATES

STUDS

STUDS

FIRST FLOOR
JOISTS

SHOE

BOX SILL

BOX SILL

SILL

SHEATHING

SUB-FLOOR (PLYWOOD)

FOUNDATION

CORNER POST

Fig. 8-3. Platform framing.

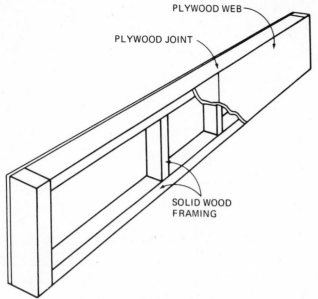

Fig. 8-4. Box beam.

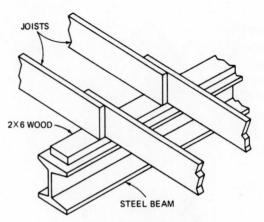

Fig. 8-5. Joists on top.

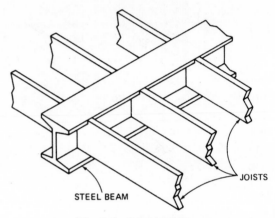

Fig. 8-6. Joists cut in.

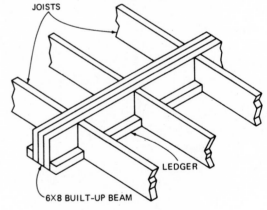

Fig. 8-7. With ledgers.

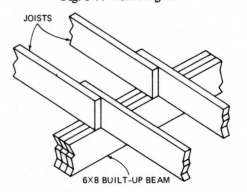

Fig. 8-8. Resting on top.

plywood nailed (sometimes glue is used) to 2 x 4 or 2 x 6 framing members (Figure 8-4).

If the girder is made of steel, there are several methods used to frame the wood joists to the steel beam. One method is to attach a 2 x 6 to the top of the girder and rest the joists on top of the 2 x 6 (Figure 8-5). Sometimes, instead, the beam is flush with the top of the floor joists and the joists are cut to rest on the flange of the beam (Figure 8-6).

The size of the built-up wood girder is given on the basement plan. Notes are included to explain how it is to be constructed. For example, a note that the girder is to be 6 x 8 means that three 2 x 8s are nailed together to make the girder. When the girder is built up in this way the end joints of the 2 x 8s should occur only at the supporting posts. The joists may be notched to rest on a *ledger*, as shown in Figure 8-7. This method is used where headroom is needed in the basement, as with a finished room. Another method commonly used is to rest the joists on top of the girder (Figure 8-8).

Still another method of attaching joists to a girder is with the use of *joist hangers*. These are specially designed steel devices which are nailed in place on the sides of the girder to form a pocket into which the joist fits. Joist hangers are available in a variety of sizes for different size joists.

Estimating Girders

The basement and foundation plan for the house shows the location and size of the girder. Solid lines with arrowheads on each end show the direction of the floor joists. The size of the floor joists and their spacing are noted just above this line. Looking at our basement plan (Figure 2-5) we find the notations for a 3-2 x 8 beam and also a 2 x 4 partition dividing the area for the recreation room. In this case, the partition would rest upon the concrete basement floor and the top of the partition would be the same as the top of the girder. Our beam notation calls for three 2 x 8s. Lumber is sold in increments of two-foot lengths up to a length of 24 feet. If you can span an area with one continuous piece of lumber, it is better than having end joints. Our basement is 22 feet wide so we would list 3-2 x 8 x 22' for our girder.

Our partition should be framed in the same way as all partitions, with a single shoe and a double plate on top. Our distance is 22 feet, so 22 × 3 = 66 linear feet of 2 x 4 is required for our plate and shoe. Our basement walls are $10\frac{1}{2}$ blocks high or 7 feet. We need 3 studs for every 4 linear feet and we should double the studs at the doorway. We multiply 22 (length of the partition) by 3, for a total of 66, which we divide by 4, for a total of $16\frac{1}{2}$ or 17 studs required. We double on each side of the doorway, so we add 2 for a total of 19. We would round this off to 20. Instead of purchasing 20-2 x 4 x 8', which means we would waste a foot for each stud, we would list 10-2 x 4 x 14'. We would now add this material to our material list.

Steel posts are used to support the framing members of the building between the foundation walls. The ends of the floor joists rest on the beam or girder. The support posts distributes the weight to the footings. Our plan calls for two supporting steel posts.

Pieces	Feet	Description	Cost	Total
2		4-inch steel post with $\frac{3}{8}''$ x 4'' x 6'' welded cap and base plates		
3		2 x 8 x 22' built-up beam		
	66	linear feet 2 x 4 plate and shoe – basement partition		
10		2 x 4 x 14' studs for basement partition – recreation room		

Our basement and foundation plan also shows stairs going from the basement up. Normally basement stairs are a straight run and are built from stock material. The stringers are made by using 2 x

10s and the treads are also constructed with 2 x 10s. If the back of the stairs are to be closed, then 1 x 8 pine is used for the risers. The following is a list of materials required to construct a plain stairway to a 7-foot basement: 2 pieces 2 x 10 x 14' for the stringers (sides); 4 pieces 2 x 10 x 12' for the treads; 1 piece 2 x 4 x 12' for the handrail; and 4 pieces 1 x 8 x 12' for the risers (if required). We would add this to our material list.

Pieces	Feet	Description	Cost	Total
2		2 x 10 x 14' stringers – basement stairs		
4		2 x 10 x 12' treads – basement stairs		
1		2 x 4 x 12' pine – hand rail		
4		1 x 8 x 12' pine – risers		

SILL SEAL AND SILL

The top of the foundation wall may not be absolutely true and even. To prevent small gaps between the foundation and the sill, a 6-inch wide by 3/4-inch thick fiberglass material called *sill seal* is used (see Figure 8-9). This is placed on top of the foundation wall and under the sill. This sill seal prevents insects and cold air from entering the basement of the house.

The sill is a single piece of wood (usually 2 x 6) laid flat on the top of the foundation wall (Figure 8-9). The first floor joists and the box sill are nailed to the sill. Usually the sill is fastened to the foundation every six or eight feet by anchor bolts that extend into the foundation.

Estimating Sill Seal and Sill

Enough sill seal and 2 x 6 (for the sill) must be ordered to cover the perimeter of the building. If a steel supporting beam is used for the main bearing and the joists rest on the top of the beam,

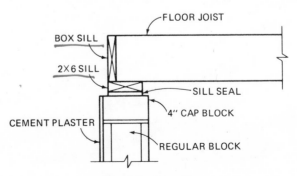

Fig. 8-9. Box sill.

then enough 2 x 6 to cover the length of the steel beam must be added. If built-up wood girders are used, it is not necessary to cover them with a sill.

The distance around the outside of the foundation is 108 linear feet. Sill seal is sold by the linear foot, so we would add 108 linear feet of sill seal to our material list. Lumber is sold by the board foot, but most lumber yards list it by the piece. Standard lengths are from eight to twenty-four feet in two-foot increments, but pieces over sixteen feet are more expensive. If the sill is 108 linear feet, nine 12-foot pieces are required (108 divided by 9 equals 12). We would add this to our material list.

Pieces	Feet	Description	Cost	Total
	108	Linear feet 6-inch sill seal		
9		2 x 6 x 12' sill		

BOX SILL

The *box sill* is the header joist placed at right angles to the ends of the floor joists (Figure 8-9). It is the same width as the floor joists and rests on top of the sill. The box sill is nailed to the other joists and toe-nailed into the sill.

Estimating Box Sill

The box sill covers the perimeter of the house, so it will also be 108 linear feet. It is the same height as the floor joists, and the plan indicates that they are 2 x 8s, so the box sill is 2 x 8s. Again this is added to our material list, and these would be the same length as the pieces we figured for the sill.

Pieces	Feet	Description	Cost	Total
9		2 x 8 x 12' box sill		

FLOOR JOISTS

The floor joists are the horizontal members that support the floor. They are usually spaced 16 inches o.c. (on center). For short spans, the joists may extend from one foundation wall to another. Over larger spans, the joists rest with one end on the foundation and the other on the supporting beam or girder. Joists sizes, like beam sizes, are dependent upon the length of the span they have to bridge and the load they have to carry. Joists should overlap for the full width of the beam that supports them, and they should be nailed together. Figure 8-10 shows the framing of floor joists.

Estimating Floor Joists

The size and direction of the floor joists is indicated on the foundation plan. A straight line with arrow-heads on each end indicates the direction of the joists and a note just above this line indicates the size

Fig. 8-10. Framing floor joists.

and spacing. For example: 2 x 8s — 16"o.c. This tells us that the joists are 2 x 8s and that they are spaced 16 inches on center; the arrowheads show the orientation of the joists. To find the number required, multiply the length of the foundation wall on which they rest by three-fourths, then add 1. This provides for 3 joists every 4 feet and 1 additional joist at the end of the wall. Also add one joist for each partition that runs in the same direction as the joists.

Our plan shows the joists running from the left and right side foundation wall to the partition and the built-up beam, and also spanning the space between the partition and the beam where the basement stairs go down. Using our formula we multiply the length of the foundation wall on which they rest by three-fourths, and then add one. Also we add one for each partition that runs in the same direction as the floor joists. Our foundation wall where the ends of the joists rest is 22 feet. 22 X 3 = 66, which divided by 4 gives $16\frac{1}{2}$ or 17, plus one more, for a total of 18 joists for the living room area. This same rule applies for the kitchen-dining area, plus we have one partition, so we will need 19 in this area. 19 + 18 = 37, so we will need 37-2 x 8 x 12' joists. Our center section shows a span of 8 feet from the center of the partition to the center of the girder. We use the same rule. 22 X 3 = 66; 66 ÷ 4 = $16\frac{1}{2}$ or 17. We also have one partition running the same direction, so we need 19-2 x 8s. We would add these to our material list.

Pieces	Feet	Description	Cost	Total
37		2 x 8 x 12' floor joists — living room, kitchen–dining area		
19		2 x 8 x 8' floor joists — hallway and stair area		

BRIDGING

Bridging is the lateral bracing of the floor joists. Wood bridging is made of short pieces of 1 x 3 nailed to the top of one joist and the bottom of the next (Figure 8-11). Bridging is also made from metal. The purpose of bridging is to keep the joists in alignment and to distribute the load over all the joists.

To determine the amount of bridging needed, multiply the number of floor joists by 3. This is the number of linear feet of bridging needed for one row. Spans of 10–14 feet usually have one

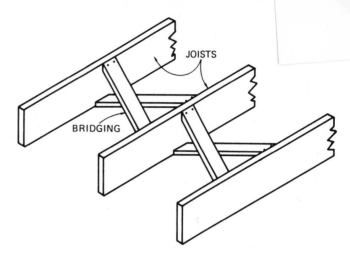

Fig. 8-11. Bridging.

row of bridging in the center of the span. If the span is over 14 feet wide, allow for two rows of bridging for each span.

We need one single row of bridging for each of our 12-foot spans, one over the living room area and the other over the kitchen–dining room area. We needed 37 joists to cover this area, so we multiply 37 by 3 for a total of 111 linear feet of 1 x 3 bridging. We would add this to our material list.

Pieces	Feet	Description	Cost	Total
	111	linear feet 1 x 3 spruce bridging		

HEADERS AND TRIMMERS

When the joists must be cut to make an opening for a stairway, chimney, or fireplace, *headers* are used to support the cut ends of the joists (Figure 8-12). The joists at the ends of the headers are called *trimmers*. Trimmers should be doubled to support the extra load when more than one joist is cut.

Estimating Headers and Trimmers

The material for headers and trimmers must be the same width as that for the floor joists. The length of the trimmers is the same as the length of the floor joists. We need headers and trimmers for the stairwell and the fireplace. Our trimmers should be 2–2 x 8 x 12' for the fireplace and 2–2 x 8 x 8' for the stairwell. We will need 1–2 x 8 x 12' header for the fireplace opening and 4–2 x 8 x 14' for headers at the stairwell. We would add this to our material list.

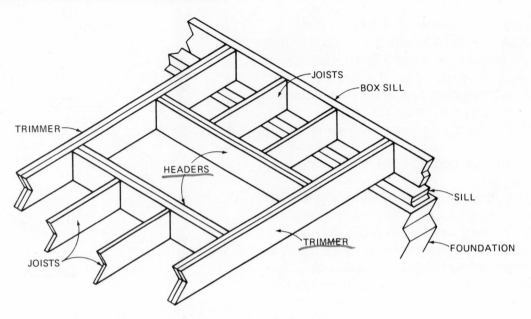

Fig. 8-12. **Headers and trimmers.**

Pieces	Feet	Description	Cost	Total
2		2 x 8 x 12' trimmers for fireplace		
1		2 x 8 x 12' header for fireplace (doubled)		
2		2 x 8 x 8' trimmers for stairwell		
4		2 x 8 x 14' headers for stairwell		

SUBFLOORING

The subfloor is the first layer of the floor. It is nailed to the floor joists and the finished flooring is applied to it. The subfloor helps keep the joists in alignment, holds the floor level, prevents the movement of dust and dirt between floors, helps deaden sound, acts as insulation, and provides a solid base to receive the finished floor. Most subflooring is done with 4' x 8' sheets of plywood (Figure 8-13).

Our first floor area is 32' x 22' or 704 square feet. Our plans and specifications call for $\frac{5''}{8}$ x 4' x 8' plywood. We take our 704 square feet of area and we divide by 32. There are 32 square feet in a 4' x 8' piece of plywood. This gives us a total of 22 sheets are required to cover the first floor area of our house. We would add this to our material list.

Pieces	Feet	Description	Cost	Total
22		$\frac{5''}{8}$ x 4' x 8' plywood — subflooring first floor		

To conserve material and eliminate the need to use underlayment, the American Plywood Association has developed new ways to construct floors using only one application of special plywood materials (Figure 8-14).

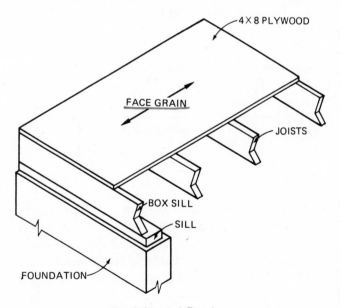

Fig. 8-13. **Subflooring.**

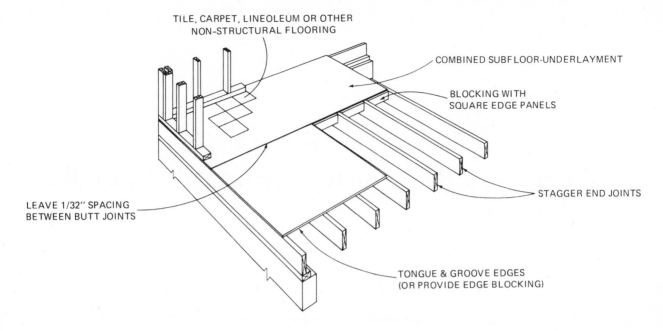

Combined subfloor-underlayment/ For direct application of tile, carpeting, linoleum or other non-structural flooring (Plywood continuous over two or more spans; grain of face plys across supports. Seasoned framing lumber is recommended.)

Plywood Grade[3]	Plywood Species Group	Maximum Support Spacing[1][2]						Nail Spacing (inches)	
		16" o.c.		20" o.c.		24" o.c.			
		Panel Thickness	Deformed Shank Nail Size[4]	Panel Thickness	Deformed Shank Nail Size[4]	Panel Thickness	Deformed Shank Nail Size[4]	Panel Edges	Intermediate
C-C Plugged Exterior	1	1/2"	6d	5/8"	6d	3/4"	6d	6	10
Underlayment with EXT. glue	2 & 3	5/8"	6d	3/4"	6d	7/8"	8d	6	10
Underlayment	4	3/4"	6d	7/8"	8d	1"	8d	6	10

Notes:

(1) Edges shall be tongue and grooved, or supported with framing.

(2) In some non-residential buildings, special conditions may impose heavy concentrated loads and heavy traffic requiring subfloor-underlayment constructions in excess of these minimums.

(3) For certain types of flooring such as wood block or terrazzo, sheathing grades of plywood may be used.

(4) Set nails 1/16" and lightly sand subfloor at joints if resilient (flooring is to be applied.)

Fig. 8-14. Combined subfloor-underlayment. One of the best ways to save money and time in subfloor construction is to use one layer of plywood as a combination subfloor-underlayment. This gives you all the advantages of plywood in both applications, yet speeds construction and custs costs. Use only the recommended grades shown in table above.

9
CARPENTRY FRAMING:
WALLS, CEILINGS, AND SECOND FLOOR

PLATES

The *plates* are the horizontal pieces, the same width as the studs (usually 2 x 4s) placed at the bottom and the top of a stud wall (Figure 9-1). The *sole plate* is a 2 x 4 nailed through the subfloor and into the joists and the box sill on the outside walls, or through the subfloor alone for the inside partitions. Sole plates are used wherever exterior walls and interior partitions are shown on the floor plan.

The top 2 x 4s, called *top plates*, are doubled. The top plates are lapped at the corners and where interior partitions and outside walls meet (Figure 9-2). The bottom 2 x 4 of the top plate is nailed to each stud and corner post. The top 2 x 4 is then nailed to the lower one.

Many carpenters lay out the work and frame the wall on the floor before tipping it up into place. The sole plate is nailed to the bottom of the studs and the top plate is nailed to the top of the studs. The wall is then tipped up and set in place and the sole plate is nailed to the subfloor and joists. The top plate is then nailed to the walls and partitions, lapping at corners and where the inside partitions meet the outside walls.

Estimating Plates

Working from the first floor plan for our house, we find the distance around the outside (perimeter) where the outside walls are to be erected. This is 108 linear feet. If this dimension is found with a scale, be sure the scale used in estimating is the same as the one used for the drawings. The scale of the drawing is given in the title box in the lower right-hand corner of the sheet. The scale for most residential house plans is $\frac{1}{4}'' = 1'0''$. If a dimension

is given, it takes precendence over measurements taken with a scale. Most plans do not show every dimension, so it is necessary to scale dimensions frequently in estimating.

Using an architect's scale, add the length of all the interior partitions. Do not deduct for door or window openings. Our house has a total of 78 linear feet for the interior partitions on the first floor. Now add the length of the interior partitions to all the length of the exterior walls: 108 + 78 = 186.

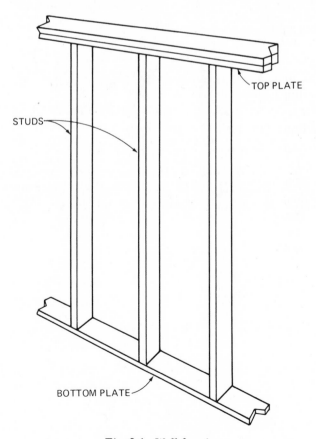

Fig. 9-1. Wall framing.

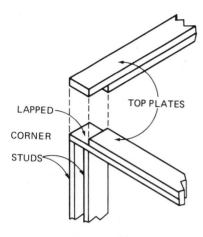

Fig. 9-2. Plates.

All walls and partitions have a single sole plate and a double top plate, so the total length of the walls and partitions is multiplied by three to find the quantity of material needed for the plates. Therefore 186 × 3 = 558 linear feet of 2 x 4s are needed for plates. We add this to our material list.

Pieces	Feet	Description	Cost	Total
	558	linear feet 2 x 4 plate and shoe — first floor		

STUDS

Studs are vertical members, usually 2 x 4s spaced 16 inches o.c., which support the weight of the ceilings, upper floors, and the roof. The studs on the exterior walls are covered with sheathing and siding. On the interior partitions they are covered with the interior wall finish.

There are two methods commonly used for fastening the studs in place. If the sole plate is first nailed to the subfloor and joists, then the studs are toenailed into the sole plate. A more common method is to nail the plates to the ends of the studs before the plates are fastened in place on the sub-

floor. When this method is used, an entire section of the wall is assembled and then it is tipped up and fastened in place.

There are several methods used for framing corners. Figures 9-3 and 9-4 show methods which are commonly used to provide a nailing surface for both the inside and the outside wall covering.

Estimating Studs

In estimating for studs, allow 1 stud for every linear foot of interior and exterior walls and partitions. Also add 2 studs for each corner. This provides an allowance for waste and double studs, which are needed in some places. From the first floor plan it can be seen that we have 15 corners (either outside corners or places where two walls meet). This means that 30 studs should be allowed for corners. Add this to our 186 (number of linear feet of walls and partitions) for a total of 216 studs required for the first floor. We add this to our material list.

Pieces	Feet	Description	Cost	Total
216		2 x 4 x 8" studs — first floor walls and partitions		

HEADERS

Whenever openings for windows or doors are cut out of the walls, *headers* (sometimes called *lintels*) are installed to carry the vertical loads to other studs. The header is always double and spaced so that its faces are flush with the faces of the studs on each side. Headers are always installed on edge, not flat. Under each header and nailed to the stud is a shorter 2 x 4 called a *jack stud*. Figure 9-5 is an illustration of the framing for a door and window opening.

The size of headers varies. They must be of sufficient size to carry the load. If the header sizes

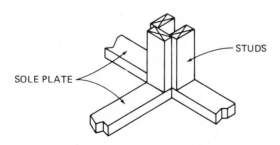

Fig. 9-3. Interior corner.

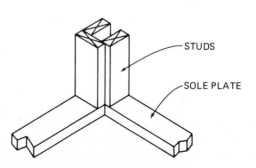

Fig. 9-4. Outside corner.

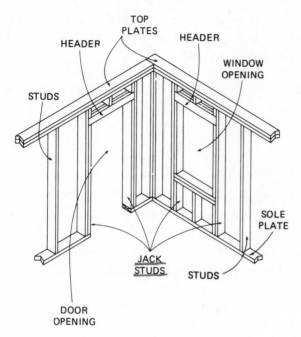

Fig. 9-5. Wall framing.

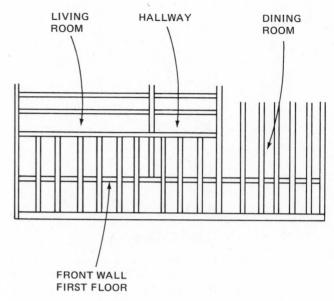

Fig. 9-6. Framing overhang.

are not given, the following sizes can frequently be used for house construction:

- spans up to 4 feet wide: 2–2 x 4s
- spans 4–5½ feet wide: 2–2 x 6s
- spans 5½–7 feet wide: 2–2 x 8s
- spans over 7 feet wide: 2–2 x 10s

Estimating Headers

To find the length of headers, measure each opening for doors or windows. According to our plan, we have no opening over three feet in width, so we will only need 2 x 4s. There are 8 windows and 2 doors in the exterior walls and there are 7 interior doors in the first floor partitions. Allow 3 feet for each opening and multiply 17 X 3 to get 51 feet of opening. Headers are double, so we double this figure and get 102 linear feet of 2 x 4 headers. We would add this to our material list.

Pieces	Feet	Description	Cost	Total
	102	linear feet 2 x 4 headers – first floor openings		

SECOND FLOOR FRAMING

This completes the framing for the first floor. We estimate the second floor the same way as the first. No sill is needed on the second floor, because the top plate of the first floor partitions provides a good

nailing surface for the joists. A box sill is required to frame the end of the second floor joists. Our second floor has a two-foot projection over the front wall, so our second floor area is 24' x 32'. In estimating our box sill we have to take the perimeter of the second floor which is 32 + 32 + 24 + 24 = 112 feet. Our first floor plan shows that the joists are to be 2 x 8s, so our box sill must be 2 x 8s. We would add the box sill to the material list.

Pieces	Feet	Description	Cost	Total
	112	linear feet 2 x 8 box sill – second floor		

We would figure the joists in the same way as we did for the first floor, taking into consideration the framing for the front projection. The joists over the living room and hallway would be the same direction. The joists over the dining room and kitchen would run from the front to the back. Framing the overhang on the front, the joists would be 4 feet long and run from the left end of the living room to the right side of the hallway (Figure 9-6).

Figuring a run of 20 linear feet from the exterior wall of the living room to the hall partition on the dining room would require 15 joists plus 1, or 16–2 x 8 x 12' over the living room and 16–2 x 8 x 8' over the hall. The front projection would be 16–2 x 8 x 4' or 8–2 x 8 x 8'. The joists over the dining room would run from the front to the back. A 12-foot run would require 9 + 1 = 10 joists. The span is 11'9" plus the 2'0" overhang, so we would need 10–2 x 8 x 14' joists for this area. Our kitchen area has the same run with a 10'3" span so we would need 10–2 x 8 x 12′ for this area. We would add this to our material list.

Pieces	Feet	Description	Cost	Total
16		2 x 8 x 12′ joists over living room		
16		2 x 8 x 8′ joists over hallway		
10		2 x 8 x 14′ joists over dining room		
10		2 x 8 x 12′ joists over kitchen		
8		2 x 8 x 8′ joists for overhang over living room and hallway		

Headers and Trimmers

We will need 1–2 x 8 x 12′ header over the living room, where we change the direction of the joists for the overhang and we will need 2–2 x 8 x 8′ for the hallway. Also we will need 4–2 x 8 x 14′ trimmers for the stairway in the hall. We would add this to our material list.

Pieces	Feet	Description	Cost	Total
1		2 x 8 x 12′ header – living room joists		
2		2 x 8 x 8′ headers – hallway joists		
4		2 x 8 x 14′ trimmers – stairwell		

Bridging

We have two areas needing bridging: living room area (second floor) and the dining-kitchen area. Our living room area had 16–2 x 8 x 12′, the dining room had 10–2 x 8 x 14′, and the kitchen area had 10–2 x 8 x 12′ for a total of 36 joists. We multiply 36 by 3 for a total of 108 linear feet of bridging required. We add this to our material list.

Pieces	Feet	Description	Cost	Total
	108	linear feet 1 x 3 bridging – second floor joists		

Subflooring

Our second floor area is 32′ x 24′, for a total of 768 square feet. Our specifications call for $\frac{5}{8}$-inch

plywood, so we divide 768 by 32 and get a total of 24 pieces required. We would add this to our material list.

Pieces	Feet	Description	Cost	Total
24		$\frac{5''}{8}$ x 4′ x 8′ plywood – subfloor second floor		

Plate and Shoe

We take the perimeter (distance around) the outside of the house on the second floor and we have 32 + 32 + 24 + 24 = 112 linear feet. We scale the interior partitions, measuring full lengths including door openings, for a total of 118 linear feet. We add the 112 linear feet (exterior walls) to our 118 linear feet (interior partitions) for a total of 230 linear feet of walls. We have a double plate on the top and a single plate (sole) on the bottom, so we would multiply 230 X 3 = 690 linear feet of 2 x 4 needed for the plates. We would add this to our material list.

Pieces	Feet	Description	Cost	Total
	690	linear feet 2 x 4 plate and shoe – second floor partitions		

Studs

Figure one stud for every foot of wall and add two for each corner. We have 230 linear feet of exterior and interior walls, and counting the corners on the second floor plan we find we have 28. We double this, getting 56, which we add to our 230, for a total of 286 studs required.

Headers

We have a total of 10 windows and 10 doors for a total of 20 openings. We figure three feet for each opening: 20 X 3 = 60. We double all headers: 60 + 60 = 120 linear feet of 2 x 4 is required for headers on second floor openings. We would add the studs and headers to our material list.

Pieces	Feet	Description	Cost	Total
286		2 x 4 x 8′ studs – second floor		
	120	linear feet 2 x 4 headers – second floor		

CEILING JOISTS

Ceiling joists support the ceiling of the top floor of the house. They also support the rafters and help secure them to the house. The ceiling joists must span from the outside wall to an interior bearing wall or the opposite exterior wall. If possible, they usually run in the same direction as the rafters. The ceiling joists are shown on the plans in the same manner as the floor joists. They would be shown on the uppermost floor plan. Figure 9-7 shows how ceiling joists are installed. Notice they tie the rafters and the roof from pushing the outside walls out.

Estimating Ceiling Joists

If possible, the ceiling joists (overlays) should run in the same direction as the rafters. Looking at our second floor plan, we see that this can easily be done except over the central hall. Scaling our plan, we find that the back bedrooms and baths will require 12-foot lengths, the front bedrooms 14-foot lengths, and the central hall 7-foot lengths. Our building is 24-feet deep. We multiply 24 by 3,

divide by 4 and add 1: $24 \times 3 = 72$, $72 \div 4 = 18$, $18 + 1 = 19$–2 x 6 x 7' ceiling joists required over the central hall. Lumber is sold in even lengths in two-foot increments, so we would have to change our central hall ceiling joists to 10–2 x 6 x 14' to avoid waste. From the left exterior wall to the center of the first partition is 12 linear feet. We multiply $12 \times 3 = 36$, divide by 4 to get 9, plus one, for a total of 10–2 x 6 x 12' ceiling joists required for the back bedroom and closet, and 10–2 x 6 x 14' ceiling joists required over the front left bedroom area. Our bath and front bedroom area is 13'3". We multiply $13 \times 3 = 39$, divided by 4 to get 10, plus one, for a total of 11–2 x 6 x 12' ceiling joists over the bath area, 11–2 x 6 x 14' ceiling joists for the front right bedroom area. We would consolidate our list, so we would have 21–2 x 6 x 12' and 21–2 x 6 x 14' ceiling joists. We would add this to our material list.

Pieces	Feet	Description	Cost	Total
21		2 x 6 x 12' ceiling joists – second floor		
21		2 x 6 x 14' ceiling joists – second floor		

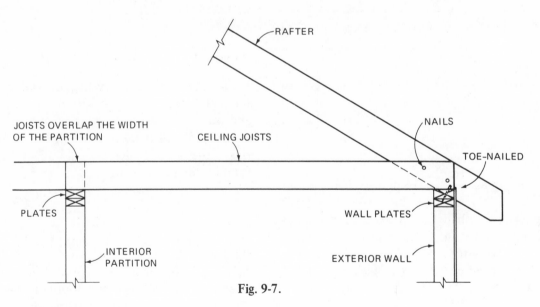

Fig. 9-7.

10
CARPENTRY FRAMING: ROOF STRUCTURE

ROOF TRUSSES

Many modern buildings are designed with roof trusses instead of rafters. A *truss* is a preassembled unit made up of framing members and fasteners. Trusses are available from most major building supply dealers, in a wide range of sizes and styles. The most common styles for roofs on residential construction are W *truss* (Figure 10-2), and *king-post trusses,* (Figure 10-1). The bottom chords of these trusses act as ceiling joists. Where no ceiling joist is desired, another type known as a *scissor truss* is used.

Trusses offer a number of advantages over conventional rafters. They are deliverd to the job site preassembled, the roof can be constructed faster, with fewer man-hours. Being built under controlled conditions in a shop, trusses are frequently stronger than on-site assemblies. Frequently, trusses are specified for a building because of their ability to span wide areas without supporting posts or partitions.

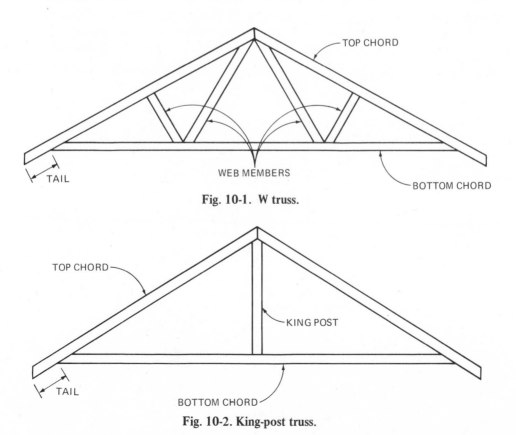

Fig. 10-1. W truss.

Fig. 10-2. King-post truss.

RAFTERS

A *common rafter* is one that extends from the ridge to a bearing plate, as in the second floor plan. For a plain roof having just two sides, all of the rafters are common rafters and the roof is referred to as a *gable roof*. Rafters, usually spaced 16 inches o.c., must be strong enough to support the weight of the roof (in some areas the roof must carry a snow load) and to resist wind pressure. The size of the rafter depends on the span and the pitch of the roof. Figure 10-3 shows the locations and names of the various members used to frame a roof. An estimator should be familiar with these.

Estimating Roof Trusses

Roof trusses used in residential construction are usually either of the king-post or W type. They are normally spaced 2 feet on centers. To determine the number of trusses required for a straight roof, divide the length of the building by 2 and add 1. If special gable end trusses are used, 2 fewer king-post or W trusses are needed. When trusses are specified, the style, size of all parts, pitch (steepness of the roof), and amount of overhang tail must be indicated. It is common practice to include truss detail drawings with the working drawings for the building. Some information is also included in the specifications.

A *hip rafter* is required where two adjacent slopes meet on a hip roof (one that slopes on all four sides). Hip rafters may be two inches deeper than the jack and common rafters to provide a full nailing surface for the jack and common rafters that are nailed to them.

Jack rafters span the space between the wall plates and the hip rafters, or the ridge and the valley rafter.

A *valley rafter* carries the ends of the jack rafters where two roof surfaces meet to form a valley. The valley rafter, like the hip rafter, may be two inches deeper than the jack rafter to afford a full nailing surface.

The *ridge board* is the top member of the roof framing and runs the length of the roof. The tops of the rafters are nailed to the ridge board. This board is usually a 2 x 8 or 2 x 10, depending upon the size of the rafters. The ridge board is usually two inches deeper than the rafters, to offer a full nailing surface. One-inch nominal thickness stock is sometimes used, but 2-inch stock is preferred because the $1\frac{1}{2}$-inch true thickness is better for holding the abutting rafters in alignment.

Estimating Rafters

To determine the number of rafters 16 inches o.c. required on each side of a gable roof, multiply the length of the ridge by $\frac{3}{4}$ and add 1. Assume jack rafters are the same size and length as the common

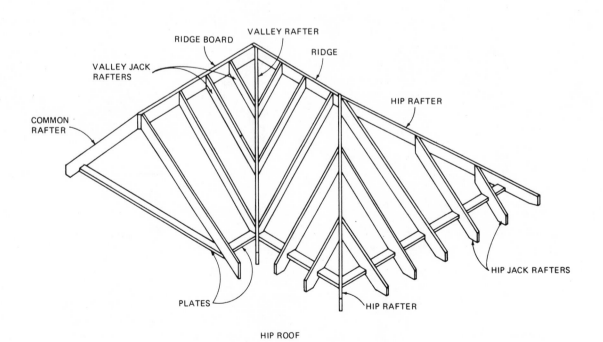

HIP ROOF

Fig. 10-3. Roof framing parts.

rafter. To find the number of jack rafters needed, take $1\frac{1}{2}$ times the width of the building and add 2. Allow 1 hip rafter for each hip and 1 valley rafter for each valley.

The following procedure can be used to find the true length of hip or valley rafters from the working drawings. Work on the elevation drawing showing the hip or valley (Figure 10-4).

Draw line CX perpendicular to line CA (the line of the level cornice of the building) passing through point B. Mark point D on line CX so that CD is the same length as AB. The distance from point A to point D is the true length of the hip or valley rafter. Be sure to use the same scale to measure this as was used in making the drawing. The amount of overhang required by the cornice must be considered when ordering hip or valley rafters. Also remember that if the length of the hip or valley is an odd number, the next higher even number must be used, because lumber is sold in even-foot sizes.

If there is a top view only showing the roof layout, as in Figure 10-5, the length of the valley or hip can still be found. To find the true length of AB, draw line AC perpendicular to AB. Draw AC equal to the height of the gable. Then draw CB which is the true length of the valley rafter AB.

Looking at our elevations, we find that our house has a straight gable roof. The plan calls for a 6/12 pitch. Using the scale on the right side elevation, we find that our rafter is 14 feet long. We can double check this length. If our building is 24 feet wide, one-half of this would be 12 feet. For every foot we go horizontally the rafter rises 6 inches, so in 12 feet our ridge height would be 6 feet. To find the hypotenuse of a right triangle, you square the base (12 feet) for 144, and then you square the height (6 feet) for 36. You add 144 and 36 for a total of 180. Take the square root of 180 and you get 13.416408 or $13\frac{1}{2}$ feet, so a 14-foot rafter would be long enough.

To determine the number of rafters required on our house, we take the length of the ridge, which is 32 feet. We multiply 32 by 3 for 96 and divide by 4, which gives 24, plus one for a total of 25 rafters needed for each side. Double this for both the front and back side: we need 50–2 x 6 x 14' rafters.

RIDGE BOARD

We need a ridge board to hold each face of the rafters. Our building is 32 feet long, so we will need

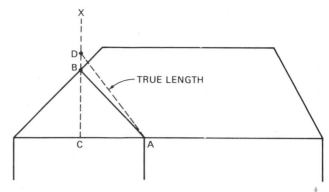

Fig. 10-4. **Finding the true length of hip and valley rafters from elevation drawings.**

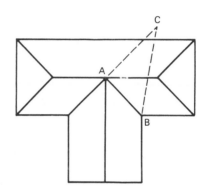

Fig. 10-5.

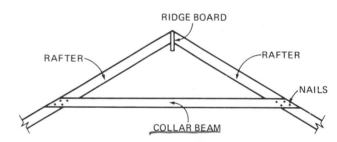

Fig. 10-6. Collar beams.

2–2 x 8 x 16' for our ridge. Figure the ridge board 2 inches wider than the rafters to provide a solid nailing surface for the rafter. We would add this to our material list.

Pieces	Feet	Description	Cost	Total
50		2 x 6 x 14' rafters		
2		2 x 8 x 16' ridge board		

COLLAR BEAMS

Collar beams (Figure 10-6) are horizontal members which tie opposing rafters together to prevent sagging. These are usually 1 x 8s, 2 x 4s, or 2 x 6s, depending upon the finish required in the area just below them. If the house is a single-story ranch house, the collar beams are probably 1 x 8s; if the attic area of a two-story house is unfinished, the collar beams are probably 1 x 8s; in a house that has an expansion attic or a shed dormer requiring a finished interior next to the rafters, the collar beams are 2 x 4s or 2 x 6s. Their size depends on the span and the ceiling load they have to carry.

Estimating Collar Beams

Occasionally one collar beam is used for each pair of rafters, but most buildings have one for every third pair of rafters. Check the plans and specifications carefully to be sure of what is required. If the collar beams are spaced 16 inches o.c., divide the number of rafters by 2 and add 1. If they are spaced 32 inches o.c., divide the number of rafters by 4 and add 1. Our specifications call for collar beams on every other set of rafters or every 32 inches. We have 50 rafters, so we divide 50 by 4 and get $12\frac{1}{2}$ or 13, plus one, for a total of 14. We would add to our material list 14–1 x 8 x 10′ shiplap.

Pieces	Feet	Description	Cost	Total
14		1 x 8 x 10′ shiplap — collar beams		

GABLE STUDS

A house with a hip roof does not have gable studs, but when a gable roof is used, the gable end must be framed (Figure 10-7). Gable studs are vertical studs, the same size as the ones used for the exterior walls and the interior partitions. These are placed between the end rafters of a gable roof and the top plate of each end wall. Gable studs are usually spaced 16 inches o.c. and are covered with the same type sheathing as the exterior walls.

Estimating Gable Studs

On a straight roof with just two gable ends it is only necessary to measure one end of the roof. The height of the gable from the top wall plates to the ridge is the length of the material that must be ordered for gable studs. The number of feet the roof spans is the number of pieces that must be ordered. Although all of the studs in the gable end are not the same length, when a stud is cut for one end of the roof, the remainder of that piece is used for a stud in the opposite end. Our house is 24 feet wide, so we will need 24 studs. The gable is 6 feet high, so instead of figuring 8-foot lengths and having waste, we will figure on using 12-foot stock. We would estimate 12–2 x 4 x 12′ studs are needed for both gable ends of our house. This we would add to our material list.

Pieces	Feet	Description	Cost	Total
12		2 x 4 x 12′ studs — gable ends		

ROOF SHEATHING

Roof sheathing is the material applied to the rafters to provide a surface for attaching shingles or other roofing material (Figure 10-8). Most often $\frac{1}{2}$ -inch plywood is used to cover rafters, which are spaced 16 inches o.c. On some buildings the sheathing may be 1 x 8 or 1 x 10 shiplap or tongue and groove.

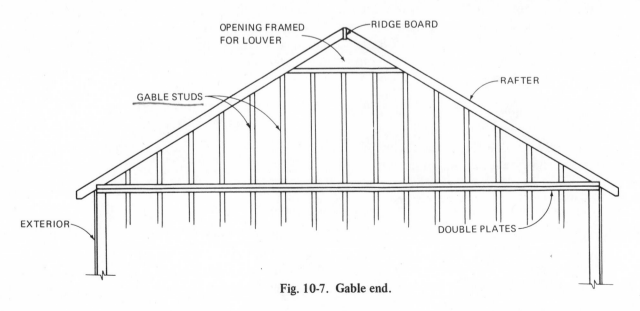

Fig. 10-7. Gable end.

Estimating Roof Sheathing

There are six different types and styles of roofs which are commonly constructed on buildings.

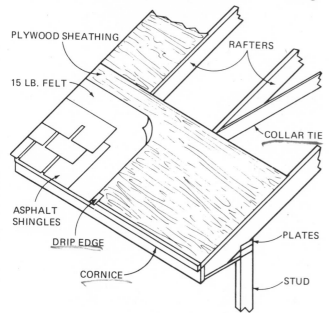

Fig. 10-8. Roof construction.

Figures 10-9–10-14 show the types of roofs and the formula for finding the area of each.

On some buildings with hip or mansard roofs, the pitch may not be the same on the ends and the sides. In this situation, the area of the sides and ends must be calculated separately and then added together.

In estimating the sheathing for our house, we have already figured the rafters and we know them to be 14 feet long. Our house is 32 feet long from one end of the ridge to the other. We multiply 32×14 for a total of 448 square feet. We double this (both sides, front and back) for a total of 896 square feet. Our specifications call for $\frac{1}{2}$-inch plywood, so we would divide 32 (4' x 8' piece) into 896 for a total of 28 sheets required to cover the rafters. We would add this to our material list.

Pieces	Feet	Description	Cost	Total
28		$\frac{1}{2}''$ x 4' x 8' plywood – roofers		

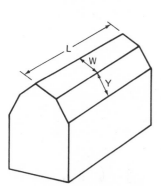

Fig. 10-9. Gable roof.
Area equals $2 \times L \times W$.

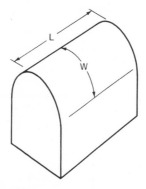

Fig. 10-10. Hip roof.
Area equals $2 \times L \times W$.

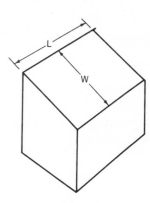

Fig. 10-11. Shed roof.
Area equals $L \times W$.

Fig. 10-12. Gambrel roof.
To find the area, add W and Y.
Multiply this sum by $2 \times L$.

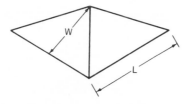

Fig. 10-13. Gothic roof.
Area equals $L \times W$.

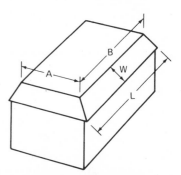

Fig. 10-14. Mansard roof.
To find the area,
add the following:
$(A \times B) + (2 \times A \times W) + (2 \times L \times W)$.

11
CARPENTRY FRAMING:
ROOF FINISHING

ROOFING

Many factors affect the selection of material to be used as roofing. It must provide protection against sun, snow, ice, wind, and sleet. It must also show a considerable ease of application, good appearance in relation to the design of the house, permanence, and low cost.

The most common roofing material for either new construction or the replacement of older roofs is asphalt roofing. Asphalt roofing material is made from a base of dry felt, which is processed from rag, wood, or other cellulose fibers and then saturated with asphalt. Additional coats of asphalt may be applied to the surface to produce a smooth-surfaced roll roofing, or mineral granules of selected colors may be rolled into the coated surface to produce mineral-surfaced roll roofing or shingles.

Saturated felt, to which no coating of asphalt has been added, is used as a building paper between the exterior sheathing and the finished siding, as underlayment between the roof sheathing and the finished roof, and between the subfloor and the finished floor. It is also used to form layers of built-up roofing.

Roofing material is available in different weights. The weight referred to is the weight of enough roofing material to cover one square (100 square feet of roof area — an area 10 feet by 10 feet, or 14,400 square inches). Mineral-surfaced roll roofing comes in 15, 30, 45, 55, and 65 pound weights. Double-coverage roll roofing is nailed and cemented down in such a way that only a 19-inch edge is exposed and all nails are covered. Double coverage roll roofing weighs 140 pounds per square. It is used primarily on low-pitched roofs. Asphalt shingles weigh 210 to 315 pounds per square.

Asphalt roof shingles (Figure 11-1), are made in a wide variety of colors and in straight or blended shades. They are manufactured so that they have a maximum of four square or three hexagonal tabs to the strip. Square-edge strip shingles are 36 inches long and 12 inches wide. They are applied with either a 4-inch or 5-inch exposure. They are packaged in bundles which cover one-third of a square each.

The second most widely used product for roofing is wood shingles. Western red cedar is the most widely used wood. Wood roofing is manufactured in three basic products: shingles, hand-split shakes, and grooved sidewall shakes (Figure 11-2).

When a roof is covered with shingles or shakes, a double course (2 layers) is applied at the eaves. Extra shingles must also be allowed for ridges, hips, and starter courses. The first course (row) of asphalt shingles is applied with the tabs pointing upward; another course is applied directly over this with the

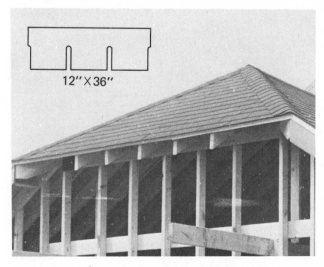

Fig. 11-1. Asphalt shingles.

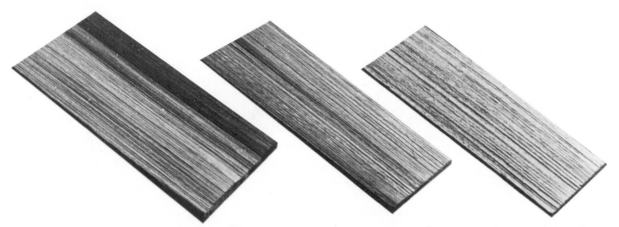

#1 Handsplit-resawn shakes. Tapered. One side heavily textured, one side sawn smooth. Thicker and heavier through the butt than other shakes.

#1 Tapersplit shakes. Medium texture and thickness. Produced by reversing the cedar block and handsplitting to achieve a natural taper.

#1 Straight-split shakes. (Also called Barn Shakes.) Medium textured. Handsplit without reserving the block, so a generally uniform thickness is achieved.

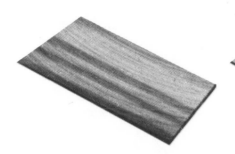  

#1 Cedar Shingles. Shingles are produced by sawing both faces to produce two smooth sides. #1 grade is 100% clear, heartwood, edge-grained.

#1 & #2 Rebutted-Rejointed shingles. Trimmed with straight edges, right-angle corners, for close tolerance application on exterior walls. Also available with smooth-sanded faces.

Grooved sidewall shakes. A #1 rebutted-rejointed shingle with grain-like grooves. Available natural or in a variety of factory-applied tones.

#2 shingle, flatgrain, limited sapwood;
#3 utility shingle, for economy applications;
#4 undercoursing, also effective on interior walls as accent covering.

Fig. 11-2. Wood shingles.

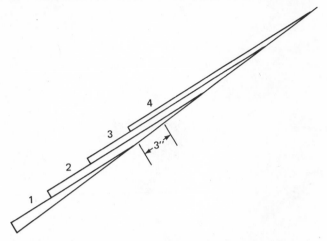

Fig. 11-3. Exposure. Third shingle overlaps the first by 3 inches. To find the exposure take the length of the shingle and subtract 3 inches, then divide by 3 to get equal portions exposed to the weather.

tabs pointing downward. Wood shingles are doubled at the starting eaves with the thick end towards the eaves. Wood shingles used in roofing are laid so that the third shingle above overlaps the first by approximately 3 inches (Figure 11-3). Wood shingles provide 240 feet of starter course per square. One bundle of shingles or shakes covers $16\frac{2}{3}$ feet of ridge or hip.

Another product that is used for finished roofing is slate. Slate roofing is expensive and is installed one slate at a time. Its advantage is its durability. Slate is manufactured in various sizes and thicknesses as well as colors. Standard slate is the quarry run of $\frac{3}{16}$ inch thickness, with considerable variations above and below $\frac{3}{16}$ inch. In figuring slate the usual lap or cover of the lowest course of slate by the upper two is 3 inches, so a square of roofing slate means sufficient number of slates of any size to cover 100 square feet with a 3-inch lap. To determine the exposed length, deduct the lap from the length of the slate and divide by 2. For example, if a 14″ x 20″ slate is used, subtract 3 from 20 which is 17 and divide by 2, for $8\frac{1}{2}$ inches exposed area; then multiply 14 by $8\frac{1}{2}$ to get a total of 119 square inches. There are 14,400 square inches in a square, so divide 119 into 14,400 to get 121 slates for each square using a 14 x 20 slate with $8\frac{1}{2}$-inch exposure.

The metals used for roofing material are tin, copper, galvanized iron, zinc, and aluminum. Tin is manufactured in two regular sizes: 20″ x 28″ and 14″ x 20″. It is also manufactured in two thicknesses: IC or 29 gauge, which weighs approximately 8 ounces to a square foot; and IX or 27 gauge, which weighs approximately 10 ounces to a

square foot. Using standing joints, a 14″ x 20″ sheet will cover 235 square inches, or 1 box (112 sheets) will cover 182 square feet. With a flat lock seam, allowing $\frac{3}{8}$ inch all around for joints, 1 box will cover 198 square feet. Depending upon the shape of the roof, you will have to add for waste and cutting and fitting for any corners, protrusions, etc.

You will also find roofs built-up of other materials. On flat roof decks, the built-up roof is the most commonly used type. This is a series or layers of felt laid one on top of the other. Each layer is nailed and cemented down with hot tar. After the finished layer is applied, the roof is then mopped with a coating of hot tar. The tar is then usually covered with a stone granual. Most built-up hot tar roofing is bonded, meaning that the contractor guarantees the roof for a number of years.

ESTIMATING ROOFING

Roofing material is measured by the square. A square is enough material to cover 100 square feet when applied according to the manufacturer's instructions. Divide the square-foot area of the roof by 100 to find the number of squares required. If asphalt shingles are being used, add $1\frac{1}{2}$ square feet for every linear foot of ridge, eaves, hip, and valley.

Our specifications (under Division 7: Moisture Protection) calls for 235-pound asphalt shingles for the roof. Our roof sheathing was 896 square feet. The house has 32 linear feet of ridge and 64 linear feet of eaves: add 32 and 64 for a total of 96 linear feet, and multiply 96 by $1\frac{1}{2}$ for a total of 144 square feet for the ridge and the eaves. We add 144 to 896 (our roof area) for a total of 1040 square feet. We divide this by 100 to get 10.4 squares. Allowing 10 percent for waste, we will need 11.44 squares of roofing. Asphalt shingles are sold 3 bundles to a square, so we will have to purchase $11\frac{2}{3}$ square of shingles for the roof. We would add this to our material list.

Pieces	Feet	Description	Cost	Total
$11\frac{2}{3}$		square asphalt shingles—roof		

Roofing Felt

When asphalt shingles are used, building felt is placed between the roof sheathing and the finished roof. This is not required for wood shingles or a metal roof. Building felt is sold in 500 square foot rolls. Add 10 percent for waste and overlapping of the edges. Our roof is 896 square feet; adding 10 percent = 89 to 896, we get a total of 985 square feet, so we will need 2 rolls of roofing felt. We add this to our material list.

Pieces	Feet	Description	Cost	Total
2		rolls 15-pound asphalt felt—roof		

Drip Edge

When asphalt shingles are used on a roof, a metal drip edge is installed along the eaves and the rake of the house. Drip edge is manufactured in 10-foot lengths. This is nailed over the top of the roofing felt along the eaves and up the rake of the cornice. The asphalt shingles go over the top of the edge of the drip edge. Measure the length of the eaves and the rakes. On our house we have 14-foot rafters: $4 \times 14 = 56$ feet for the rake. The eaves across the front and back measure $32 + 32 = 64$ feet. We add $64 + 56$ for a total of 120 linear feet. We divide 120 by 10 (length of each piece of drip edge) to get 12 pieces required.

Pieces	Feet	Description	Cost	Total
12		galvanized drip edge 10-foot lengths—roof		

12
CARPENTRY FRAMING: CORNICES

CORNICE CONSTRUCTION

A *cornice* is the assembly of boards and moldings used in combination with each other to provide a finish to the ends of the rafters which extend beyond the face of the outside walls (Figure 12-1). On a hip roof or a mansard roof, the cornice runs horizontally around the house. When the house has a gable, the cornice runs up each end rafter to meet at the ridge. This is referred to as the *rake* (Figure 12-2).

The rake cornice can extend beyond the end rafters, in which case it is called a *boxed rake* (Figure 12-1). Otherwise, it is a *tight rake* (Figure 12-3). With either type, when the house has a gable end, the cornice running up and down the rafters meets the level end of the cornice at the eaves. Where the level cornice at the eaves turns the corner at the end of the house, it is called a *cornice return* (Figure 12-4). The length of the cornice return depends upon the style of the cornice and the amount of overhang at the eaves.

Both the material used in the house and the type of architecture influence the design of the cornice. Most cornices include a frieze, bed mold, soffit, and crown mold. A simple cottage cornice, an elaborate colonial cornice, and a wide sweeping cornice are all made from the same basic parts.

On some houses the rafters bear on a *rafter shoe* which is nailed to the top of the ceiling joists (Figure 12-5). This is done to allow more overhang without the cornice being too low. It also makes the cornice high enough to come even with the top of the window casing.

Study the various styles of cornices (Figures 12-5–12-7) to become acquainted with the names of the various styles and their parts. Before estimating the materials, study the working drawings to learn what type cornice is needed. This is usually found on a special detail or on the wall section view detail.

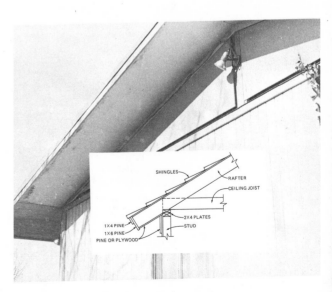

Fig. 12-1. Cornice.

Fig. 12-2. Rake cornice.

Fig. 12-3. Tight rake.

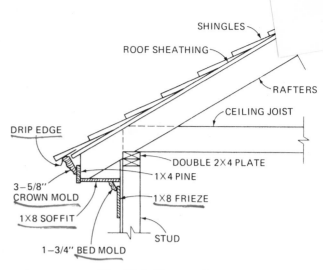

Fig. 12-6. Box cornice.

Fig. 12-4 Cornice return.

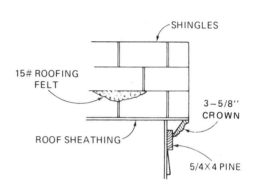

Fig. 12-7. Tight rake cornice.

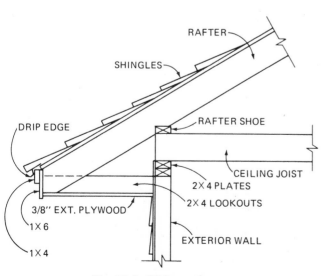

Fig. 12-5. Wide overhang.

ESTIMATING CORNICES

Our house drawings show a tight cornice on the rake and a box cornice at the eaves. Our wall section detail and the side elevations show that the rake is made up using a 5/4″ x 6″ and a 1″ x 3″. The cornice at the eaves is made up using a 1 x 6, 1 x 3, 1 x 8, and a 1 x 2. To determine the amount of each needed, remember that our rafters are 14 feet long, so we will need 4 pieces of 5/4 x 6 for the rake. The cornice returns are 2′6″ each, so we will need 1–5/4 x 6 x 10′ for the returns. At the eaves we need 64 linear feet of 1 x 6 for the fascia, 64 linear feet of 1 x 8 for the soffit. Our frieze returns each corner approximately one foot, so we will need 68 linear feet of 1 x 8 for the frieze. We will need 68 linear feet of 1 x 2 where the soffit and frieze meet. We need 64 linear feet of 1 x 3 to go over the fascia. We would consolidate each by size and add to our material list.

Pieces	Feet	Description	Cost	Total
4		5/4 x 6 x 14' pine cornice—rake		
1		5/4 x 6 x 10' pine cornice—returns		
	64	linear feet 1 x 6 pine cornice—eaves		
	132	linear feet 1 x 8 pine cornice—soffit, frieze, and returns		
	134	linear feet 1 x 3 pine cornice		
	74	linear feet 1 x 2 pine cornice		

13
CARPENTRY FRAMING: OUTSIDE WALLS

SHEATHING

Exterior sheathing is the material applied to the outside face of the studs. Sheathing strengthens the frame, acts as insulation, and provides a surface on which the outside finish of the building may be fastened. Sheathing may be plywood, fiberboard, or shiplap (Figure 13-1).

Plywood is commonly used for subflooring, roof sheathing, and wall sheathing. It is a wood material usually made up of an odd number of thin veneer sheets, cross laminated and bonded together under high pressure with glue. It is available as exterior plywood with waterproof glue and interior plywood with water-resistant glue. It is generally classified according to the quality of wood that is used on the face panel. It is graded according to the quality and appearance of the face and back panels.

All plywood (except sheathing grades) is sanded. Panels which are $\frac{3}{8}$ inch thick and less have a minimum of three plys. Those that are $\frac{1}{2}$ inch to $\frac{3}{4}$ inch have a minimum of five plys. Most plywood is manufactured in 4' x 8' sheets, but other sizes may be ordered.

Plywood which must be exposed to water should always be made with exterior glue. Several types of plywood are manufactured for marine use, building construction, or with special surfaces for use as siding. Figures 13-2 and 13-3 shows the characteristics of most common types of plywood.

Estimating Sheathing

The specifications (under Division 6: Carpentry and Millwork, C-7) state that the wall sheathing and roof sheathing shall be $\frac{1}{2}''$ x 4' x 8' plywood APA grade C-D exterior. To find the sheathing required, multiply the perimeter of the house by the height. The height can be scaled from the elevations, by measuring the distance from the top of the foundation to the eaves. Our foundation is 108 linear feet. On our house we have a 2-foot projection on the front, so the second floor is larger than the first. In this case, figure each floor separately. Scaling from the top of the foundation to the projection is 9 feet. 108 (perimeter) times 9 equals 972 square feet on the first floor. Our second floor perimeter is 112 linear feet and our height is again 9 feet: $112 \times 9 = 1,008$ square feet. Our gable is 24 feet wide and 6 feet high: $24 \times 6 = 144$ square feet. We add $972 + 1,008 + 144$ for a total of 2,124 square feet. We divide this by 32 for a total of $66\frac{3}{8}$ or 67 sheets of 4 x 8 plywood is needed for

Fig. 13-1. Plywood and fiberboard sheathing.

Table 1: Guide to Engineered Grades of Plywood

Specific grades and thicknesses may be in locally limited supply. See your dealer for availability before specifying.

Interior Type

Use these terms when you specify plywood	Description and Most Common Uses	Typical Grade-trademarks	Veneer Grade Face	Back	Inner Plies	Most Common Thicknesses (inch) (1)				
C-D INT-APA (2) (3)	For wall and roof sheathing, subflooring, industrial uses such as pallets. Also available with intermediate glue or exterior glue. Specify intermediate glue for moderate construction delays; exterior glue for better durability in somewhat longer construction delays and for treated wood foundations.	C-D 32/16 APA INTERIOR PS 1-74 000	C	D	D	5/16	3/8	1/2	5/8	3/4
STRUCTURAL I C-D INT-APA and STRUCTURAL II C-D INT-APA	Unsanded structural grades where plywood strength properties are of maximum importance: structural diaphragms, box beams, gusset plates, stressed skin panels, containers, pallet bins. Made only with exterior glue.	STRUCTURAL I C-D 24/0 APA INTERIOR PS 1-74 000 EXTERIOR GLUE	C (6)	D (7)	D (7)	5/16	3/8	1/2	5/8	3/4
UNDERLAYMENT INT-APA (3) (2) (9)	For underlayment or combination subfloor underlayment under resilient floor coverings, carpeting in homes, apartments, mobile homes. Specify exterior glue where moisture may be present, such as bathrooms, utility rooms. Touch-sanded. Also available in tongue-and-groove.	UNDERLAYMENT GROUP 1 APA INTERIOR PS 1-74 000	C Plugged	D	(8) C & D	1/4		3/8	1/2	5/8	3/4
C-D PLUGGED INT-APA (3) (2) (9)	For built-ins, wall and ceiling tile backing, cable reels, walkways, separator boards. Not a substitute for Underlayment, as it lacks Underlayment's punch-through resistance. Touch-sanded.	C-D PLUGGED GROUP 2 APA INTERIOR PS 1-74 000	C Plugged	D	D	5/16	3/8	1/2	5/8	3/4
2·4·1 INT-APA (2) (5)	Combination subfloor underlayment. Quality base for resilient floor coverings, carpeting, wood strip flooring. Use 2·4·1 with exterior glue in areas subject to moisture. Unsanded or touch-sanded as specified.	2·4·1 GROUP 1 APA INTERIOR PS 1-74 000	C Plugged	D	C & D	(available 1-1/8″ or 1-1/4″)				

Exterior Type

Use these terms when you specify plywood	Description and Most Common Uses	Typical Grade-trademarks	Veneer Grade Face	Back	Inner Plies	Most Common Thicknesses (inch) (1)				
C-C EXT-APA (3)	Unsanded grade with waterproof bond for subflooring and roof decking, siding on service and farm buildings, crating, pallets, pallet bins, cable reels.	C-C 42/20 APA EXTERIOR PS 1-74 000	C	C	C	5/16	3/8	1/2	5/8	3/4
STRUCTURAL I C-C EXT-APA and STRUCTURAL II C-C EXT-APA	For engineered applications in construction and industry where full Exterior type panels are required. Unsanded. See (9) for species group requirements.	STRUCTURAL I C-C 32/16 APA EXTERIOR PS 1-74 000	C	C	C	5/16	3/8	1/2	5/8	3/4
UNDERLAYMENT C-C Plugged EXT-APA (3) (9) C-C PLUGGED EXT-APA (3) (9)	For Underlayment or combination subfloor underlayment under resilient floor coverings where severe moisture conditions may be present, as in balcony decks. Use for tile backing where severe moisture conditions exist. For refrigerated or controlled atmosphere rooms, pallets, fruit pallet bins, reusable cargo containers, tanks and boxcar and truck floors and linings. Touch-sanded. Also available in tongue-and-groove.	UNDERLAYMENT C-C PLUGGED GROUP 2 APA EXTERIOR PS 1-74 000 / C-C PLUGGED GROUP 3 APA EXTERIOR PS 1-74 000	C Plugged	C	C (8)	1/4		3/8	1/2	5/8	3/4
B-B PLYFORM CLASS I & CLASS II EXT-APA (4)	Concrete form grades with high reuse factor. Sanded both sides. Mill oiled unless otherwise specified. Special restrictions on species. Also available in HDO.	B-B PLYFORM CLASS I APA EXTERIOR PS 1-74 000	B	B	C				5/8	3/4

(1) Panels are standard 4x8 foot size. Other sizes available.
(2) Also available with exterior or intermediate glue.
(3) Available in Group 1, 2, 3, 4 or 5.
(4) Also available in STRUCTURAL I.
(5) Available in Group 1, 2 or 3 only.
(6) Special improved C grade for structural panels.
(7) Special improved D grade for structural panels.
(8) Ply beneath face a special C grade which limits knotholes to 1 inch or in Interior Underlayment D under Group 1 or 2 faces 1/6 inch thick.
(9) Also available in STRUCTURAL I (all plies limited to Group 1 species) and STRUCTURAL II (all plies limited to Group 1, 2 or 3 species).

Fig. 13-2. Plywood grade–use chart—engineered grades.

Table 2:   Guide to Appearance Grades of Plywood[1]

For strength properties of appearance grades, see "Plywood Design Specifications," Y510

Interior Type

Use these terms when you specify plywood [2]	Description and Most Common Uses	Typical Grade-trademarks	Veneer Grade			Most Common Thicknesses (inch) [3]					
			Face	Back	Inner Plies						
N-N, N-A, N-B INT-APA	Cabinet quality. For natural finish furniture, cabinet doors, built-ins, etc. Special order items.	N-N G1 INT-APA PS 1.74	N	N.A. or B	C						3/4
N-D INT-APA	For natural finish paneling. Special order item.	N-D G3 INT-APA PS 1.74	N	D	D	1/4					
A-A INT-APA	For applications with both sides on view. Built-ins, cabinets, furniture and partitions. Smooth face; suitable for painting.	A-A G-4 INT-APA PS 1.74	A	A	D	1/4		3/8	1/2	5/8	3/4
A-B INT-APA	Use where appearance of one side is less important but two smooth solid surfaces are necessary.	A-B G-4 INT-APA PS 1.74	A	B	D	1/4		3/8	1/2	5/8	3/4
A-D INT-APA	Use where appearance of only one side is important. Paneling, built-ins, shelving, partitions, and flow racks.	A-D GROUP 1 APA INTERIOR PS 1.74 000	A	D	D	1/4		3/8	1/2	5/8	3/4
B-B INT-APA	Utility panel with two smooth sides. Permits circular plugs.	B-B G-3 INT-APA PS 1.74	B	B	D	1/4		3/8	1/2	5/8	3/4
B-D INT-APA	Utility panel with one smooth side. Good for backing, sides of built-ins. Industry: shelving, slip sheets, separator boards and bins.	B-D GROUP 3 APA INTERIOR PS 1.74 000	B	D	D	1/4		3/8	1/2	5/8	3/4
DECORATIVE PANELS INT-APA	Rough sawn, brushed, grooved, or striated faces. For paneling, interior accent walls, built-ins, counter facing, displays, and exhibits.	DECORATIVE GROUP 2 APA INTERIOR PS 1.74 000 GROUP 1 FACE	C or btr.	D	D		5/16	3/8	1/2	5/8	
PLYRON INT-APA	Hardboard face on both sides. For counter tops, shelving, cabinet doors, flooring. Faces tempered, untempered, smooth, or screened.	PLYRON INT-APA PS 1.74			C & D				1/2	5/8	3/4

Exterior Type [7]

A-A EXT-APA [4]	Use where appearance of both sides is important. Fences, built-ins, signs, boats, cabinets, commercial refrigerators, shipping containers, tote boxes, tanks, and ducts.	A-A G3 EXT-APA PS 1.74	A	A	C	1/4		3/8	1/2	5/8	3/4
A-B EXT-APA [4]	Use where the appearance of one side is less important.	A-B G-1 EXT-APA PS 1.74	A	B	C	1/4		3/8	1/2	5/8	3/4
A-C EXT-APA [4]	Use where the appearance of only one side is important. Soffits, fences, structural uses, boxcar and truck lining, farm buildings. Tanks, trays, commercial refrigerators.	A-C GROUP 4 APA EXTERIOR PS 1.74 000	A	C	C	1/4		3/8	1/2	5/8	3/4
B-B EXT-APA [4]	Utility panel with solid faces.	B-B G-1 EXT-APA PS 1.74	B	B	C	1/4		3/8	1/2	5/8	3/4
B-C EXT-APA [4]	Utility panel for farm service and work buildings, boxcar and truck lining, containers, tanks, agricultural equipment. Also as base for exterior coatings for walls, roofs.	B-C GROUP 2 APA EXTERIOR PS 1.74 000	B	C	C	1/4		3/8	1/2	5/8	3/4
HDO EXT-APA [4]	High Density Overlay plywood. Has a hard, semi-opaque resin-fiber overlay both faces. Abrasion resistant. For concrete forms, cabinets, counter tops, signs and tanks.	HDO AA G1 EXT-APA PS 1.74	A or B	A or B	C or C plgd		5/16	3/8	1/2	5/8	3/4
MDO EXT-APA [4]	Medium Density Overlay with smooth, opaque, resin-fiber overlay one or both panel faces. Highly recommended for siding and other outdoor applications, built-ins, signs, and displays. Ideal base for paint.	MDO BB G-4 EXT-APA PS 1.74	B	B or C	C		5/16	3/8	1/2	5/8	3/4
303 SIDING EXT-APA [6]	Proprietary plywood products for exterior siding, fencing, etc. Special surface treatment such as V groove, channel groove, striated, brushed, rough sawn.	303 SIDING 16 oc GROUP 1 APA EXTERIOR PS 1.74 000	(5)	C	C			3/8	1/2	5/8	
T 1-11 EXT-APA [6]	Special 303 panel having grooves 1/4" deep, 3/8" wide, spaced 4" or 8" o.c. Other spacing optional. Edges shiplapped. Available unsanded, textured, and MDO.	303 SIDING 16 oc T 1-11 GROUP 1 APA EXTERIOR PS 1.74 000	C or btr.	C	C					5/8	
PLYRON EXT-APA	Hardboard faces both sides, tempered, smooth or screened.	PLYRON EXT-APA PS 1.74			C				1/2	5/8	3/4
MARINE EXT-APA	Ideal for boat hulls. Made only with Douglas fir or western larch. Special solid jointed core construction. Subject to special limitations on core gaps and number of face repairs. Also available with HDO or MDO faces.	MARINE AA EXT-APA PS 1.74	A or B	A or B	B	1/4		3/8	1/2	5/8	3/4

(1) Sanded both sides except where decorative or other surfaces specified.
(2) Available in Group 1, 2, 3, 4 or 5 unless otherwise noted.
(3) Standard 4x8 panel sizes, other sizes available.
(4) Also available in Structural I (all plies limited to Group 1 species) and Structural II (all plies limited to Group 1, 2 or 3 species).
(5) C or better for 5 plies, C Plugged or better for 3 ply panels.
(6) Stud spacing is shown on grade stamp.
(7) For finishing recommendations, see Form V307.

Fig. 13-3. Plywood grade–use chart–appearance grades.

our sheathing. We would add this to our material list.

Pieces	Feet	Description	Cost	Total
67		$\frac{1}{2}''$ x 4' x 8' C-D APA exterior plywood–sheathing		

SIDING

Wood siding is available in various patterns and in several widths and thicknesses. The most common sizes and types of wood siding are: $\frac{1}{2}$ x 6, $\frac{1}{2}$ x 8, $\frac{3}{4}$ x 10 bevel siding, 16-inch shingles, random-width vertical siding, and shakes. Other common sidings are brick or stone veneer, aluminum siding, and vinyl siding.

Wood shingles and wood shakes may be applied to either single courses or double courses (a course is a layer). When the double-course method of application is used, first-grade shingles or shakes are applied in each course so that they project about $\frac{1}{2}$ inch below the underlayer of third-grade shingles. In many cases, asphalt-impregnated undercover or backer boards are used as the undercourse.

Wood shingles are also remanufactured into processed shakes with square butts and vertical edges finely machined to obtain tight, invisible joints. Processed shakes have a face side that has a combed surface which makes them look like hand-split shakes. These shakes are packed so that two bundles cover 100 square feet (one square).

Figure 13-4 shows the wood products that are most commonly used for siding on houses. Some of these may be used in combination with each other, such as hand-split shakes and colonial brick, or stone and vertical board and battens. There is no limit to the combinations that can be used. Additional information is available through siding manufacturers' specifications and most lumber yards. It is important for an estimator to be familiar with as many types of siding as possible.

ESTIMATING SIDING

The sheathing should be covered with building paper before the siding is applied. Kraft building paper is sold in 500-square-foot rolls. Our house has a total of 2,124 square feet, so five rolls of building paper are needed. Add this to the material list.

Pieces	Feet	Description	Cost	Total
5		rolls kraft building paper		

If the material to be used for the siding is sold by the square, divide the total square foot area of the walls by 100. This is the number of squares needed to cover the walls. Many siding materials

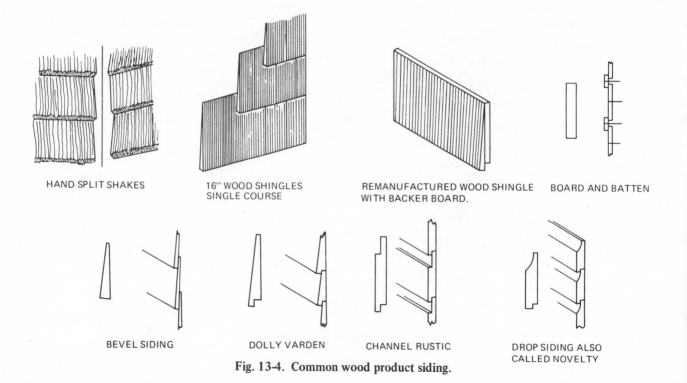

HAND SPLIT SHAKES   16" WOOD SHINGLES SINGLE COURSE   REMANUFACTURED WOOD SHINGLE WITH BACKER BOARD.   BOARD AND BATTEN

BEVEL SIDING   DOLLY VARDEN   CHANNEL RUSTIC   DROP SIDING ALSO CALLED NOVELTY

Fig. 13-4. Common wood product siding.

are sold by the board foot. The following should serve as a guide to estimating materials:

- Wood shingles. Allow 3 bundles for each square, if they are to be installed with 12-inch exposure.
- 6-inch beveled siding with a $4\frac{3}{4}$-inch exposure. Add 30 percent to the wall area to be covered and order by the board foot.
- 8-inch beveled siding with $6\frac{3}{4}$-inch exposure. Add 25 percent to the wall area to be covered and order by the board foot.
- 10-inch bevel siding with $8\frac{3}{4}$-inch exposure. Add 20 percent to the wall area to be covered and order by the board foot.
- Brick veneer. Allow $6\frac{1}{2}$ brick for every square foot of wall area.
- Stone veneer. Allow 20 percent for waste and order by the ton.

Our plans show that we have brick veneer on the lower half of the front elevation and the remainder of the house has $\frac{1}{2}''$ x 8″ bevel siding. Our total square foot area is 2,124 square feet. We subtract the area where the brick veneer is applied so we multiply 32 (distance across the front of the house) by 10 (height where the veneer work is installed) for a total of 320 square feet. We subtract 320 from 2,124 for a total square foot area of 1,804 for our bevel siding. To determine the amount of siding required, we take the square foot area to be covered and add 25 percent for 8-inch bevel siding with a $6\frac{3}{4}$-inch exposure. 25 percent of 1,804 is 451, which we add to 1,804 for a total of 2,255 board feet of $\frac{1}{2}$ x 8 bevel siding. We would add this to our material list.

Pieces	Feet	Description	Cost	Total
	2,255	$\frac{1''}{2}$ x 8″ bevel siding		

14
CARPENTRY EXTERIOR FINISH

CORNER BOARDS

According to the elevation drawings the outside corners have corner boards. The corner boards are usually made of 5/4-inch material. The reason for this is so the bevel siding can be fitted against it without the bevel part protruding beyond the face of the corner board. Our plans call for 5/4 x 5 pine corner boards. We have four corners and they are approximately 18 feet high. We need two boards for each corner except the front lower section, where the brick veneer work is installed. We multiply 18 feet (height) by 8 (four corners doubled) for a total of 144 linear feet, less 20 feet for the brick front; this means we will need 124 linear feet of 5/4 x 5 pine for our corner boards. We would add this to our material list.

Pieces	Feet	Description	Cost	Total
	124	linear feet 5/4 x 5 pine—corner boards		

OVERHANG

Our plans show the second floor protruding two feet beyond the face of the first floor across the front of the house. The front elevation shows a 1 x 8 with a drip cap on top of it but does not show the finish under the overhang, so we find the wall section detail of our working drawings, which shows how this area is to be finished. The detail shows the underside of the overhang finished with $\frac{3}{8}$-inch exterior plywood. Our side elevations show how the end returns are to be finished. Our house is 32 feet across the front and the projection is two feet wide. Plywood is sold in 4' x 8' sheets. Dividing 8 into 32, we get 4. With a 2-foot projection, we should get a total of 16 linear feet, 2-feet wide,

from each sheet, so we will need 2 sheets of $\frac{3}{8}$" x 4' x 8' exterior plywood for the bottom of the overhang. We will need 32 linear feet of 1 x 8 plus the return on each end, for a total of 36 linear feet of 1 x 8 and 36 linear feet of drip cap. We would add this to our material list.

Pieces	Feet	Description	Cost	Total
2		$\frac{3}{8}$" x 4' x 8' exterior plywood—overhang soffit		
	36	linear feet 1 x 8 Pine—second floor projection		
	36	linear feet Wood drip cap—second floor overhand		

LOUVERS

It is important to allow air flow over insulation to reduce the humidity level before condensation builds up. As a general rule, attics should have one square foot of free-flow ventilation area for every 300 square feet of attic floor space. This ventilation is normally provided by louvers installed in the gables and soffits. Our specifications (Division 10, Section C) calls for 3 wood louvers with aluminum wire screen backing. The louvers are to be 1'6" x 2'0", or $1\frac{1}{2} \times 2 = 3$ square feet of free-flow air per louver. Three louvers means we have 9 square feet. Our second floor area is 768 square feet, so our ventilation is adequate. Looking at our side elevations, we see where these louvers are installed. One is placed on each side of the chimney to balance the appearance. We would add these to our material list.

Pieces	Feet	Description	Cost	Total
3		wood louvers No. 600-L 1'6" x 2'0"—gables		

ENTRANCE ROOF

There are as many possible styles and methods of construction for porch entrances and entrance roofs as there are for houses. It is necessary to study the drawings and the specifications for each building that is estimated to determine how the entrances and roofs are to be constructed. Our plans show a roof over the side entrance door. This can be seen on the right side elevation and the front and rear elevations. The side entrance shows a wood framed roof supported by wood brackets protecting the side entrance door. We have to list this material on our estimate. The framing is 3 feet out and 4 feet wide. We determine this by using our architect's scale. We will need 2–2 x 6 x 10' for the bottom beam. Our ceiling joists and rafters can be framed with 2 x 4s, with the rafters bearfooting on the ceiling joists (Figure 14-1).

We will need 3–2 x 4 x 12' for the ceiling joists and rafters. We have 24 square feet, so we will need 1 piece $\frac{1}{2}''$ x 4' x 8' plywood for roof sheathing and 1 bundle of asphalt shingles for the roof. Our rake and eaves totals 14 linear feet, so we will need 2 10-foot lengths of galvanized drip edge. For the finish wood on the entrance, we will need 2 pieces of 1 x 6 x 10' pine for the beam, 1–1 x 6 x 8' for the rake, 20 linear feet 1 x 3 for the rake and eaves,

1–1 x 2 x 10'. For the finish ceiling we will need 1–$\frac{3}{8}''$ x 4' x 8' exterior plywood; the remainder will be used for the front gable part of the entrance roof. Our brackets will require 2–2 x 4 x 8' pine. We also need wrought iron railings for both the side and front entrances. We would list this material on our material list.

Pieces	Feet	Description	Cost	Total
2		2 x 6 x 10' beam–roof over side entrance		
3		2 x 4 x 12' ceiling joists and rafters–side entrance		
1		$\frac{1}{2}''$ x 4' x 8' plywood sheathing		
1		bundle asphalt shingles–entrance roof		
2		galvanized drip edge, 10 foot		
2		1 x 6 x 10' pine–beam		
1		1 x 4 x 10' pine–beam		
1		1 x 6 x 12' pine cornice–eaves		
1		1 x 6 x 8' pine cornice–rake		
	20	linear feet 1 x 3 pine		
1		1 x 2 x 10' pine		
2		2 x 4 x 8' pine brackets		
1		$\frac{3}{8}''$ x 4' x 8' exterior plywood–ceiling and gable		
2		wrought iron railings–side entrance 3 feet long		
2		wrought iron railings–front entrance 4 feet long		

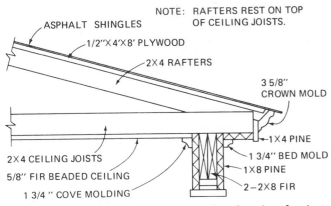

NOTE: RAFTERS REST ON TOP OF CEILING JOISTS.

ASPHALT SHINGLES
1/2"X4'X8' PLYWOOD
2X4 RAFTERS
3 5/8" CROWN MOLD
1X4 PINE
1 3/4" BED MOLD
1X8 PINE
2–2X8 FIR
2X4 CEILING JOISTS
5/8" FIR BEADED CEILING
1 3/4" COVE MOLDING

Fig. 14-1. Entrance cornice detail with rafters bearfooting on ceiling joists or overlays.

15
CARPENTRY INTERIOR WORK: WALLS

INSULATION

The basic function of insulation is to resist the flow of heat. Insulation is installed in the sidewalls, ceilings, roof, and floors of buildings to reduce the flow of heat to the outside in the winter and to the inside in the summer.

Insulation may be made from various materials, such as wood fibers, plastic foam, rock wool, fiberglass, aluminum foil, and vermiculite. It is sold in loose form, in batts, and attached to paper. It is chemically treated to be vermin-proof and flame-resistant. The estimator should be acquainted with the types most often used.

Moisture Control

Household moisture from daily activities, like cooking and washing, tends to build up in the air. Warm air holds more moisture than cold air; so when warm, humid, inside air comes in contact with cold surfaces, the air releases this moisture in the form of condensation. If the moisture present in the warm air is allowed to pass through the walls, it may condense on or under the cold exterior siding, causing the surface paint to peel. A vapor barrier (material that prevents the movement of water vapor) is needed between the interior of the house and the outer surface of the house walls. Most insulation sold in batts or rolls has a vapor-resistant facing. This vapor barrier, usually a specially coated kraft paper or aluminum foil, should always be installed toward the heated side of the wall or ceiling (Figure 15-1).

To help prevent condensation, it is important to have adequate ventilation in kitchens, bathrooms, and attics.

Sizes

It pays to use good insulation in both the sidewalls and the ceilings of a house. Ceilings, walls, thermal windows, and insulation all have *thermal resistance values* or *R values*. The higher the R value of a material, the higher is its resistance to the flow of heat. For example, $3\frac{1}{2}$-inch, R-11 fiberglass insulation has the same thermal resistance as a 9-inch wooden wall, a $4\frac{1}{2}$-foot brick wall, or an 11-foot stone wall.

Fig. 15-1. **Fiberglass insulation between wall studs.**

Insulation is manufactured in widths suitable for inserting between the framing members of the building. It is usually 15 inches wide for framing on 16-inch centers and 23 inches wide for 24-inch centers. It has a reinforced paper flange along each edge for nailing or stapling to the studs, joists, or rafters.

Estimating Insulation

Because there are so many types and thicknesses of insulation available, the first step in estimating insulation is to read the specifications and drawings. Division 7, Section A-4 of our specifications calls for the ceiling of the second floor to be 6-inch fiberglass; all sidewalls $3\frac{1}{2}$-inch fiberglass.

To determine the amount of insulation required for the ceiling, figure the square foot area to be covered. Our second floor area is 32 feet by 24 feet, for a total of 768 square feet. The 6-inch fiberglass insulation is sold in batt form. The batts are 15″ x 48″ and come 10 pieces to the bag. Each bag covers 50 square feet. Don't forget that the wall section detail also shows insulation where the front overhang protrudes beyond the front wall. This area is 32 feet long by 2 feet wide, for a total of 64 square feet. We add 768 to 64 for a total of 832 square feet. We divide 832 by 50 square feet per bag for a total 16.64 or 17 bags needed for the ceiling.

In estimating the sidewall area to be insulated, do not include the gable ends. Multiply the distance around the house and multiply by the height. Our first floor perimeter is 108 linear feet times the 8-foot ceiling height, for a total of 894 square feet. Our second floor exterior wall is 112 linear feet times the 8-foot ceiling height, for a total of 896 square feet. We add 864 to 896 for a total of 1760 square feet. $3\frac{1}{2}$-inch fiberglass is sold in 70-square-foot rolls, so divide 1760 by 70 for 26 rolls needed. We would now add the insulation to our material list.

Pieces	Feet	Description	Cost	Total
17		bags 6-inch fiberglass insulation — ceiling		
26		rolls $3\frac{1}{2}$-inch fiberglass insulation — sidewalls		

WALLBOARD

Before the interior walls are covered, all work must be completed inside the walls. The rough plumbing, wiring, and heating equipment must be in place and inspected. Then the insulation is installed. With this done, the house is ready for wood paneling, lath and plaster, or gypsum wallboard. The most commonly used wall covering is gypsum wallboard, sometimes referred to as *drywall*.

Gypsum wallboard is made by encasing a core made of gypsym rock and other ingredients between two sheets of paper. It is manufactured in widths of 4 feet, and lengths from 6 to 16 feet. The long edges are reinforced and tapered. Its most popular thicknesses are $\frac{3}{8}$ inch and $\frac{1}{2}$ inch. The tapered edge allows the joints to be treated in such a way that they are entirely concealed. This treatment consists of using a special cement, reinforced with a perforated tape which fits into the recess formed at the joints by the boards (Figure 15-2).

Gypsum wallboard is usually applied directly to the wood framing (Figure 15-3). The ceilings are applied first, then the sidewalls. The boards should be accurately cut and positioned, but not forced together. Horizontal application, with the long edges at right angles to the framing members, is preferred because it minimizes joints and strengthens the wall or ceiling. Outside corners that are not protected by millwork are reinforced with steel corner bead. Annular nails are recommended for wallboard application.

Estimating Gypsum Wallboard

The first step in estimating interior walls is to read the specifications to determine what kind of material is to be used. This is found under Division 9, Section B which states that our house is to have $\frac{3}{8}$-inch thick gypsum wallboard.

The quantity needed can be found in two ways. One is to lay out each room by making a rough sketch of each wall and ceiling, showing the dimensions. From these sketches, the number of pieces and their length can be determined. Because some contractors apply the wallboard horizontally and

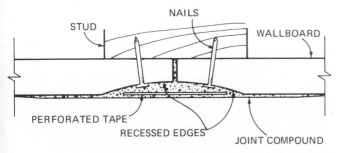

Fig. 15-2. Joint finishing.

Fig. 15-3. Gypsum wallboard getting nailed in place.

other contractors apply the wallboard vertically, making a square foot measurement may be more practical. This allows the contractor to order the needed lengths when the wallboard is to be installed. The price will be the same whether it is listed by the piece or by the square foot.

The amount of wallboard needed for the ceilings is the same as the area of the floor. We have our second floor area of 768 square feet. The ceiling of the first floor is 32 x 22 or 704 square feet. We just figured our exterior wall area for the insulation, which was 1760 square feet. Our interior partitions on the first floor were 78 linear feet and the interior partitions on the second floor were 118 linear feet. We add: 118 + 78 = 196 linear feet. We double this (finish wallboard on each side) for a total of 392.

We then multiply this by 8 (our ceiling height) for a total of 3,136 square feet. We now add 768 square feet (second floor ceiling) plus 704 square feet (ceiling of first floor) plus our exterior wall area of 1,760 square feet and our interior wall area of 3,136 square feet. This gives us a total of 6,368 square feet of wallboard needed for this house. We add this to our material list.

Pieces	Feet	Description	Cost	Total
	6,368	$\frac{3}{8}$-inch gypsum wallboard — ceilings and walls		

JOINT SYSTEM

Joint compound is a premixed, vinyl-based cement that may be used directly from the container. It is used to apply the paper tape to the wallboard joints and for concealing nails (spotting).

The tape that goes over the seams in the wallboard comes in 250- and 500-foot rolls. Joint compound comes in 5- and 1-gallon cans. Allow one 250-foot roll of tape and 6 gallons of compound for every 1,000 square feet of wallboard.

Our house has 6,368 square feet of wallboard, so we will need 7 rolls of 250-foot tape, and 8 five-gallon cans of joint cement. We would add this to our material list.

Pieces	Feet	Description	Cost	Total
7		250-foot rolls Perfa-tape — wallboard finish		
8		5-gallon cans joint cement — wallboard finish		

PLASTER

Plastered walls are not as common as gypsum wallboard, because of the easy application of wallboard and the long drying time involved with the use of plaster. Plaster contains a lot of moisture and the building must be allowed to dry out before finished woodwork is taken into the building. However, some buildings do have plastered walls, so as an estimator you should be familiar with this material.

Gypsum lath has a gypsum core encased between two sheets of paper. It is manufactured in two sizes, 16″ x 32″ and 16″ x 48″; and in two thicknesses, $\frac{3}{8}$ inch which is standard, and $\frac{1}{2}$ inch, which is recommended for use where the studs are more than 16 inches o.c. but not more than 24 inches o.c. It is manufactured either plain or perforated. The perforated type is the same size as the plain except that it has one $\frac{3}{4}$-inch hole for every 16 square inches of surface. This provides a mechanical key for the plaster coat. The amount of gypsum lath needed for a building is found the same way as the amount of wallboard is found by the square foot method.

To find the amount of plaster required for a house, first find the number of square yards to be covered. To do this divide the number of square feet of lath by 9. For every 100 square yards to be plastered the following materials are required:

Neat plaster for brown coat	850 pounds
Gauging plaster	125 pounds
Hydrated lime	250 pounds

Plaster is sold in 100-pound bags and hydrated lime is sold in 50-pound bags.

Our house has 6,368 square feet of wall and ceilings. We divide 9 into 6,368 for a total of 707.56 or 708 yards. We would estimate our plaster based on 750 square yards. Multiplying the quantities for 100 square yards by $7\frac{1}{2}$ gives the following quantities required for our house:

64 bags of neat plaster
10 bags of gauging plaster
50 bags of hydrated lime.

MOLDINGS

The interior millwork or woodwork includes all of the finished trim and moldings. This may vary with each individual building, so it is important for the estimator to check the working drawings and specifications carefully before beginning to estimate.

The inside of each window and each side of the door jambs are trimmed with *casing*. *Stop molding* is the small piece that stops the door from swinging through the opening when it is closed. Windows also use stop molding to hold the sash in place. In addition, the bottom of the window is trimmed with the *stool*, which forms the finished window sill, and the *apron*, which is placed against the wall directly below the stool. These door and window parts are all considered interior millwork, but they are estimated with the doors and windows. A variety of moldings are illustrated in Figure 15-4.

Cove Molding

The corner where the sidewalls and the ceiling meet is often covered with a decorative molding. This corner is called a *cove molding*, but crown molding and bed molding are also used here (Figure 15-5). In some cases, combinations of two or more kinds of moldings may be used. To find out how the corner next to the ceiling is to be finished, look at the drawings. This information is found on the drawing of the half section view, the cornice detail, or a special detail of the cove. The wall section detail is usually drawn to the scale $\frac{3}{4}'' = 1'0''$. It shows the construction of the house from the bottom of the footings up to the rafters at the eaves. The wall section of our house calls for $1\frac{3}{4}$-inch cove molding.

ESTIMATING COVE MOLDING

Molding is estimated by the linear foot. The distance around the outside of the house on each floor is known. We have 108 linear feet on the first floor and 112 linear feet on the second floor. The total length of the interior partitions on both floors is 196 linear feet. We have molding on each side of the interior partitions so we double 196 for a total of 392. We now add: 108 + 112 + 392 = 612 linear feet of $1\frac{3}{4}$ cove needed. We add this to our material list.

Pieces	Feet	Description	Cost	Total
	612	linear feet $1\frac{3}{4}$-inch cove mold — ceilings		

BASEBOARD AND CARPET STRIP

These are the finished moldings at the bottom of a wall where the wallboard or plaster and the finished floor meet. The carpet strip (sometimes called a *shoe mold*) is nailed to the bottom of the baseboard and covers the crack formed where the base meets the finished floor. This allows for any movement between the hardwood floor and the wall.

ESTIMATING BASEBOARD AND CARPET STRIP

There are two ways that the length of the baseboard can be found. One is to find the perimeter of each room and deduct the width of the door openings. Another is to deduct twice the width of the interior door openings and the width of the exterior door openings from the total length of the ceiling cove moldings. The width of the interior door openings and the width of the exterior door openings on the first floor totals 40 feet and the second floor totals 60 feet. We add: 60 + 40 = 100, which we subtract from the 612 linear feet of our $1\frac{3}{4}$-inch cove mold, for a total of 512 linear feet of base and carpet strip required on our house. We would add this to our material list.

Pieces	Feet	Description	Cost	Total
	512	linear feet base and carpet strip – both floors		

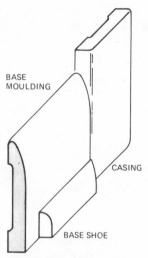

BASE MOULDING

CASING

BASE SHOE

BASE: Applied where floor and walls meet, forming a visual foundation. Protects walls from kicks and bumps, furniture and cleaning tools. Base may be referred to as one, two or three member. The base shoe and base cap are used to conceal uneven floor and wall junctions.

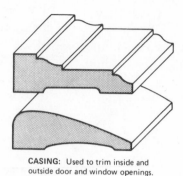

CASING: Used to trim inside and outside door and window openings.

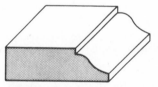

STOOL: A moulded interior trim member serving as a sash or window frame sill cap.

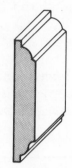

MULLION CASING: The strip which is applied over the window jambs in a multiple opening window. Sometimes called a panel strip, used for decorative wall treatments.

CHAIR RAIL: An interior moulding usually applied about one third the distance from the floor, paralleling the base moulding and encircling the perimeter of a room. Originally used to prevent chairs from marring walls. Use today is as a decorative element or a divider between different wall covering such as wallpaper and paint or wainscoting. A key decorative detail in traditional and Colonial design.

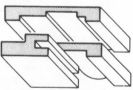

JAMB: The top (header) and two sides (legs) of a door or window frame which contacts the door or sash. Flat jambs are of fixed width while split jambs are adjustable.

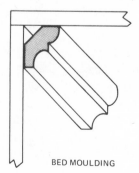

STOP: In door trim, stop is nailed to the faces of the door frame to prevent the door from swinging through. As window trim, stop holds the bottom sash of a double-hung window in place.

Fig. 15-4. Wood moldings.

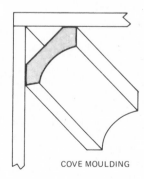

COVE MOULDING

BED MOULDING

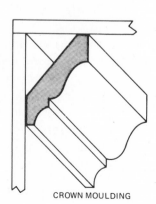

CROWN MOULDING

Fig. 15-5. Moldings used between ceilings and walls.

16
CARPENTRY INTERIOR WORK: FLOORS

Most homes have a different kind of floor covering on the kitchen, bathroom, and entrance floors than on the other floors of the house. Kitchen floors are usually tile or vinyl sheet material. Bathroom floors are usually vinyl sheet, vinyl tile, or ceramic tile. Entrances are usually of a masonry material such as ceramic tile, quarry tile, slate, brick, or marble. You should check the plans and specifications carefully to find out what is called for in these areas.

UNDERLAYMENT

All the floors should be laid on the same level to eliminate the possibility of tripping on their edges. Different materials are manufactured in various thicknesses. For example, hardwood flooring is $\frac{25}{32}$ inch thick, but vinyl is $\frac{1}{8}$ inch thick. To overcome this difference, *underlayment* is nailed down on top of the subfloor before the finish floor material is installed.

Underlayment can be any material that will give a smooth surface to receive the finish flooring. Often plywood that has a smooth sanded surface on one side is used. Other materials such as hardboard and tempered particle board are also used. The thickness of the underlayment varies according to the thickness of the material used for the finished floor. The thickness of the underlayment plus that of the finished flooring material should be the same as the floor it will butt against.

In some entrance and bathroom floors where masonry material is set in a bed of mortar to form the finished floor, the subflooring is cut to rest flush with the top of the floor joists instead of on top of the joists. This allows for a thicker bed of mortar. However, because of the high strength of modern adhesives, most ceramic floors are cemented in place with adhesives instead of mortar. Where a masonry floor comes next to finished wood floors, a marble threshold is placed in the doorway. In entrance halls, the common practice is to have the masonry installed, and then the finished carpet or wood floor fitted against it. Figure 16-1 shows how the subfloor may be installed on ledgers to allow a thicker bed of mortar, and Figure 16-2 shows how a marble threshold is used between a masonry floor and a wood floor.

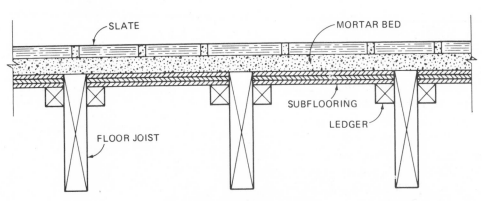

Fig. 16-1. Subfloor on ledgers.

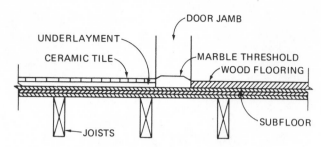

Fig. 16-2. Marble threshold.

Estimating Underlayment

The house we are estimating calls for $\frac{5}{8}$-inch plywood APA underlayment grade (Division 6, Section C-12). To find the quantity needed, take the square-foot area of the kitchen, bathrooms, and entrance hallway. Then divide this total area by 32 to find the number of pieces of underlayment needed. Our first and second floor plans give us the size of the rooms and the material for the finish floor. Our plans show slate on the entrance hall, linoleum on the kitchen and half bath, and ceramic tile floors on the second floor bathrooms. Our entrance hall is 8' x 22' including the closet and half bath. The kitchen is approximately 10' x 12'. The two baths on the second floor are approximately 93 square feet. The first floor central hall is 8 X 22 = 176 square feet. The area of our stairwell (4 X 10 = 40) is subtracted from this: 176–40 = 136 square feet. The kitchen is 10 X 12 = 120 square feet. We add 93 square feet for the second floor baths: 136 + 120 + 93 = 349 square feet area for underlayment. We divide 349 by 32 and get a total of 11–$\frac{5}{8}$ x 4 x 8 plywood underlayment. We would add this to our material list.

Pieces	Feet	Description	Cost	Total
11		$\frac{5''}{8}$ x 4' x 8' plywood APA underlayment		

RESILIENT FLOORING

Kitchen floors are usually covered with either carpeting or some type of resilient flooring. Indoor/outdoor carpeting is popular because it resists staining and may be easily cleaned. Resilient flooring is also popular because of the ease with which it may be cleaned. New types of resilient flooring are now available with a "no wax" finish. It is available in tile form or in sheets. Standard inlaid linoleum is manufactured in 12-foot wide rolls. Our specifications (Division 9, Section G) calls for inlaid linoleum in the kitchen and half bath on the first floor.

Estimating Resilient Flooring

Find the area over which the resilient floor is to be installed and divide this total by 9. This gives the number of square yards required. Look at our first floor plan: the kitchen floor is approximately 10 x 12 and the half bath is 3 x 5. Using 12-foot wide material you have $1\frac{1}{4}$ square yards of material for every linear foot. We do not need inlaid flooring underneath the kitchen cabinets, but we do under the range and refrigerator. Our half bath can be covered with the piece cut out where the cabinets are installed by adding 1 more linear foot. We would need a piece of inlaid linoleum 11-feet by 12-feet to cover our area. We multiply 11 by $1\frac{1}{4}$ = 13 $\frac{3}{4}$ yards, rounded off to 14 yards. We would add this to our material list.

Pieces	Feet	Description	Cost	Total
	14	square yards inlaid linoleum — kitchen and half bath		

SLATE FLOORING

There are a great number of materials which are commonly used for floors. The most common materials are hardwood, carpet, ceramic tile, vinyl tile, vinyl asbestos tile, and slate. The use for which the room is intended is an important factor in deciding which material to use. Figure 16-3 shows the in-

Fig. 16-3. Resilient flooring.

Fig. 16-4. Slate floor.

stallation of a resilient flooring like our floor for the kitchen, and Figure 16-4 shows a sample of a slate floor for our hallway.

Although hardwood flooring is a popular material for many rooms, it is not normally used in entrances. Hardwood may be stained by the mud and water which are tracked into the entrance, and heavy traffic can cause hardwood to show wear. The most common materials for entrance floors are brick, quarry tile, slate, stone, and vinyl tile. These are durable and are not affected by water. Division 9, Section F of the specifications for our house indicates that the front entrance and hallway is to be slate. Slate (Figure 16-4) is set in much the same manner as ceramic tile: individual pieces of slate are set in adhesive, then the joints are grouted.

Estimating Slate Flooring

Slate flooring is sold by the square foot. If the slate is set with adhesive, you will have to determine the quantity and also the grout for the joints.

We already have the square-foot area of the entrance hall and the half bath: 8 × 22 = 176 square feet. We subtract the area of the stairwell (40 square feet). Our half bath is 3' x 5' or 15 square feet. We add 15 to 40 for a total of 55 square feet, which we subtract from 176 square feet; this means we will need 121 square feet of slate. We need 1

gallon of adhesive for every 100 square feet, so we will need 5 quarts of adhesive. It takes approximately 5 pounds of white cement grout for every 100 square feet, so we should figure 6 pounds of grout for our slate entrance. We would add this material to our material list.

Pieces	Feet	Description	Cost	Total
	121	square feet multicolored slate — hallway		
5		quarts special slate adhesive — hallway		
6		pounds white cement grout — hallway		

HARDWOOD FLOORING

The hardwoods most commonly used for flooring are oak, maple, beech, birch, and pecan. Oak is the most plentiful and is by far the most used. Hardwood flooring may be classified as strip flooring, plank flooring, or parquet. Strip flooring is the type most extensively used. It is tongue and grooved on the ends as well as the edges, so that each piece joins the next when it is laid. Strip flooring is manufactured in a variety of sizes. The most popular is $\frac{25}{32}$ inches thick and $2\frac{1}{4}$ inches wide.

The principal American producers of hardwood flooring have adopted uniform grading rules and regulations. Every bundle of flooring produced by a member of either the National Oak Manufacturers' Association or the Maple Flooring Manufacturers' Association is identified as to grade. Usually the manufacturer's name and mill mark or identification are found on each bundle.

Standard oak flooring grades have been established. Quarter sawed is clear or select, and plain sawed is select, clear, No. 1 common, or No. 2 common. Hardwood flooring comes in random lengths.

Strip flooring is usually nailed through the building paper directly to the subflooring and joists. The nails are driven through the base of the tongue of the flooring at a 45-degree angle (Figure 16-5) so that the nail heads do not show. Staples made for use with a flooring hammer may be used instead of nails.

When a wide-plank flooring is installed, the width of the boards make blind nailing impractical. In this case, the flooring is drilled and counterbored so that the floor can be fastened to the subflooring with screws. The heads of the screws are then cov-

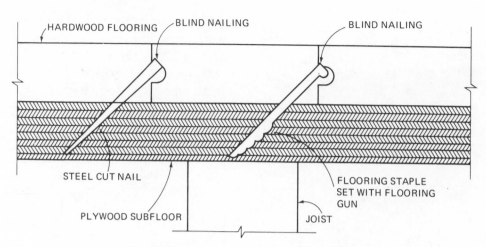

Fig. 16-5. Blind nailed hardwood flooring.

ered with wood plugs and finished the same as the rest of the floor. Where it is necessary to face nail the flooring near the walls, the nail heads are set and the holes filled with wood putty and sanded the same as the floor.

Estimating Hardwood Flooring

Division 9, Section C states that the hardwood floor is to be 1 x 3 select oak. All subfloors are to be broomed cleaned and covered with deadening felt before the finished flooring is laid. The wood flooring, where scheduled, is to be 1 x 3 T&G and end-matched select oak. Flooring is to be laid evenly and blind nailed every 16 inches without tool marks.

To figure the quantity needed, first find the area to be covered. All the floors in our house except the entrance hallway, kitchen, baths are to be hardwood. We already have determined the square-foot area of our house: 32 × 24 = 768, second floor plus 32 × 22 = 704, first floor, for a total of 1,472 square feet. Our entrance hallway, half bath, kitchen and bathrooms on the second floor totals 326 square feet. We subtract this from 1,472 for a total of 1,146 square feet to be covered with hardwood flooring. We have to allow one-third for matching. One third of 1,146 is 382. We add: 382 + 1,146 = 1,528 square feet of hardwood required for the job.

One roll of deadening felt covers 500 square feet. We have 1,472 square feet for both floors so we will need 3 rolls. We add this to our material list.

Pieces	Feet	Description	Cost	Total
	1,528	1 x 3 select oak flooring		
3		rolls deadening felt		

CERAMIC TILE

There are many materials which may be used on the bathroom walls and floors. The most common one is ceramic tile (Figure 16-6 and 16-7), because it is durable and easy to clean. Other materials that might be used on the bathroom walls are plastic laminates, vinyl-coated wallpaper, and paint. Carpeting or resilient flooring might be used on the floors. The plans and specifications indicate the material to be used. Our specifications call for ceramic tile in the tub area and the finish floors. We have two bathrooms with tubs. A standard 5-foot tub is 2'6" wide and 5'0" long. This gives us a total

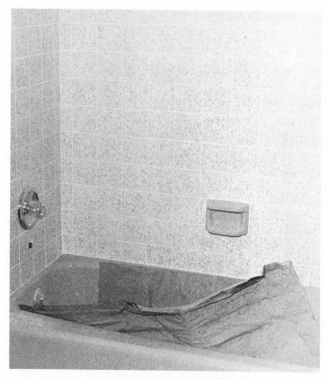

Fig. 16-6. Ceramic tile.

the door of each bath meets the hardwood floor we will need a marble threshold.

Also the specifications call for a vitreous china grab bar and soap dish. We will also need adhesive for the wall tile and grout for the wall tile. Adhesive covers 100 square feet to the gallon, so we will need 5 quarts of adhesive and 6 pounds of wall grout. We would now add this material to our material list.

Pieces	Feet	Description	Cost	Total
	70	square feet 1 x 1 ceramic tile flooring – baths		
1		gallon ceramic tile floor adhesive		
5		pounds white floor grout – tile work		
2		marble thresholds, 2'4" long		
6		$\frac{1}{2}$" x 4' x 8' waterproof plywood – tub area		
	120	square feet $4\frac{1}{4}$ x $4\frac{1}{4}$ ceramic tile – tub areas		
	44	linear feet 2" x 6" ceramic cap tile		
4		inside corner caps		
4		outside corner caps		
2		vitreous china soap and grab		
2		vitreous china grab bars		
5		quarts ceramic wall adhesive		
6		pounds white ceramic wall grout		

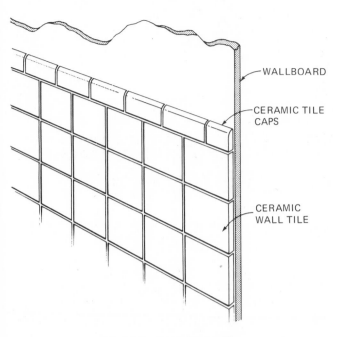

Fig. 16-7. Ceramic cap tile.

of 10 linear feet of tile for each tub. Our specifications state the tile is to be set on waterproof plywood. Each sheet is 4 feet wide, so we will need three sheets for each tub area.

The tile is to be applied to a height of 6 feet above the tub. Multiply this height by the linear feet to be covered, and we find that 60 square feet of wall tile will be required for each tub area.

Tile caps are used to finish the exposed edges of the tile (Figure 16-7). These are 2 inches by 6 inches with a round edge. We need 22 linear feet of cap (6 feet for each end and 10 feet for top edge). We will need two inside corners and two outside corners for each tub. Our floor tile is to be 1 x 1 ceramic tile. Our baths have approximately 35 square feet of finished floor in each, so we will need 70 square feet of ceramic floor tile. We will need 1 gallon of adhesive and 5 pounds of grout. Where

17
CARPENTRY MILLWORK AND FINISHES: DOORS AND CLOSETS

Doors are classified as either exterior or interior, depending on whether they are hung in exterior walls or interior partitions. They can also be classified as either *flush* (with a smooth surface) or *panel* doors (with decorative panels).

EXTERIOR DOORS

Exterior doors are usually $1\frac{3}{4}$ inches thick, solid pine or fir. Flush doors usually have a hardwood veneer on the surface. They may have either a solid wood or a hollow center. All exterior doors are glued with a waterproof glue.

A great assortment of styles of doors is available. A colonial door, for example, is constructed with various size panels and moldings. Some have glass panels and others have all solid wood panels. A colonial door is constructed so that the four top panels form a cross (Figure 17-1).

Doors and windows are always designated with the width first, the height second, and the thickness last. Most main entrance doors in residential construction are 3'0" x 6'8" x $1\frac{3}{4}$". This means that they are 3 feet wide by 6 feet 8 inches high by $1\frac{3}{4}$ inches thick. Other entrance doors, such as a back or side door, are usually 2'8" x 6'8" x $1\frac{3}{4}$". These sizes may be varied, but they have almost become standard for residential construction throughout the building industry.

INTERIOR DOORS

Interior doors are manufactured in a variety of styles and sizes. They may be $1\frac{3}{4}$ inches, $1\frac{3}{8}$ inches, or $1\frac{1}{8}$ inches thick. Most interior doors are $1\frac{3}{8}$ inches thick; however, narrow louvered doors are $1\frac{1}{8}$ inches thick.

Flush interior doors are usually hollow-core doors which are covered with hardwood veneer. The top, bottom, and side rails are made of solid wood, as is the area where the hardware is installed. The remainder of the inside of the door is filled with a lightweight material, such as honeycombed paper. This makes a lightweight door that is attractive and serviceable. Flush doors are usually finished with a light stain or clear finish. This is why manufacturers use select veneers for the finished surface.

Flush doors are manufactured in 6'0", 6'4", 6'8", and 7'0" heights. They are made in widths from 9 inches to 3 feet.

Panel doors are manufactured in various styles with from two to eight panels (Figure 17-2). The most popular style panel door is a 6-panel colonial door which has two small panels at the top and four larger panels directly below.

Door heights may vary according to the available headroom where they have to be installed. It is common practice to have door heads and window heads the same height from the finished floor. The most common height for both interior and exterior doors is 6'8". This allows the casing for the top of

Fig. 17-1.

Fig. 17-2. Panel doors.

the doors and windows to be the same height, keeping the appearance in balance.

PREHUNG DOORS

A prehung door is a complete unit with the door hung at the factory. These units are assembled at the factory with the door, jambs, stop, and casing all in place. The door jamb slides apart so that the two halves can be slid into the opening. With the unit in place, the casings are nailed through the wall and into the studs. In ordering a prehung door, the size, style of the door, style of the casing and hardware, and the swing of the door must all be listed. To determine the *swing* of a door, stand on the side toward which the door opens. If the knob is on the right side, it is a *right-hand door*. If the knob is on the left side, it is a *left-hand door*.

Prehung door units are available in bypass sliding units and folding units. Any style door can be prehung, either exterior or interior.

Louver doors (Figure 17-3) are popular in such places as the laundry, family room, closets, and kitchens. These doors are manufactured in various widths, heights, and styles.

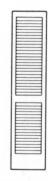

Fig. 17-3.

ESTIMATING DOORS

List the exterior doors first. The specifications list the size and style of the exterior doors on our house. If the information is not available in the specifications, it can be found on the drawings or a special schedule of doors.

When job-hung doors are used instead of prehung, the parts must be listed separately. Exterior door frames may be assembled by a local lumber dealer. In this case the frame is delivered to the job site with the jambs and outside casing installed. The oak threshold is sent separately and the frame and sill are installed at the site.

We first list the exterior door frames and sills, the doors by size and model number. Our front door is 3'0" x 6'8" and the side door is 2'8" x 6'8". We list them with the sill.

Pieces	Feet	Description	Cost	Total
1		exterior door frame with sill, 3'0" x 6'8"—front door		
1		exterior door frame with sill, 2'8" x 6'8"—side door		
1		exterior door, 3'0" x 6'8" x $1\frac{3}{4}$" M-108—front door		
1		exterior door 2'8" x 6'8" x $1\frac{3}{4}$" No. 233—side door		

The door stop and the interior door casing are listed separately. Estimate the casing on the interior side only, since the exterior casing is installed when the frame is assembled. It takes approximately 17 linear feet of casing for each side of a door. We have two exterior doors, hence 2 × 17 = 34 linear feet of casing. To prevent the door from swinging through the doorway, molding called *door stop* is installed around the inside of the jambs. The door closes against the door stop. Each door requires approximately 17 linear feet of door stop; hence 2 × 17 = 34 linear feet of stop is required. We would add this to our material list.

Pieces	Feet	Description	Cost	Total
	34	linear feet casing style No. 37—exterior doors		
	34	linear feet $1\frac{3}{4}$" door stop No. 93—exterior doors		

We next list all the interior doors on the first floor according to the specifications and the first floor plan. All interior doors are to be birch flush doors except where noted. We would list them as follows.

Pieces	Feet	Description	Cost	Total
2		set door jambs 3'8" x 6'8"—cased opening, hallway		
3		set jambs 2'6" x 6'8"—dining room, kitchen, cellar stairs		
1		set jambs 2'0" x 6'8"—half bath		
1		set jambs 1'8" x 6'8"—closet		
2		louver doors 1'3" x 6'8" x $1\frac{1}{8}$ M-510—dining to kitchen		

Pieces	Feet	Description	Cost	Total
1		2'6" x 6'8" x $1\frac{3}{8}$" birch flush door—cellar stairs		
1		2'0" x 6'8" x $1\frac{3}{8}$" birch flush door—half bath		
1		1'8" x 6'8" x $1\frac{3}{8}$" birch floor door—closet		

We would next figure the door casing and the door stop. Each door will require casing on each side and only the doorways with doors will require door stop. We estimate 17 linear feet of casing for each side of a door and 17 linear feet of stop for each door. We would now list this material on our material list.

Pieces	Feet	Description	Cost	Total
	238	linear feet casing style No. 37—first floor doors		
	68	linear feet door stop $1\frac{3}{8}$" No. 93		

We now refer to the second floor plan and again list each door and/or cased opening according to size. We have 3 bedroom doors 2'6" x 6'8", 2 closet doors 2'6" x 6'8", 2 bathroom doors 2'4" x 6'8", 1 sliding bypass door 4'0" x 6'8", 1 linen closet door 1'8" x 6'8", and 1 closet door in the large bathroom 1'3" x 6'8". We would list each door by size, listing jambs, doors, casing, and stop.

Pieces	Feet	Description	Cost	Total
5		sets jambs 2'6" x 6'8"—second floor		
2		sets jambs 2'4" x 6'8"—bathrooms, second floor		
1		set jambs 1'8" x 6'8"—linen closet, second floor		
1		set jambs 1'3" x 6'8"—towel closet, second floor		
1		set jambs 4'0" x 6'8"—closet, master bedroom		
5		birch flush doors 2'6" x 6'8" x $1\frac{3}{8}$"—bedrooms		
2		birch flush doors 2'4" x 6'8" x $1\frac{3}{8}$"—bathrooms		
1		birch flush door 1'8" x 6'8" x $1\frac{3}{8}$"—linen closet		
1		birch flush door 1'3" x 6'8" x $1\frac{3}{8}$"—towel closet		
2		birch flush doors 2'0" x 6'8" x $1\frac{3}{8}$"—closet, master bedroom		
1		set sliding door hardware 4'0"—master bedroom closet		

374	linear feet casing style No. 37— second floor doorways			
187	linear feet $1\frac{3}{8}''$ door stop No. 93— second floor doors			

We will need adjustable closet rods for each closet and we will also need some 1 x 3 pine to fasten the closet rods on, and also we will need 1 x 12 pine shelving for a shelf in each closet. The rods are adjustable, so we only need list the number. We have to scale the working drawings to get the total length of 1 x 3 and 1 x 12 pine for the closets.

The 1 x 3 should go across the back and on each side. The shelves just across the back. We would now list this material on our material list.

Pieces	Feet	Description	Cost	Total
4		adjustable closet rods—1 first floor; 3, closets, second floor		
	70	linear feet 1 x 3 pine closet rod supports		
	30	linear feet 1 x 12 pine—closet shelves		

18
CARPENTRY MILLWORK AND FINISHES: WINDOWS

WINDOWS

There are too many sizes and styles of windows available to list them all in this book. The following is a brief description of the windows used most commonly in house construction.

Double hung windows (Figure 18-1) consists of two sash (the frame holding the glass) which slide up and down in the jambs (the side members of the window unit). The cross member where the two sash meet is called the meeting rail or check rail. The sash lock may be locked in the closed position by a sash lock installed on the check rail. Double hung windows have spring balances which help relieve the weight of the sash for ease in opening the window. *Sliding windows* are similar to double hung windows, except that they slide from side to side instead of up and down. Because there is no weight to overcome in opening sliding windows, they do not normally have balance mechanisms. *Awning windows* (Figure 18-2) open out from the bottom sash. They may be operated by either a push bar or a cranking mechanism. *Casement windows* (Figure 18-3), are similar to awning windows except that they open from the side instead of the bottom.

Some manufacturers produce units which can be installed as either casement or awning windows.

The various types of movable sash are often used in combination with fixed sash (those which cannot be opened) to produce picture windows, bow windows, bay windows, and mullion windows. A *picture window* (Figure 18-4) has a large fixed sash in the center and usually some type of movable sash on the sides for ventilation. *Bow windows* (Figure 18-5) curve out from the face of the house with a series of small fixed or awning windows or casement windows. *Bay windows* (Figure 18-6) are similar to bow windows except they do not form a uniform curve. They consist of three windows,

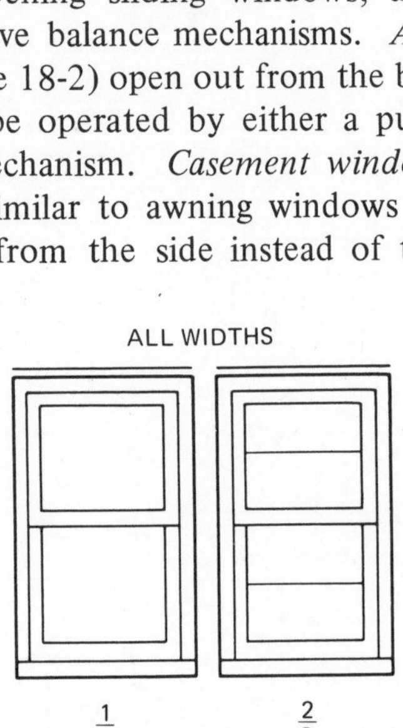

ALL WIDTHS

Fig. 18-1. Dougle hung.

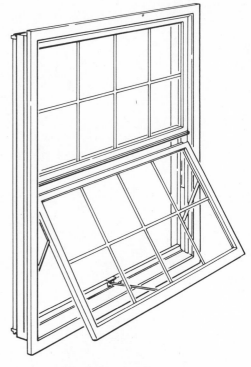

Fig. 18-2. Awning.

usually double hung, at 30-degree or 45-degree angles. *Mullion windows* (Figure 18-7) consist of two or more windows mounted side by side. Where the jambs of each window are joined, the space is covered with a molding called mullion casing on the inside and also the outside.

Windows come from the manufacturer with weather stripping and sash installed. Normally double hung and sliding windows are supplied with sash locks installed as well. The glass may be either standard glazing or insulated glazing, two pieces of glass with space between them, (Figure 18-8). Insulated windows eliminate the need for a storm sash.

WINDOW SIZES

Windows and doors are always listed with the first dimension shown being the width and the second dimension being the height. For example, a 3'0" x 4'6" window is 3 feet wide and 4 feet 6 inches high.

Fig. 18-3. Casement.

Fig. 18-4. Picture window.

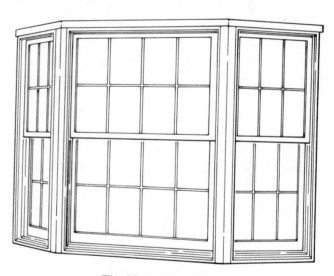

Fig. 18-6. Bay window.

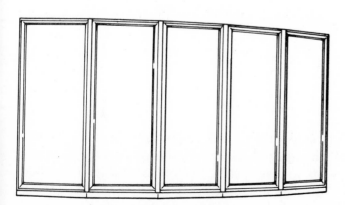

Fig. 18-5. Bow window.

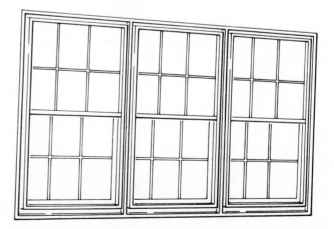

Fig. 18-7. Mullion windows.

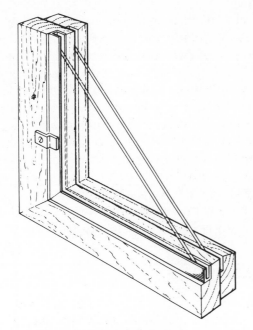

Fig. 18-8. Insulated glazing.

The symbol used on the drawing indicates whether it is a single, mullion, or triple unit.

Usually the window is referred to by its nominal size, that is, including everything but the casing. For example, double-hung windows may be available in sizes ranging from 1'8" x 2'6" to 3'8" x 4'6". The sizes increase in multiples of two inches in width and four inches in height.

Sometimes the size of a window is referred to by its glass size. To convert from glass size to nominal size; double the glass size (for a double hung window) and add six inches for the height. For example, if the glass size is shown as 24/24 the window is 24 + 24 + 6 or 54 inches. Therefore 24-inch high glass means a 4'6" height. To find the width, take the glass size and add 4 inches. A 24-inch glass would be a 28-inch window or 2'4". Therefore, a 24/24 glass size window is 2'4" x 4'6". These figures are intended only as a guide. Each manufacturer's catalog should always be checked for sizes and styles of windows.

In addition to size and types of windows, the estimator must know the number of lights (panes of glass) to be installed in each window. On double hung windows this is given as a fraction with the top number being the number of lights in the top sash and the bottom number being the number of lights in the bottom sash. They would show a picture of the window and show 1/1, 2/1, 6/1 or 6/6, or 8/8, depending upon the size of the window.

Estimating Windows

The number and size of the windows are shown on the first and second floor plan. The specifications indicate that they are to be double hung windows with removable grilles. Wood combination storm and screen are to be furnished with each window. This is found in Division 8, Section F of our specifications. Looking at our floor plans we would list our windows according to size.

Pieces	Feet	Description	Cost	Total
6		double hung windows, single 3'0" x 4'6", 6/6—first floor		
1		double hung window, single 3'0" x 3'2", 6/6—first floor kitchen		
1		double hung window, single 2'0" x 3'2", 6/6—first floor bath		
8		double hung windows, single 3'0" x 4'6", 6/6—second floor		
1		double hung window, single 2'4" x 3'10", 6/6—second floor bath		
1		double hung window, single 2'0" x 3'10", 6/6—second floor bath		

We should now list the combination storm sash and screen that is required for each window according to our specifications. We would group these according to size.

Pieces	Feet	Description	Cost	Total
14		combination storm and screen windows 3'0" x 4'6"		
1		combination storm and screen window 3'0" x 3'2"—kitchen		
1		combination storm and screen window 2'0" x 3'2"—half bath		
1		combination storm and screen window 2'4" x 3'10"—bath		
1		combination storm and screen window 2'0" x 3'10"—bath		
2		exterior door weatherstripping 3'0" x 6'8"		

WINDOW TRIM

Window units come complete, except for the interior trim. After the window is installed, the stool, apron, casing, stop, and mullion casing (if required) must be installed; Figure 18-9 shows the parts of a double hung window and the moldings.

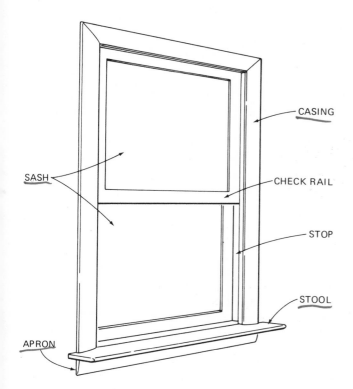

Fig. 18-9.

Casing and Stop

We need casing from the top of the window stool up each side and across the top of the window. We have 14 windows 3′0″ wide and 4′6″ high; one window 3′0″ wide and 3′2″ high; one 2′0″ wide and 3′2″ high; one 2′4″ wide and 3′10″ high; and one 2′0″ wide and 3′10″ high. Measuring the distance up one side, across the top and down the other side we find that we need 206 linear feet of casing. We add 10% for waste so we will need 227 linear feet of casing.

We will need the same amount of window stop as casing because all our windows are double hung, single units. We would also add to our material list 227 linear ft. of No. 90 window stop.

Window Stool and Apron

For each window we need a window stool and apron to finish the inside of the window. The stool ex-

tends approximately $2\frac{1}{2}$ inches on each window. The apron rests underneath the comes even with the outside edge of th window casing. We have 15 windows 3′0″ wide and 3 windows approximately 2′6″ wide, giving us a total length of 60 feet (including 5 inches extra per window for the stool). We add 10% for waste, so we will need 66 linear feet. We add this material to our material list.

Pieces	Feet	Description	Cost	Total
	227	linear feet window casing style No. 37 – windows		
	227	linear feet window stop 7/16 x 7/8 No. 90 – windows		
	66	linear feet window stool No. 71-PL 11/16 x $2\frac{1}{2}$ – windows		
	66	linear feet window apron No. 35 11/16 x $2\frac{1}{4}$ – windows		

Under Division 10, Section B-2, our specifications call for a fireplace mantel. We would add this to our material list.

Pieces	Feet	Description	Cost	Total
1		fireplace mantel M-1448		

Our elevations show that some windows have blinds on each side. This is found on the front elevation, the right elevation and the left elevation. Also a louver door is placed on each side of the front door. We would list them according to size.

Pieces	Feet	Description	Cost	Total
11		pair blinds 3′0″ x 4′6″ M-500 – windows on front and sides		
1		pair blinds 2′4″ x 3′10″ M-500 – bath window on right side		
2		louver doors 1′6″ x 6′8″ x $1\frac{1}{8}$″ M-510 – each side front door		

19
CARPENTRY MILLWORK AND FINISHES: STAIRWORK

STAIRS

All houses, except those with only one story and no basement, have a stairway. Normally, when the cellar is unfinished the stairs from the first floor to the cellar are of rough construction with open risers. This means that there is no piece covering the opening from one step (tread) to the next. The *stringers* (pieces at the sides supporting the stairs) are commonly made of 2 x 10s on this type of stairs.

Each part of the stairway has a specific name and purpose. In order for an estimator to discuss stairs and list the materials to construct them, it is important to know the name and purpose of each part.

Tread

The *tread* (Figure 19-1) is the part one steps on when using the stairs. Treads are usually made of $1\frac{1}{16}$-inch thick oak or pine and may be purchased

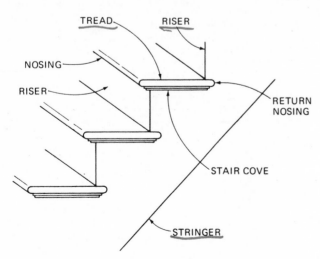

in several widths and lengths. The front edge of the tread, called the *nosing,* is rounded. The width of the tread is measured from the face of one riser to the face of the next and does not include the nosing. The nosing projects $1\frac{1}{8}$ inches beyond the face of the riser. This means that if a stairway has 10-inch treads, the actual width of the tread is $11\frac{1}{8}$ inches.

Risers

The space at the back of the treads is enclosed by the *risers* (Figure 19-1). The risers are usually constructed of clear pine. Do not confuse the riser with the *rise of the stairs:* the rise is the vertical distance from the top of one tread to the top of the next.

A $\frac{5}{8}$-inch by $\frac{3}{4}$-inch cove molding, called a *stair cove,* is nailed to the bottom of the tread nosing and the face of the riser. It is for appearance only and has no structural purpose.

The *stringers* are the inclined pieces that support the treads. They are frequently made of 1 x 10 pine. A *housed stringer* (Figure 19-2) has dados (grooves) cut $\frac{1}{2}$ inch deep on the inside to receive the treads and risers. These grooves are tapered, so wedges can be glued into them after the treads and risers are in place. This assures a tight fit.

Types of Stairs

Stairs may be either closed, half open, or open. They may be straight, have landings, or turn with winders. *Straight stairs* are those which do not turn from top to bottom. A *landing* is a platform partway up the stairs and is generally used where the stairs

Fig. 19-1. Stair parts.

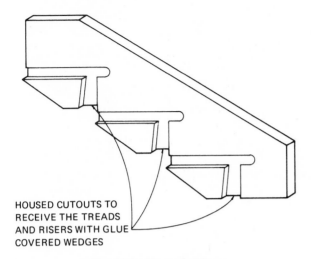

Fig. 19-2. Housed stringer.

HOUSED CUTOUTS TO RECEIVE THE TREADS AND RISERS WITH GLUE COVERED WEDGES

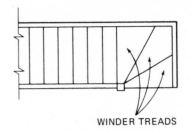

WINDER TREADS

Fig. 19-3. Winders.

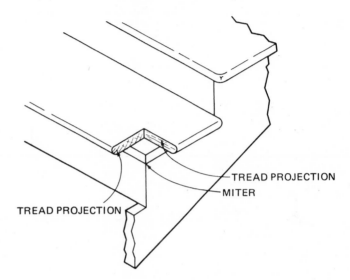

TREAD PROJECTION
MITER
TREAD PROJECTION

Fig. 19-4. Open tread.

make a turn. *Winders* (Figure 19-3) are tapered treads, used to make a turn without a landing. *Closed stairs* are those with a partition or wall at both sides, concealing the ends of the treads. *Open stairs* have one or both ends of the treads exposed to view, and *half open stairs* are those which are open for part of the run. On half-open stairs two to five treads, not counting the starting step, are left open.

Half-open and open stairways require more parts than do closed stairways. The stringer is mitered where the riser is joined to it (Figure 19-4), and the tread nosing is continued around the open end of the tread. On stairs with a housed stringer, such as closed stairs, the cove molding extends from one stringer to the other. On open stairs the cove molding follows the return nosing around the end of the tread.

Open stairs frequently use a special *starting step* to which the *newel post* is attached. Starting steps are available in several styles, including *circle end, quarter circle,* and *scroll.* They are reversible, so they can be used for either right open or left open stringers. The vertical pieces which support the handrail at each tread are called *balusters.* There are usually two or three balusters on each tread and they are doweled into the tread and the handrail. Balusters are available in 30, 33, 36, and 42-inch lengths. If two balusters are used on each tread they should be 30 inches and 33 inches long. If three are used they should be 30 inches, 33 inches, and 36 inches long. Balusters under volutes should be 36 inches long; under goosenecks they are 42 inches long; and under the railing at the second floor they are 33 inches long. These dimen-

sions will allow the handrail to be 30 inches high above the treads and 34 inches high on the level.

If the stairway includes a landing, a *landing nose* must be installed at its edge. Landing nosing is $1\frac{1}{16}$ inches thick by $3\frac{1}{2}$ inches wide with a rabbet to allow it to match 13/16-inch flooring.

Figure 19-5 shows suggested starting sections for half open or open stairways. Notice the various styles and starting steps, newel posts, volutes, easements, and rail.

ESTIMATING STAIRS

Information pertaining to the stairs is included in Division 6, Section D-3 states what is called for. The first floor plan shows the stairs to be the Abbington M-844.

Most builders purchase factory-built stairs from a millwork manufacturer. In this case it is only necessary to list the size and style of the stairs. Figure 19-6 shows the M-844 Abbington stairway to be used in our house.

Looking at our stair detail and the first and second floor plans we find the stairway is a half

M-847
Newel M-786 with
Rail M-820 and
Balusters M-840,
Tread and Riser

M-849
Newel M-800 with
Starting Step M-830,

Rail M-820,
Balusters M-840

M-859
Newel M-769 with
Tread and Riser

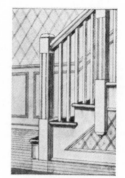

M-853
Newel M-885 with
Starting Step M-896,
Landing Newel M-890,
Balusters 1⅛" S4S

M-856
Volute M-719 with
Starting Step M-780
or
Volute M-722 with
Starting Step M-779V
Newel M-765 for either

M-857
Turnout M-723 with
Starting Step M-779T,
Newel M-895

M-858
Easement M-725 with
Tread and Riser,
Newel M-766A

M-761
Vertical Volute M-750
with Newel M-753,
Tread and Riser

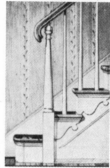

M-762
Rail Drop M-751
with Newel M-754
Starting Step M-896

Fig. 19-5. Suggested starting stair sections.

open stair from the first floor to the second. At the second floor we find another newel post and a straight handrail with balusters returning against the stairwell partition. We have 14 risers and 13 treads. Notice that the stairs are half open but the treads are closed, so we will figure housed stringers for the total length. We would list 1 set of housed stringers with 14 risers and 13 treads. We need a newel post at the start and at the top of the stairs. The balusters are spaced approximately one to a tread. With our return on the second floor we should figure 11 balusters 36 inches long. We need 26 linear feet of handrail, 26 feet of subrail, and 26 linear feet of fillet. We list our 13 treads giving size, and the 14 risers giving size. Our landing nosing goes across the floor at the top of the stairs and also returns to the stairwell partition on the second floor. We need cove molding and handrail brackets to support the handrail along the closed section of the stairs.

We would now add the stair material to our material list.

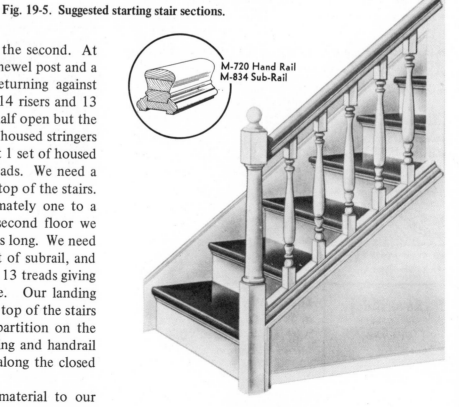

M-720 Hand Rail
M-834 Sub-Rail

Fig. 19-6. Stairway.

Pieces	Feet	Description	Cost	Total
1		set housed stringers with 14 risers — 13 treads — pine		
2		newel posts M-769 3″ x 3″ x 4′2″		
11		balusters M-836 33″ — birch		
	26	lineal feet M-720 handrail — birch		
	26	lineal feet M-834 subrail — birch		
	26	lineal feet M-835 fillet		
13		oak treads 11/16″ x $10\frac{1}{2}$″ x 3′6″		
14		oak risers $\frac{3}{4}$″ x $7\frac{5}{8}$″ x 3′6″		
1		oak landing nosing $1\frac{1}{16}$″ x $3\frac{1}{2}$″ x 8′0″		
	60	lineal feet oak cove molding No. 8062		
4		handrail brackets		
50		pine wedges — stairwork		
1		pint white glue — stairwork		

20
PAINTING AND DECORATING

Before subcontractors can bid on a job they must estimate the cost of labor and materials for their part of the work. In many cases the general contractor relies on the subcontractor to estimate that portion of the work. However, this is not always true and the estimator should have knowledge of all phases of the construction process. Division 9, Section H covers the painting for our house.

It is quite common for specifications to indicate the name of a paint manufacturer and specific grades of materials to be used. This is done to indicate the quality expected, not to restrict the contractor from using other brands. Most manufacturers supply paint, varnish, and other finishing materials with approximately the same characteristics.

PAINT AND OTHER FINISHES

Paint used in building construction is an opaque (not transparent) material which is applied in a thin film for protection or decoration. Paint consists of two main parts: the *pigment*, which supplies the color, and the *vehicle*, which carries the pigment to the surface being painted. The vehicle may also contain a *drier* to quicken the drying process. There are two general classifications of paint, depending upon the vehicle used.

Oil-Based Paint

Oil-based paints use either linseed oil, soybean oil, or tung oil for the vehicle. Soybean oil is the most widely used, because of its nonyellowing characteristics. Mineral spirits or turpentine is normally used to thin oil-based paints and to clean equipment which is used with these paints. The manufacturer will have printed on the label of each can its contents and the percentage of each: for example,

Pigment 53% (listing the pigments) and Vehicle 47% (listing the ingredients in the vehicle).

Water-Based Paint

These paints use a combination of water and synthetic resin for the vehicle. Latex was the resin used in the first water-based paints; therefore, they are sometimes referred to as *latex paint*, even though other resins, such as acrylic, are more often used. These paints use pigments which are similar to those used in oil-based paints. One of the chief advantages of water-based paint is that it can be cleaned up with water, yet still provides a tough, water-resistant surface when dry. The following is an example of the contents of a water-based paint;

Pigment-36%		Vehicle-64%	
Titanium		Acrylic resin	10%
dioxide	22%	Additives	2%
Silicates	12%	Water	44%
Tinting colors	2%	Ethylene glycol	8%

Primer

Some surfaces hold paint better than others, so special *primer* may be used for the first coat. For example, wood grows more rapidly in the spring than in the summer, so the summer wood is less porous and does not hold paint as well; but if the entire surface is painted with a coat of primer, the surface coat of paint has a uniform surface on which to adhere. Most paint manufacturers make primer especially for use with their paint. It is generally best to use primer and paint made by the same manufacturer.

Frequently, wood products contain knots and sap pockets which release sap or pitch and bleed through the finished paint (Figure 20-1). This can

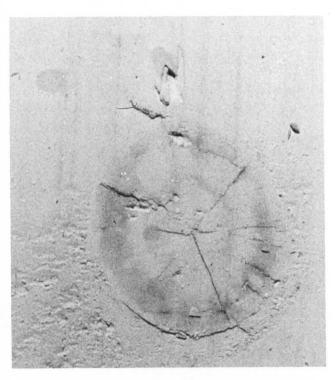

Fig. 20-1. Knot.

be prevented by painting knots and sap pockets with shellac before the wood is painted. This seals the surface and prevents the pitch from bleeding through into the paint. Shellac can be thinned and cleaned up with denatured alcohol.

Stain

Frequently wood siding, shingles, and trim are stained a darker color. This is done by applying *wood stain*, which is similar to paint in that it contains a vehicle and pigment, but is not opaque. Stain darkens the wood while allowing the grain to remain visible. When woodwork is stained, it is normally varnished afterwards.

Varnish

Varnish is a combination of resin, oil, thinner and drier which makes a durable, transparent coating. Most varnish does not contain pigment, although additives may be included to make satin varnish. Satin varnish is transparent, but does not provide the high gloss that would result if no additives were included.

Some specifications call for the use of *polyurethane*, which is sometimes referred to as varnish. Polyurethane is a clear, plastic coating material which provides the appearance of varnish, but is even more durable than varnish. Most polyurethane and varnish can be cleaned up with turpentine or mineral spirits.

Coverage

Most paints and finishing materials cover about 450 square feet per gallon if applied to unpainted wood, and about 600 square feet per gallon if applied to a previously painted surface. This is only an approximation; more accurate coverage rates can be found by reading the manufacturer's label on the container. Before using any finishing material, it is important to read the manufacturer's directions carefully.

ESTIMATING PAINT

The area of the exterior walls can be determined from the number of sheets of plywood estimated for sheathing. Each sheet covers an area of 32 square feet. Our front is brick veneer halfway up, so we multiply 32 × 10 for 320 square feet. 320 divided by 32 equals 10 sheets. Our total of 65 sheets of plywood for sheathing was estimated earlier. We subtract 10 from 65 for a total of 55 sheets. We multiply: 55 × 32 = 1,760 square feet. Allowing 450 square feet per gallon for primer, it would require 4 gallons. Our specifications call for siding and trim to be primed on the back, so we will need 8 gallons of exterior primer for the siding.

The specifications indicate the house is to be painted with two coats of house paint. We have figured primer, so the finished coat should cover 600 square feet per gallon. Our total area is 1,760 sq. ft. We divide 600 into 1760 for a total of 3 gallons for one coat, so we will need 6 gallons for two coats.

Cornice

The box cornice is made up of a 1 x 3, a 1 x 6, a 1 x 8, a 1 x 2, and a 1 x 8 frieze. Adding all of these together the total width of the cornice is approximately two feet, so for every linear foot of cornice there are two square feet to be painted. The length of the cornice on the front and back is 64 linear feet; 64 × 2 = 128 square feet to be painted.

The rake is made up of a piece of 5/4 x 6 and a 1 x 3, which adds up to approximately 6 inches when installed. The rafters of our house are 14 feet long, and we have two gables: 28 + 28 = 56 linear feet. We should also figure our cornice returns; these are 2'6" long and we have four of them: 2'6" × 4 = 10 linear feet. We add 10 to 56 for a total of 66 linear feet for the rake and returns. We determined the rake to be approximately 6 inches per foot, so we would divide 66 by 2 for a total of 33, which we add to our 128 square feet at the eaves, for a total of 161 square feet of cornice.

This is to be primed on the back as well: 161 × 2 = 322 square feet; 1 gallon of primer covers 450 square feet, so we will need 1 gallon of primer. One gallon of house paint is enough to apply two coats to the outside of the cornice and rake.

Windows and Doors

As a rule of thumb, the sash and frame of one side of an average size window is equal to about 25 square feet. Most doors are also about 25 square feet. We have 18 windows and two exterior doors for a total of 500 square feet. Most manufacturers prime their windows and doors unless requested not to when ordering. Our specifications call for trim and shutter paint on all exterior woodwork of windows and doors. We will need two gallons of paint for this work.

Interior Walls and Ceilings

All rooms except the kitchen and baths are to have two coats of flat wall paint. Our total wallboard area was determined to be 6,368 square feet. We subtract the area of the walls and ceilings in the bathrooms and kitchen from the total wallboard area and we have the area to be covered with flat wall paint.

Our kitchen and bathroom ceilings and walls total 1,158 square feet. We subtract this from the wallboard total: 6,368 – 1,158 = 5,210 square feet, so we will need approximately 10 gallons for 1 coat, or need 20 gallons of flat wall paint in all.

Kitchen and bathrooms have a total of 1,158 sq. ft. Our specifications call for satin finish enamel on these walls. We already have the primer figured, so we will need 2 gallons for the finish satin.

Interior Trim

Paint for the interior trim is estimated the same way as paint for the exterior trim. The cove molding at the ceiling need not be included because this is normally painted with the ceiling. We have the exterior doors and windows already estimated at 500 square feet. We have 14 interior doors which must be finished on both sides, so this area is 700 square feet (14 × 25 = 350; 350 × 2 = 700). The total area of doors and windows inside the house is 500 + 700 = 1,200 square feet. In addition, we have three sets of jambs and trim with no doors. Allow 20 square feet for each opening with no door: 20 × 3 = 60 square feet.

The baseboard and carpet strip are about 4 inches wide, so allow 1 square foot for every 3 linear feet. Our baseboard has already been estimated at 512 linear feet, so we have approximately 171 square feet of base and carpet strip. We add all of the interior trim and get a total of 1,431 square feet (1,200 + 60 + 171 = 1,431). Our specifications call for enamel undercoat and enamel finish, so we will need 3 gallons of each.

Hardwood Flooring

Our specifications call for 1 x 3 select oak flooring. Oak flooring is normally sanded smooth, filled with a wood filler, then varnished. The area of the hardwood floors can be estimated by subtracting the allowance made for matching (one-quarter the total) from the total amount of flooring included on our material list. We estimated 1,528 sq. ft. of flooring: $\frac{1}{4}$ × 1,528 = 382; 1,528 – 382 = 1,146 square feet of floor area. Wood filler covers approximately 600 square feet per gallon, so 2 gallons are required. Varnish covers approximately the same, so two coats will require 4 gallons.

You should also figure in thinners, brushes, roller, cloths, sandpaper, etc. To cover these miscellaneous expenses make an allowance of $100.00. We would now add our paint material to our material list.

Pieces	Feet	Description	Cost	Total
9		gallons exterior primer		
7		gallons exterior house paint — nonchalking		
2		gallons trim and shutter exterior paint		
20		gallons flat wall paint — interior		
2		gallons satin finish paint — kitchen bathrooms		

3		gallons enamel undercoat		
3		gallons enamel paint		
2		gallons paste wood filler — hardwood floors		
4		gallons varnish — hardwood floors		
		miscellaneous paint articles — allowance		100.00

21
CARPENTRY MILLWORK AND FINISHES: CABINETS

The kitchen cabinets and bathroom vanities for a job may be site constructed (built on the job by the carpenter, Figure 21-1), or factory built and finished. The most common practice is to use factory-built cabinets, except when special requirements are needed. Information about the cabinets may be found in several places. The floor plan of the kitchen and bathrooms usually indicates where cabinets are to be placed in the room. Most sets of working drawings also include special detail drawings or cabinet elevations to show more precise information. These special details or cabinet elevations usually indicate the size of each cabinet. Most often, the manufacturer, style, and finish is included in the specifications for the house.

TYPES OF CABINETS

Wall Cabinets

All the kitchen cabinets that are mounted on the wall above the countertop are considered *wall cabinets*. Although wall cabinets are manufactured in a variety of sizes to suit the needs of the kitchen, they are standardized in some ways. Usually the tops of all of the wall cabinets are positioned 84 inches from the floor. The space above the cabinets is usually enclosed and finished with the same material as the kitchen walls (Figure 21-2). The height of the cabinet is varied according to the kitchen layout. Where the cabinet is mounted directly above the counter, the standard cabinet height is 30 inches.

Usually cabinets which are mounted above a range or cooking surface are 18 to 21 inches high. Frequently a metal hood with an exhaust fan is mounted below these cabinets (Figure 21-3). The hood is fastened to the bottom of the wall cabinet and the exhaust fan is ducted through the wall or up through the cabinet. Above a refrigerator, the wall cabinets are 12 or 14 inches. All wall cabinets are usually 12 inches deep, but their width varies in 3-inch increments. They most frequently range

Fig. 21-1. Site-constructed cabinets.

Fig. 21-2. Enclosed soffit.

Fig. 21-4. Manufactured countertop with backsplash.

Fig. 21-3. Exhaust hood.

from 9 inches wide to 48 inches wide. In addition, blank filler pieces are available to allow for minor adjustments.

Base Cabinets

All of the kitchen cabinets that rest on the floor are considered *base cabinets*. Base cabinets vary in width only. The standard height for base cabinets is $34\frac{1}{2}$ inches from the floor to the top of the cabinet. The standard depth is 24 inches. Like wall cabinets, base cabinets are available in widths from 9 inches to 48 inches, in 3-inch increments.

Countertops

The countertop may be either site constructed or factory built, regardless of how the cabinets are constructed. The surface is usually made of plastic laminate. At the rear of the countertop is a 4-inch *backsplash* (a raised lip to prevent water from running down the back of the cabinets). Most countertops are 25 inches wide, including the backsplash. Manufactured countertops are usually $1\frac{1}{2}$ inches thick and have a rounded edge and backsplash (Figure 21-4).

Site-constructed countertops are usually made of $\frac{3}{4}$-inch plywood, with plastic laminate glued to the surface. The front edge of a site-built counter-

top is usually built up by adding a $\frac{3}{4}$-inch thick piece of wood to the bottom edge.

In some kitchens the space between the base cabinets and the wall cabinets is covered with the same plastic laminate as the countertop. In this case the backsplash may be eliminated.

Lavatories

Lavatories are frequently installed in a base cabinet called a *vanity*. Vanities are normally $29\frac{1}{2}$ inches high by 21 inches deep and vary in length from 12 inches to 60 inches. The top may be a laminated plastic countertop, similar to that used in a kitchen, or a single cast piece, including the lavatory.

ESTIMATING CABINETS

The manufacturer and style of the cabinets is indicated in the specifications. We find ours in Division 10, Section E. The sizes and quantities of kitchen cabinets can be determined from the kitchen cabinet plan and the cabinet elevations. The numbers of these drawings indicate the size and type of cabinet. On wall cabinets, the first two digits indicate the width in inches, and the last two digits indicate the height in inches. For example, a 2418 wall cabinet is 24 inches wide and 18 inches high. All base cabinets are the same height, so only two digits are shown, to indicate the width. The letters preceding the numbers indicate the type of cabinet. These letters may vary from one manufacturer to another, so the estimator should have the manufacturer's catalogs available. Some of the most common cabinets are shown in Figures 21-5 – 21-8.

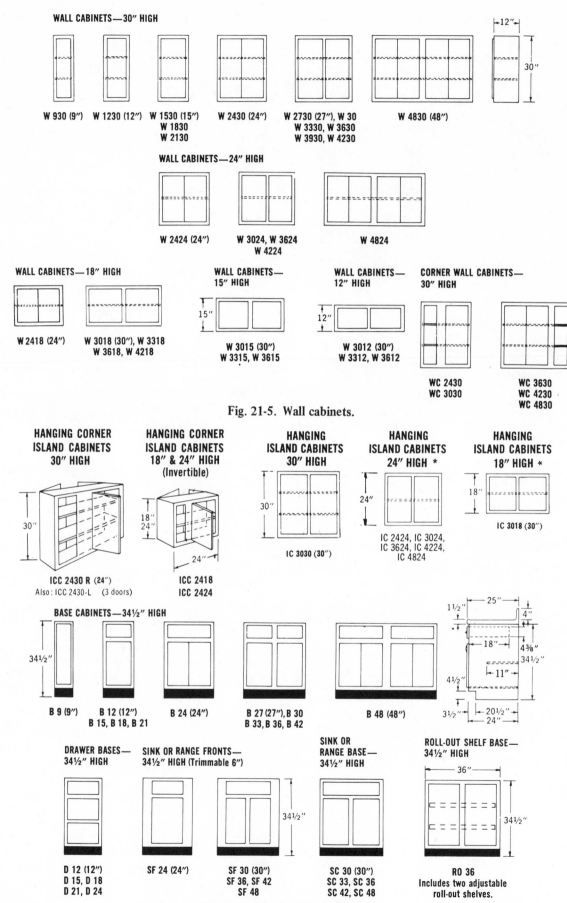

Fig. 21-5. Wall cabinets.

Fig. 21-6. Island and base cabinets.

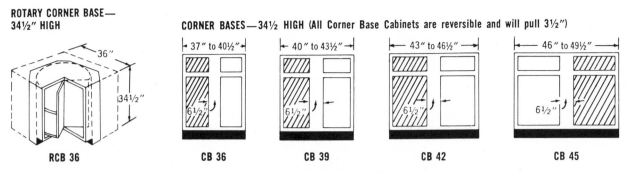

Fig. 21-7. Base corner units.

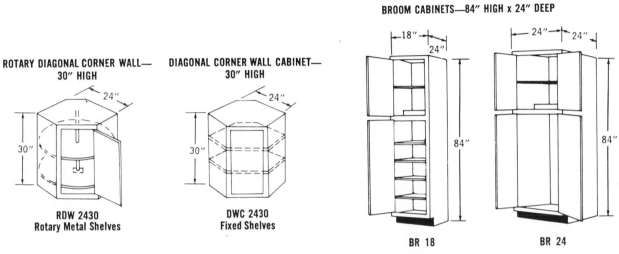

Fig. 21-8. Wall corner cabinets and utility cabinets.

We would list our kitchen cabinets from the first floor plan, which gives us the size and location. Our vanities we find on the second floor plan in each bathroom. The manufacturer and style is in our specifications. We list by size, starting with the wall cabinets.

Pieces	Feet	Description	Cost	Total
1		Oakmont wall cabinet 3018		
1		Oakmont wall cabinet 1830		
1		Oakmont wall cabinet 3330		
1		Oakmont wall cabinet WC-2430		
1		Oakmont wall cabinet 1530		
1		Oakmont wall cabinet 3315		

1		Oakmont base cabinet B-18		
1		Oakmont base cabinet SF-42		
1		Oakmont base cabinet D-18		
1		Oakmont base corner CB-45		
	13	linear feet laminated plastic counter-top		
1		bath vanity RV-36 complete with top — bath No. 1		
1		bath vanity RV-30 complete with top — bath No. 2		
1		$\frac{1}{4}''$ silver plated mirror 36" x 30" — bath No. 1		
1		$\frac{1}{4}''$ silver plated mirror 30" x 30" — bath No. 2		

22
CARPENTRY MILLWORK AND FINISHES: HARDWARE AND NAILS

HARDWARE

Finish hardware is generally considered to be all of the hardware which is exposed to view in the completed house. This includes such things as door lock sets, hinges for swinging doors, push plates, kick plates and door stops.

A *lock set* is an assembly consisting of the latching mechanism, lock, and knob for a door. Lock sets are available in a variety of styles and with a variety of locking or latching mechanisms, for use in different areas of the house (Figure 22-1).

Exterior lock sets have a cylinder which can be locked from the outside with a key. These are used on exterior doors such as the front or side door. Interior lock sets include bathroom lock sets, bedroom lock sets, and passage latch sets. Bathroom and bedroom lock sets have a turnbutton or some other means for locking the door from the inside, but do not have cylinders for keys. The only difference between the two is that bedroom lock sets usually have a brass finish and the bathroom lock set has a chrome finish on one side that is installed on the bathroom side. A passage latch set has no provision for locking the door from either side.

Butt hinges (sometimes called butts) are the hinges on which the doors swing. They may be either *loose pin* (Figure 22-2), meaning that the pin may be slipped out to separate the two leaves, or *fast pin*, meaning that the pin cannot be removed.

Loose-pin hinges are normally used in residential construction. Butts are available in sizes ranging from the very small ones used on minature furniture to the large ones for exterior doors. Usually $1\frac{3}{8}$-inch thick doors are hung on one pair of $3\frac{1}{2}''$ x $3\frac{1}{2}''$ butts and $1\frac{3}{4}$-inch thick doors are hung on $1\frac{1}{2}$ pairs (3 hinges) of $4\frac{1}{2}''$ x $4\frac{1}{2}''$ butts. Butt hinges

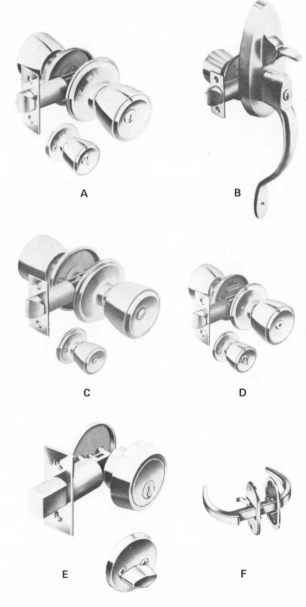

A   B

C   D

E   F

Fig. 22-1. Locksets; (A, B) entrance locks; (C, D) passage latch sets; (E) dead bolt; (F) screen door.

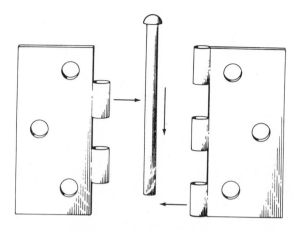

Fig. 22-2. Loose-pin butt.

are available with a variety of finishes, but most are either dull brass or polished brass.

Finish Hardware Allowance

Because of the great variety of styles and grades of hardware available for residential construction, it is impractical to estimate the cost of finish hardware for a house. Instead, it is customary to specify an allowance for these items. This allowance should be sufficient to purchase all of the finish hardware required. If the owner selects more or less expensive hardware, the difference is either deducted from or added to the contract price.

If prehung doors are specified the hardware is installed at the factory, but the style must be indicated when the doors are ordered. The hardware allowance for our house is found in Division 10, Section D.

NAILS

Due to the many sizes and types of nails used in the construction of a house, it is impractical to include nails on an estimate. Normally an amount of money is allowed for the purchase of all the nails for the building. This is unlike a finish hardware allowance in that the owner does not make up any difference in cost. Although the estimator is not required to list each type of nail, some knowledge of the types and sizes used is desirable.

Nail sizes are listed by the old English term *penny* (abbreviated d). The penny size of a nail refers to its length. For example, a 2d nail is one inch long and a 16d nail is $3\frac{1}{2}$ inches long. Figure 22-3 shows the most common sizes of nails. Most nails are made of steel, but other materials may be used for special-purpose nails. Nails for aluminum siding and trim are made of aluminum and nails for copper flashing are made of copper. Nails which are ex-

posed to the weather, where they might rust and stain the surrounding paint, are usually galvanized.

Kinds of Nails

There are a great many types of nails for general construction and special purposes. Some have smooth shanks and some have screw shanks, ring shanks, or barbed shanks for greater holding power. Only the most commonly used nails are discussed here. Figure 22-4 shows some of the most common type of nails used in house construction.

Common nails and *box nails* have smooth shanks and flat heads. The difference between the two is the diameter of their shanks. Both are used for general building construction, such as framing; however, box nails bend easily, so they are usually used only where splitting is a problem. Most framing members, such as studs, joists, beams, and sills, are nailed with 16d common nails. Subflooring, wall sheathing, and roof sheathing are usually nailed with either 6d or 8d common or box nails.

Casing and *finishing nails* have very small heads. Casing nails are slightly larger in diameter than finishing nails. These nails are used for nailing trim, such as casing, cornices, and siding, where the nailheads are concealed. They may be driven slightly below the surface of the wood with a nail set, so that the head can be covered with putty or plastic wood. The sizes used most often are 4d, 6d, and 8d.

Roofing nails are special nails used for applying roofing materials. They are available in varying lengths to suit the thickness of the roofing material. Roofing nails are normally galvanized to prevent them from rusting.

Drywall nails are special nails used for applying gypsum wallboard. Most drywall nails have ring shanks to prevent *nail popping*. Nail popping is a term used to describe the loosening of nails in gypsum wallboard, causing the nailhead to show through the joint compound.

Underlayment and *flooring nails* are special nails for use on those flooring materials. Underlayment nails have ring shanks to prevent them from loosening. For $\frac{3}{8}$-inch underlayment, 3d underlayment nails should be used; on $\frac{5}{8}$-inch underlayment, 4d nails should be used. For hardwood strip flooring, 8d hardened nails are used.

Duplex-headed nails have two heads so that they may be removed easily. These are useful for building scaffolding which is to be taken apart after it is used.

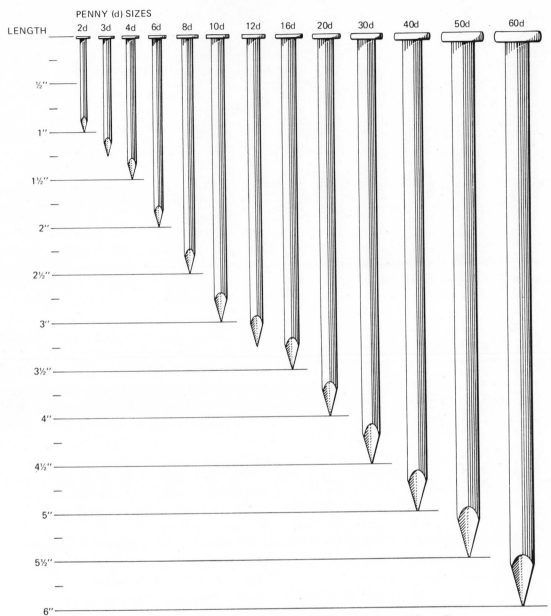

NOTE: All "d" sizes are same length. Only diameter changes between Common and Box nails.

Fig. 22-3. Nail chart.

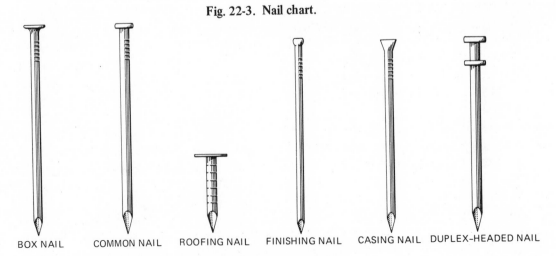

BOX NAIL   COMMON NAIL   ROOFING NAIL   FINISHING NAIL   CASING NAIL   DUPLEX–HEADED NAIL

Fig. 22-4. Common types of nail.

23
MECHANICAL: PLUMBING

As with some of the other highly specialized phases in the construction of a house, the plumbing work is usually performed by a subcontractor. The plumbing contractor will make an accurate list of materials before beginning the job. However, in order for the general contractor to estimate the overall cost of the project and act as the coordinator of all of the work, general knowledge of the basics in plumbing is necessary.

ROUGH PLUMBING

The plumbing is installed in two operations. The first operation is the installation of rough plumbing: piping for fresh water and waste, up to the wall or floor surface where the fixtures are attached. When the fresh water (supply) pipes are installed, all openings are capped or plugged and pressure is applied to them. In this manner the piping is tested for leaks before it is concealed in the structure.

Bathtubs and shower bases are fixtures, but they are installed when the rough plumbing is done. This is because the wall and floor covering must be installed up to the edge of the tub or shower.

Shower bases are available in a wide variety of sizes, so the drawings and specifications must be consulted to determine the size required. Bathtubs may be ordered in several sizes, but 5'0" long by 2'6" wide is considered a standard size. Bathtubs are available with a finish return at either end, or with no returns at the ends (the tub must be enclosed between two walls). They are available with either a left-hand drain or a right-hand drain. There are also steel and fiberglass assemblies, which include the tub and the walls in the tub area. Figure 23-1 shows a return and Figure 23-2 shows a one-piece fiberglass tub enclosure.

The waste plumbing for a house includes the pipes and fittings that carry the waste water from the fixtures to the house sewer, which runs from the building to the municipal sewer or the house septic system, and the pipes and fittings make up the vent system. Figure 23-3 shows several of the most common types of fittings.

Fig. 23-1. Tub return.

Fig. 23-2. Fiberglass enclosure.

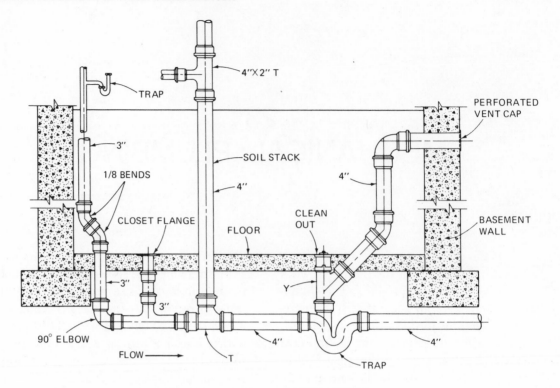

(A) COMMON FITTINGS FOR WASTE PLUMBING

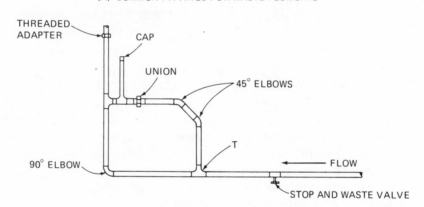

(B) COMMON FITTINGS FOR SUPPLY PLUMBING

Fig. 23-3. Common fittings. (A) waste lines, usually cast iron or plastic; (B) supply, usually copper, galvanized, or plastic.

The piping from a fixture to the point in the plumbing system where it joins piping from other fixtures is called a *branch*. The large vertical pipe in the main house drain, into which the branches run, is called a *stack*. The *vent* runs vertically up through the roof to allow atmospheric pressure to enter the system and prevent vacuum from building up as the waste water is discharged. *Traps* are included at each fixture and at the point where the sewer leaves the house. A trap is a U-shaped fitting (Figure 23-4) which prevents sewer gas from entering the house. As water passes through the system, the trap remains filled, so that gasses cannot pass backward through it.

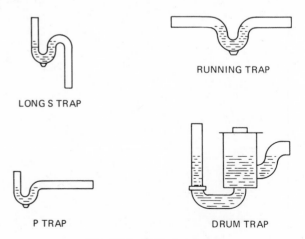

Fig. 23-4. Common traps.

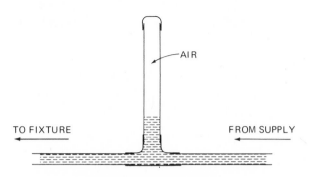

Fig. 23-5. Air chamber.

Normally, each branch of the supply includes a valve to shut off the water. This valve may be a *stop valve* (a simple valve to turn the flow on and off) or it may be a *stop and waste* valve (turns flow on and off and includes a means to drain the system). Usually, a stop and waste valve is used at a low point in the system, such as the main shutoff for the building. The individual shutoffs at the fixtures are usually stop valves which come with chrome-plated supply set, installed between the wall or floor and the fixture.

Each branch of the supply also includes an air chamber (Figure 23-5). The air chamber is a short section of pipe, capped at the end to trap air. When a sudden surge of pressure occurs in the line, the trapped air acts like a shock absorber to prevent the pipes from hammering.

FINISHED PLUMBING

Once the rough plumbing is installed, and the walls or floors are covered with whatever finish is to be used, the finished plumbing is installed. Finished plumbing includes fixtures, such as lavatories, sinks, water closets (toilets); and trim, such as faucets. Figure 23-6 shows how fixtures are indicated on a floor plan.

Water closets are available in a variety of types, and most manufacturers make more than one type.

The specifications must be consulted to determine which type is required. *Lavatories* (wash basins) may be either wall hung or mounted in a built-in vanity cabinet. Many wall-hung lavatories require supporting legs. These legs may be included with the lavatory, or it may be necessary to order them separately. Lavatories vary in size from 12″ x 12″ to 21″ x 24″. The *water heater* is basically a tank for storing water. It includes a device for heating the water with electricity, oil, or gas. Once the water is heated to the desired temperature, a thermostat shuts off the heating device. A relief valve is installed on the water heater, to release the pressure in the event that either the temperature or the pressure becomes dangerously high.

In addition to the above fixtures, the finished plumbing includes *sill cocks* (hose bibs), shower heads, faucets for all fixtures requiring them, tub and lavatory drains, and any other exposed hardware which is connected to the plumbing. Usually one or more sill cocks are included. A sill cock is a faucet mounted on the outside of the building to which a hose may be attached.

There are several materials which are commonly used for pipes and fittings. At one time, most fresh water plumbing was done with galvanized iron, and waste plumbing was done with cast iron soil pipe. Copper has replaced galvanized iron as the most common supply piping material. Plastic pipe and fittings are also available for supply plumbing, but copper is more generally used. When the plumbing is done with copper, fittings made of copper or brass are soldered to the pipe. The method of soldering used in plumbing is called *sweat* soldering and the fittings are called *sweat fittings*.

Waste plumbing may be done with any one of three materials. Cast iron pipe, called *soil pipe*, is still used frequently, but plastic and copper are also common. Cast iron soil pipe is manufactured in 5-foot lengths and is joined with oakum (a fibrous,

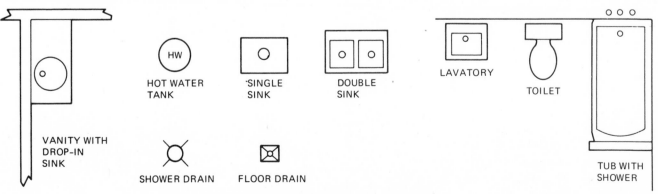

Fig. 23-6. Plumbing fixtures on a floor plan.

ropelike material) and molten lead or a man-made material called plastic lead. Plastic pipe is available in 10- or 20-foot lengths and is joined with a special adhesive. Copper waste plumbing is joined by sweat soldering, the same as copper supply plumbing. Usually all underground waste plumbing, such as the house sewer, is cast iron.

Many areas have plumbing codes which specify what material can be used for the plumbing. The estimator should be familiar with local and state plumbing codes.

ESTIMATING PLUMBING

Some architects include elevations and plans of the plumbing with the working drawings. If these are not included with the drawings, the estimator will want to make a sketch of the plumbing system. The easiest way to sketch the supply plumbing on one sheet, and the waste plumbing on another. The estimate discussed here is for the plumbing within the house itself. Normally, the materials necessary to connect houses to the municipal sewer are included in the estimate, but these vary depending on the setback of the house and the location of the municipal sewer.

With the sketches drawn, the estimator can measure the length of the pipes, count the necessary fittings, and list the finished plumbing. The specifications indicate the material and size of the piping, and the manufacturer and model of each item of finished plumbing. Figure 23-7 shows a typical water supply layout and Figure 23-8 shows a typical waste layout; from which an estimator can list the various materials for piping.

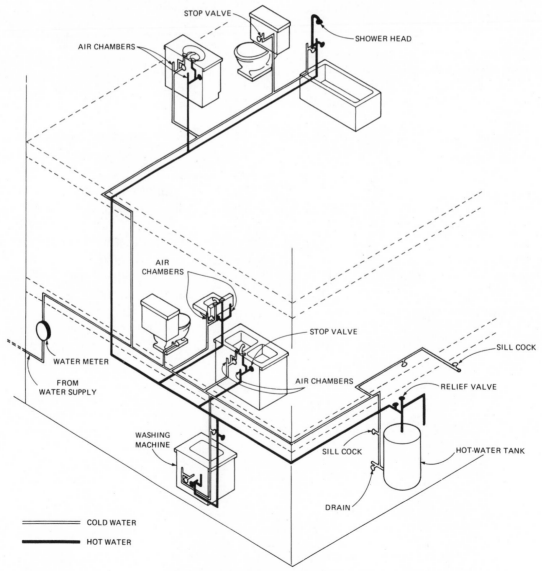

Fig. 23-7. Typical water supply layout.

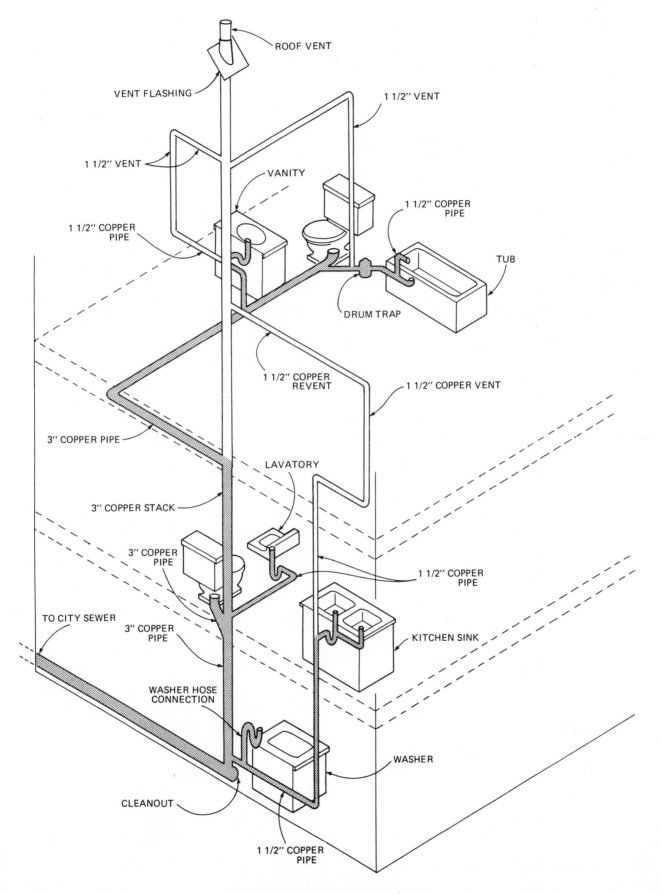

Fig. 23-8. Typical waste piping layout.

24
MECHANICAL: HEATING AND AIR CONDITIONING

The design of an efficient heating system involves more than an elementary knowledge of air handling. The heating system is usually designed by a specialized heating contractor. However, the estimator should understand the principles involved in order to interpret the information provided by the engineer.

All heating systems are based on the warming of air in the living space. This warming may be accomplished by passing the air through a chamber heated by the combustion of oil, coal, or natural gas, air passing through radiators and convectors to which hot water has been piped, or by the use of electricity to heat the air.

Heat flows from the warm objects to colder ones. As a house loses its heat to the colder outdoors, persons inside the house lose their body heat to the colder air. The basic function of a heating system is to supply heat to the building as quickly as it is lost to the outside.

Warm air is not as heavy as cold air. For this reason, the heated air rises to the ceiling and then slowly descends to the floor as it becomes cooler. As the air passes cold windows, it is chilled rapidly and becomes heavier, causing it to descend faster. This causes a draft and a layer of cold air on the floor. The heating system can be designed to overcome this by the placement of heating outlets on the outside walls and under windows. As the warm air leaves the outlet, it forms a curtain of warm air in front of the windows and the exterior walls. This reduces the rapid cooling of air and eliminates drafts.

TYPES OF SYSTEMS

There are three basic types of heating systems in common usage: central furnace, which heats air to be directed into the living space; baseboard units, which are located in the living spaces; and radiant heat, which is provided by heat lamps, electrical units embedded in the building structure. Figure 24-1 shows a heating outlet under a window.

The source of energy for any heating system may be electricity, natural gas, fuel oil, coal, or wood. The type of heating system selected depends on the local climate, the size of the house, the availability of fuel, the cost of installation, the cost of operation, and the owner's preference.

Hot Air Heating

In hot air heating, the cold air is passed over a surface heated by electricity or the combustion of

Fig. 24-1. Heating outlet under a window.

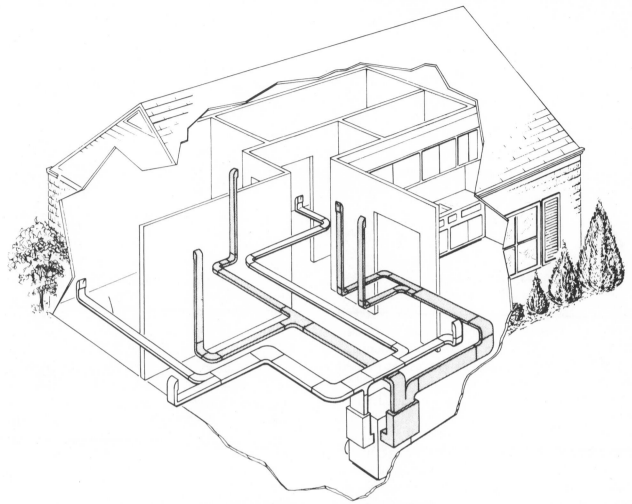

Fig. 24-2. Hot air heating system — ductwork.

fuel oil, gas, coal, or wood. This heated air is carried by ducts to outlets at various points throughout the house. The heated air is generally forced through the ducts by a fan in the furnace, but the system may rely on the force of gravity acting on cold air instead. As the air in the room cools, it is drawn into a cold air return, which carries it back to the furnace to be reheated.

Hot air systems frequently incorporate a humidifier to replace some of the moisture which has been removed from the air in the heating process. Humidified air provides a more comfortable atmosphere for the occupants of the building and protects the woodwork from excessive drying. Central air conditioning is also easily incorporated in a hot air heating system. The cooled air is forced through the same ductwork as the heated air. Figure 24-2 shows typical hot air heating system ductwork.

Hot Water Heating

In hot water heating a boiler uses either gas or fuel oil to heat the water, which is then circulated through pipes to *radiators* or *convectors*. A convector is a section of pipe with metal fins attached (Figure 24-3), housed in a special cover. The fins are heated by the hot water and they, in turn, transfer the heat, by a process called convection, to the surrounding air. In a radiator, the heat is simply transferred from the part containing the water to the surrounding air by radiation.

The heated water is circulated through the system by an electrically operated circulating pump. This pump is usually on the return side of the system.

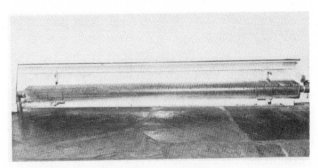

Fig. 24-3. Convector.

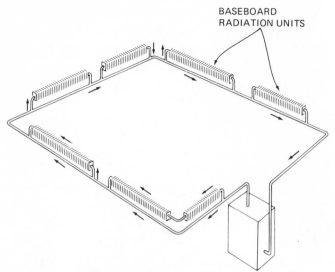

Fig. 24-4. One-pipe system.

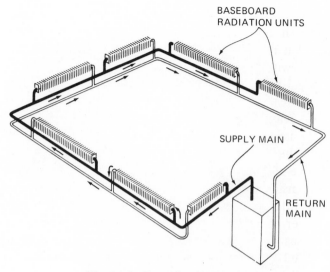

Fig. 24-5. Two-pipe system.

That is, it draws the water from the radiators or convectors back to the boiler to be reheated. Hot water may use either a one-pipe or a two-pipe system. In a *one-pipe system* a main line carries the heated water throughout the house and returns it to the boiler (Figure 24-4). Branch lines at each heating outlet carry the hot water to each heating outlet and back to the main. A heat control valve is installed at the inlet side of each outlet. This allows the heat to be controlled at each outlet without affecting others in the system. In a *two-pipe system* one pipe carries heated water to all of the outlets and a second pipe returns the cooled water to the boiler (Figure 24-5). The outlets are fed by a branch pipe from the main supply, and the cooled water is returned to the return line by another pipe.

Hot water systems frequently have zoned heat. In zoned heating, the temperature in various parts of the house can be controlled separately. In residential construction it is common for the bedrooms to be zoned separately from the living areas, so that the temperature can be kept at a lower setting in the bedrooms. Each zone of a zoned system has a separate thermostat which regulates electrically operated valves at the outlets.

Electric Heating

A variety of electric heating devices are available for residential construction. The most common is electric baseboard heat (Figure 24-6). This system uses an electrical resistance element, similar to that found on an electric cooking range or toaster, to heat aluminum fins which transfer the heat to the surrounding air. Some electric heating outlets include a fan to circulate the air.

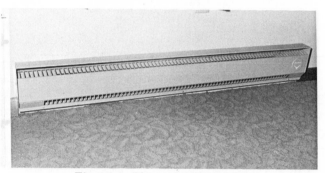

Fig. 24-6. Electric baseboard unit.

Another type of electric heat uses special resistance cable embedded in the floor or ceiling. This type provides uniform heat from the floor to the ceiling without the need for additional equipment on the walls.

Heat Pumps

As its name implies, a heat pump is a device for transferring heat from one point to another. All air contains some heat. In the cold months the heat pump transfers some of the heat from the outside air to the inside of the building. In warm weather it transfers the excess heat from the inside of the building to the outside air. A heat pump is not practical as a total heat supply in cold climates, because of the lack of heat in the outside air during winter months. It is, however, a useful means of reducing fuel consumption, when used as an aid to another heat source.

Thermostats

A thermostat is a switch which is activated by changes in temperature. As the temperature drops below that for which the thermostat is set, the switch closes and activates the heating device to which it is connected. As the temperature rises, the switch opens.

All heating systems use thermostats. With electric heat, the thermostat simply turns the power to the heating outlet. With hot air heating, thermostats are used to control the flow of fuel to the furnace. As the temperature drops, the thermostat opens a valve allowing fuel to enter the burner. When the air inside the furnace reaches a preset temperature, an automatic switch activates the blower. In a hot water heating system the thermostat opens and closes valves to control the flow of hot water. In a zoned system each thermostat controls the valves on the radiators in their respective zones.

HEATING LOADS

Thermal Conductivity and Resistance

Architects and heating engineers measure heat by Btus (British thermal units). A Btu is the amount of heat required to raise the temperature of 1 pound of water 1 degree Fahrenheit. The rate of heat transfer is expressed as Btuh (British thermal units per hour).

All materials conduct heat. The rate at which a material conducts heat is its *K factor*.

This is the number of Btuh that is conducted by 1 square foot of the material 1 inch thick with a difference in temperature of 1 degree Fahrenheit on either side. For example, consider a 1-inch thick piece of gypsum wallboard with one side 1 degree

colder than the other. If one Btu passes through a 1-square-foot area every hour, the K factor is 1.

Buildings are composed of an assortment of materials, so the K factor is not a practical way to measure the amount of heat that is lost through a wall, ceiling, or floor. The combination of all of the K factors in a section of a building is the *U factor*. Figure 24-7 lists approximate U factors for some common types of construction.

The purpose of thermal insulation is to resist the flow of heat. The resistance of a material to the flow of heat is the R value of that material. The R value is the reciprocal of the U value (R = 1/U).

The difference between the inside and outside temperatures greatly affects the amount of heat that is lost through the shell of the building. The difference between the lowest probable outside temperature and the desired inside temperature (design temperature) is called the *design temperature difference*. This design temperature difference must be known in order to determine the heating requirements of a building.

Determining the Thermal Resistance of a Building

Properly installed insulation is vital to comfortable, economical heating. Although it is impossible to completely stop the flow of heat through a building section, insulation can greatly reduce the flow of heat, depending on type density, and other characteristics. For this reason insulation is specified according to its R value, rather than its thickness.

A building section consists of the materials provided to support the structure, the inside and outside surface materials, and whatever material is used to minimize heat transfer and air infiltration. Figure 24-8 shows a building section and the ther-

Type of Building Section	U Value
Wood frame with plywood sheathing, wood siding, and 1/2 inch drywall, no insulation.	0.24
Wood frame with plywood sheathing, wood siding, and 1/2 inch drywall, 3 1/2 inch (R-11) insulation.	0.07
8-inch concrete block	0.25
Single-glazed window	1.1
Double-glazed window	0.6

Fig. 24-7. U factors for some typical building sections.

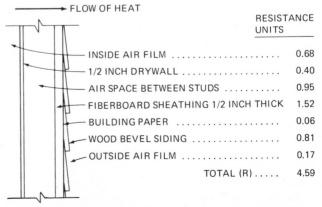

Fig. 24-8. A typical building section, showing thermal resistance, without insulation.

mal resistance of each component. Notice that the total R value for the section is 4.59. Fiberglass insulation has a thermal resistance of approximately 3.7 resistance units per inch of thickness. If all $3\frac{1}{2}$ inches of available stud space were filled with with such insulation, 12.95 resistance units would be substituted for the 0.95 units provided by the air space. This would increase the total resistance of the building section to an R value of 16.59.

It can readily be seen that the original resistance is less than one-quarter of the insulated resistance. With substantial insulation, slight variations in the resistance of the structural and finish components of the building have a minor effect on the overall resistance. In wall sections, the total resistance can be assumed to be the resistance of the insulation plus three units for the structural and finish components. Similar reasoning can be applied to floor and ceiling construction to arrive at values for these sections. Uninsulated ceilings can be assumed to have a resistance of approximately $1\frac{1}{2}$ units.

Windows and doors offer much less resistance to the flow of heat than do other building materials, so the values mentioned above do not apply to them. A single layer of window glass has an R value of approximately 0.88. However, trapped air offers substantial resistance to the flow of heat, so double glazing increases the R value to approximately 1.67. The resistance values of several common building sections is shown in Figure 24-9.

Type of Building Section	R value
Wood Frame Walls	3 plus the R value of the insulation used.
Floors above unheated spaces	2 plus the R value of the insulation used.
Single-glazed windows	1 1/2 plus the R value of the insulation.
Double-glazed windows	0.88
Doors with glass	Use R value of the glass for the entire door.
Doors without glass Less than 1″ thick or metal	0.88
Over 1″ thick nonmetal	1.67

Fig. 24-9. R values for common building sections.

The amount of heat lost through the various sections of the entire building can be found from the R value for each section and the area of each exposed section. By dividing the resistance (R) into the area, the heat transmission load (the heat that is lost through the building materials) is found in Btuh per degree Fahrenheit of temperature difference. The following three steps are used to find the building transmission load:

1. Find the square-foot area of each exposed outside section.
2. Divide the R value listed in Figure 24-9 into the section area to find the heat transmission for that section.
3. Add all section transmission loads per degree Fahrenheit.

Infiltration Losses

In addition to the heat that is lost by transmission through the various building sections, heat is also lost through *infiltration*; that is, heat is lost as air enters the building through cracks and small openings. To replace the heat lost in this manner, an *infiltration factor* is used. The infiltration factor is based on changing all of the air in the building periodically. The number of Btuh required to reheat the infiltrated air in 1 cubic foot of space is the infiltration factor. For example, if the air is changed once in two hours, the infiltration factor is 0.0088 Btuh. To find the total infiltration heat loss at this rate, multiply 0.0088 times the number of cubic feet in the building.

Example. Find the infiltration heat loss for a building 40 feet long by 24 feet wide, with 8-foot ceilings. Use an infiltration factor of 0.0123 Btuh (0.7 air change per hour). Volume = 40′ × 24′ × 8′ = 7,680 cubic feet. Heat loss per degree Fahrenheit = 7680 × 0.0123 = 101 Btuh.

The sum of all transmission losses and the infiltration loss is the *building load per degree Fahrenheit*. Multiply the building load per degree Fahrenheit by the design temperature difference to find the *total building load*. This total building load is the number of Btu per hour that the heating system must be capable of replacing or the required output of the heating system.

Estimating Heating Loads

The floor plan to be estimated in this example is shown in Figure 24-10. This house is used instead

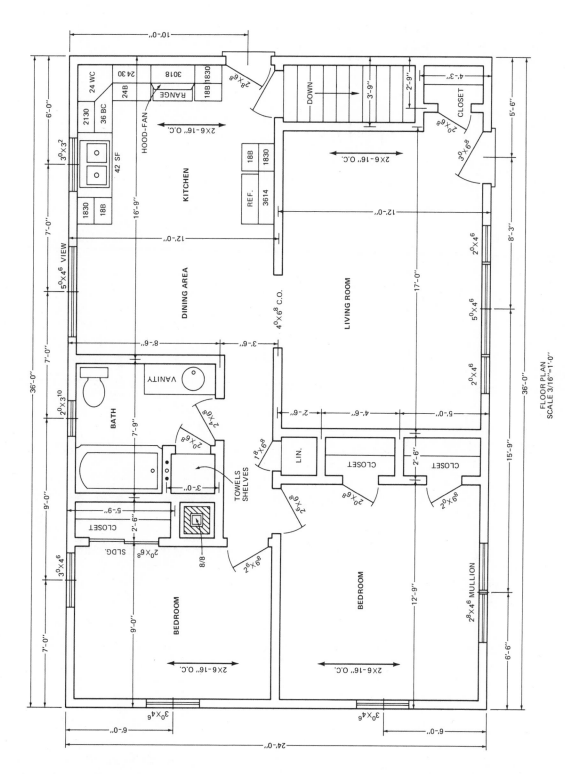

Fig. 24-10. Sample house for heating estimate.

of the two-story house we have been listing to avoid time-consuming calculations. All calculations for this house are based on 0.7 air change per hour infiltration rate (0.0123 Btuh) and a design temperature difference of 80 degrees Fahrenheit. This difference is appropriate for an outside temperature of minus 10 degrees Fahrenheit and an inside (or design) temperature of 70 degrees Fahrenheit. For the purposes of this estimate, assume that the house is of standard frame construction with R-19 insulation in the ceiling, R-11 insulation in the walls, and R-11 insulation in the floors. All windows are to be doubled glazed.

First find the square-foot area of the outside walls of the building. The house is 36 feet long and 24 feet wide, so the perimeter is 120 feet. The ceilings are 8 feet high, so the wall area is 120 × 8 = 960 square feet.

The sizes of the windows are indicated on the floor plan. The window areas are as follows: Picture window, 40 square feet; bedroom mullion window, 27 square feet; three windows 3'0" x 4'6", 40 square feet; bathroom window, 8 square feet; dining area window, 23 square feet; kitchen window, 9 square feet; front door, 21 square feet; and kitchen door, 20 square feet, for a total of 188 square feet.

The area of the ceiling is found by multiplying the width of the house by the length of the house. The house is 36' x 24', so the ceiling area is 864 square feet. The area of the floor is the same as the ceiling.

To find the infiltration load, it is necessary to know the cubic volume of the building. This is found by multiplying the floor area by the ceiling height: 864 × 8 = 6,912 cubic feet.

Using this information, the heat loss can be found as shown in Figure 24-11. Heat load in Btuh (80

degree F design temperature difference) is 432 × 80 = 34,560 Btuh. The heating system for this house must be capable of supplying 34,560 Btuh.

AIR CONDITIONING

There are two general types of air conditioning. One is the *individual room unit*, which has a basic function of cooling, dehymidifying, filtering, and circulating air. These are the small units that can be set in a window or in some cases cut through an opening in an outside wall. The other type is the *central air conditioning system*, which cools the entire house. With a hot air heating system, the central air conditioning system is installed so that it uses the ductwork of the heating system.

The main function of air conditioning is to keep the air properly dry. Where comfort is concerned, there is a close relationship between the apparent temperature and humidity. The quicker the evaporation of perspiration, the cooler the skin feels. Almost all types of air conditioning units either single room window units or a central cooling system are controlled by a thermostat. Most central air conditioning systems installed use what is called a split system. This system has the condensing unit located outside of the house and is connected by an insulated copper pipe to the evaporator cooling coil, which is located at the furnace, usually installed in the plenum.

The design and size of a central air conditioning system should be left to the engineer who specializes in this field.

With air conditioning, as with heating, the Btu unit is used in calculations. In air conditioning the Btu measurement indicates the amount of heat the system will remove. Most manufacturers of air conditioning units rate the units as so many Btus, as a measure of their cooling capacity. *Tonnage* is another method of rating air conditioner performance: 1 ton of cooling is comparable to the cooling effect achieved by melting 1 ton of ice in 1 hour. As a rule of thumb, an air conditioner will provide 12,000 Btus of cooling in 1 hour for every ton at which it is rated.

A rough way of estimating air conditioning needs is to figure one ton of cooling for every 500 square feet of floor space. This is only a rough estimate and should not be used to design a system; as with heating, many factors must be taken into consideration in adequately designing an air conditioning system.

SECTION	DESIGN DATA	LENGTH AREA OR	UNIT HEAT LOSS	HEAT LOSS Btuh per DEGREE
Glass and Doors	Double Glass	188 sq. ft.	R = 1.67	188/1.67 = 113
Walls	R 11 Insulation	960 sq. ft.	R = 11 + 3 = 14	960/14 = 69
Ceilings	R 18.5 Insulation	864 sq. ft.	R = 18.5 + 1.5	864/20 = 43
Floors	R 7.4 Insulation	864 sq. ft.	R = 7.4 + 2	864/9.4 = 92
Infiltration	.7 air change/hr.	6,912 cu. ft.	0.0123	6,912 x 0.0123 = 115
			Total	432

Fig. 24-11.

25
ENERGY CONSERVATION

Solar energy is not something new. The earliest man was aware of the value of the sun and its warming rays. In 1774, Joseph Priestley found that by concentrating the rays of the sun onto mercuric oxide he obtained a gas which caused a candle to burn brighter than it did in air; thus, oxygen was discovered. In 1881, a French scientist, Jacques d'Arsonval, experimented with the use of the thermal energy of the ocean as a source of energy (see below).

Solar energy technology is classified into six major areas. (1) *Solar climate control* describes solar radiation collection systems utilizing energy-absorbent materials located on roofs or walls of buildings or homes. (2) In *solar energy focusing systems*, the solar energy received by a large area (acres rather than the square feet of solar climate control systems) is focused upon energy-absorbent materials for transfer to a fluid medium, such as water. (3) In *direct solar energy conversion systems*, thermoelectric devices (such as solar cells) and thermionic devices receive a solar energy input and provide an electric power output. (4) In *indirect solar energy conversion systems*, the captured solar energy is not immediately transferred to a fluid medium or converted to electricity for immediate transmission, but rather is converted to some other form of energy that is easy to store. (5) In *ocean thermal energy conversion*, advantage is taken of the temperature difference between the upper layer of the ocean, which is heated by the sun, and the deeper, colder layers. (6) Finally, *wind power* is also classified as a solar approach because wind arises from atmospheric thermal differences.

Solar radiation is measured by calorie (heat) of radiation energy per square centimeter. Solar radiation per minute (1 calorie per square centimeter per minute) is equivalent to 221 Btus per square foot per hour.

Scientists have determined that approximately 43 percent of the radiation that reaches the earth from the sun is changed to heat. Clouds have a strong influence on the amount of energy that reaches the earth's surface. It is estimated that a typical cloud reflects back into space approximately 75 percent of the sunlight that strikes it. This means that on a cloudy overcast day only 25 percent of the sun's energy reaches the earth's surface.

The solar energy reaching the earth's surface is absorbed and reflected in varying degrees depending upon the surface it strikes. Bright objects, such as shining aluminum, glass mirrors, bright steel, white snow, and to some degree water, will reflect about 75 percent of the sunlight they receive; green grass and forests absorb anywhere from 80 to 95 percent of the solar energy they receive and change it to heat.

The earth's axis and its rotation around the sun, along with the earth's atmosphere, determine the amount of solar energy reaching the earth's surface at different places and times. This in turn accounts in part for the differences in temperature and climate among various sections of the earth's surface.

Scientists and engineers have been working for over half a century to find a way to absorb, store, and distribute this vast source of energy in an economical system. Solar radiation is so low in intensity that collectors of large surface area are required to get enough energy concentrated in small workable units. Solar energy is not yet suitable for use in cold, cloudy regions, or in large cities, where the acres of sunlight are not enough to supply the needs. Solar energy is being used and is suitable for use in sunny rural areas.

Solar energy used in residential construction can be classified into two categories, active and passive. An *active solar system* entails the use of solar collectors or panels attached to the outside of a structure which collects solar radiation and transfers it into the building by using either a fluid or air, pumps, a storage tank, controls, and a heat exchanger. A *passive solar system* uses solar energy by collecting it through glass on the side of the building which receives the most sunlight. In its purest definition, a passive solar house would use no energy to circulate air, relying instead on the simple principle of convection to regulate the flow of heat within the building.

An active solar heating system is very simple to understand. The sun's rays heat up a blackened absorber plate, which is usually covered with a sheet or two of glass or plastic and is insulated on the back and sides. A liquid circulates through tubes attached to the absorber, or the liquid is allowed to trickle down the absorber surface, or air is ducted to the absorber. The rays of the sun heat the liquid or the air, which is carried where it is needed: either to a hot water tank or to the living space, or to a heat storage space to be used at a later time.

There are many types of systems used to store the energy. Water, sand and pebbles, rocks, antifreeze liquids, etc. The size of the storage unit depends upon the number and size of the collectors. Most manufacturers of solar systems recommend using $\frac{1}{2}$ cubic foot of rock for every square foot of collector.

A comparison of heat capacity on a volume basis between various storage media shows that water can store 62.5 Btu per cubic foot per degree Fahrenheit, while rocks, brick, and gravel can store approximately 36 Btu per cubic foot per degree.

SOLAR HOT WATER HEATING

Active Hot Water Heating

There are many solar water heating systems in use today. The simplest system, with no energy input, is the *thermosiphon* or *natural-circulation* system, which requires that the storage tank be located above the collectors. In this system, the denser cold water (or antifreeze liquid in freezing climates) falls and absorbs the heat in the collectors, and rises to the storage tank.

It is estimated that each person uses approximately 25 gallons of hot water a day. The collectors are usually sized so that each collector can handle 25 gallons of hot water a day. The storage capacity also should be 25 gallons per person. A family of 4 would require 4 collectors and a 100-gallon storage tank. Figure 25-1 is a diagram of a simple thermosiphon system.

Solar Swimming Pool Heating

This system uses the same principle as the domestic hot water solar heating system: a pump to circulate the pool water through uninsulated collectors, where it absorbs the heat. A special valve is placed in the pool's existing filtration plumbing and is

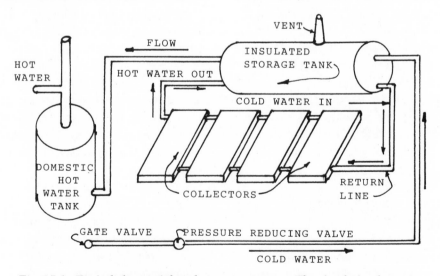

Fig. 25-1. Typical thermosiphon hot water system. The circulation between solar collector and the storage tank is affected by thermal gradients. The hot water, being less dense, rises to the storage tank at a higher level. If this is not possible, a pump can be used for forced circulation. Collectors are tilted at an angle to receive their optimum position at noon in March and September. In most climates an auxiliary heating unit is necessary.

controlled by a differential thermostat, which compares collector and water temperature. When heat can be added to the pool, the valve closes and the water is diverted to the collectors. This system can also be connected to an existing pool heater to lower the cost of heating the pool with conventional fuels.

SOLAR SPACE HEATING

Active Space Heating

The basic elements of the active air heating system exclusive of pumps, valves, and controls are (a) a solar collector, (b) an auxiliary heating device, (c) thermal storage unit, and (d) heater element fan and air duct system. This system requires more solar collectors than that called for to heat the domestic hot water. Depending upon the requirements for heating the house (Btuh loss), more and larger collectors will be required as well as a larger storage unit, usually three to five day's supply.

The air is circulated through the collectors, where it is heated. The cold air enters the bottom of the collectors and it is heated by the rays of the sun so that it rises to the top of the collectors. An air handler with blowers and dampers directs the air flow through the system. The heated air is forced into a storage bin, usually filled with pebbles and rocks. The size of the bin depends upon the required number of collectors. The storage bin is well insulated and is usually made of concrete or concrete products.

When the heat is stored, air flows from the top of the bin to the bottom, so the coldest air is returned to the collectors. When the system is heating from the storage, the flow is from the bottom to the top, so that the hottest air goes to the living space.

Passive Space Heating

In passive solar heating systems, the building itself is the collector, storage device, and heat distributer. Passive systems must be carefully designed to take advantage of the changing position of the sun. In its purest definition, a passive solar house would use no energy to circulate air, relying instead on the simple principle of convection to regulate the flow of heat within the building.

Heat flows from hot to cold. As sunlight warms the surface of a mass such as a wall, heat will begin to flow to the cooler wall interior. When the room temperature is below the temperature of the wall surface, heat will flow from the wall into the room. Heat flow will stop when the temperature within a room equals the temperature within the walls.

The rate at which a mass conducts heat is directly related to its ability to absorb heat. Passive solar buildings utilize the facing of windows and skylights to bring in the warmth of the sun. Passive buildings must be carefully oriented in response to the seasonal and daily movements of the sun to maximize solar heat gain in the winter and to minimize solar heat gain in the summer.

Today the more effective passive solar houses utilize fans to increase and direct the air flow, and have very large thermal storage units. One of the most significant characteristics of a passive solar building is the use of *thermal masses*: materials which have the ability to absorb and reradiate large amounts of energy. Passive design measures also include use of good insulation, careful entrance location with regard to winter wind, use of air-lock vestibules, and careful consideration of natural ventilation and natural light.

Passive measures such as cross ventilation, exhaustion of hot air by convection, evaporation, and absorption of heat by a thermal mass, can provide up to 100 percent of a building's cooling needs in summer.

A passive solar heating system for a house incorporates the passive solar design with a large internal thermal mass and interlocking air handling system. The most important feature of a passive solar house is its insulation envelope. The insulation envelope has a sufficient thermal integrity to ensure proper utilization of all available energies. The greatest source of energy for a passive solar house comes naturally from the sun through glass. All windows are double glazed, with a minimal amount of windows on the side where the sun does not shine on the house.

Excess energy from the sun and other sources should be stored within the structure, not only to increase its thermal performance but to prevent the house from experiencing large temperature swings. It is extremely important that the storage system be sized to properly transfer energy in and out of the storage medium.

The thermal storage units are usually underground sand, concrete, and crushed stone beds. The bed is completely enclosed in insulation and is an integral part of the building. These beds vary in size de-

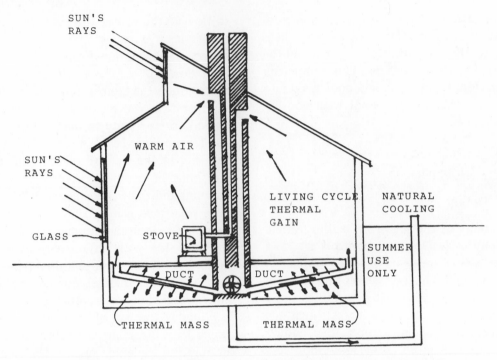

Fig. 25-2. Passive solar heating system. This interlocking thermal storage bed and air handling system is continually charged and discharged with Btus. The storage bed can accept Btus from woodburning, solar collectors, a wind generator, as well as the sun and the daily living cycle.

pending upon the size of the building. They average 150 to 200 tons under the first level of the building (Figure 25-2).

Within the mass storage bed are thermal transfer ducts connected to registers at the perimeter of the building from a central plenum. Air is continually drawn from the highest place down through the internal vertical mass (chimney). After passing through the storage bed it is circulated throughout the building. One or two small fans power this thermal loop. When the air in the building is warmer than the mass, the mass absorbs the excess heat, which acts to cool the air and increase the mass temperature. When the building begins to lose en-

ergy at night or during cloudy days the warmed mass releases its stored energy to the air flowing through it back into the building.

Solar heating systems are sized to replace a certain proportion of the heat that is lost from the house. There are well established formulas for determining heat-loss characteristics. They are complicated and should be left to a solar engineer. There are many factors to consider in solar heating. The size of the house, its location, the style (architecture), the cost of the solar system, the ability of the system to compensate for climate differences, etc. Most solar heating systems will require an auxiliary heating system for backup.

26
ELECTRICAL WIRING

The estimator for a general contractor is not usually called upon to prepare an itemized list of materials for the electrical work in a house. Electrical wiring is a highly specialized trade and the specific equipment required is usually estimated by the electrical contractor. However, the general estimator should know how the material list is prepared, so that the estimates received from the electrical contractor can be checked.

In most localities, electrical wiring is governed by the *National Electrical Code.* Many communities have local codes in addition to the National Electrical Code. In addition to the requirements set forth in the various codes, special building considerations may affect the design of the electrical installation. Because of all these variables, the information presented here should be considered as a guide only; the appropriate codes should always be consulted.

CONDUCTORS

An electrical wire is frequently referred to as a *conductor,* because it conducts electricity. Conductor sizes are listed by American Wire Gauge (AWG) numbers. The higher the AWG number is the smaller the diameter of the wire. The size, or gauge, determines the current capacity of a wire. According to the National Electrical Code, a 14-gauge copper conductor can be used for circuits carrying up to 15 amperes; a 12-gauge conductor can be used for up to 20 amperes; and 4/0-gauge conductors can be used for up to 200 amperes. Because aluminum has a slightly higher electrical resistance than copper, a larger size conductor must be used for aluminum wiring.

SERVICE

The wiring and equipment required to deliver electricity from the utility company's pole to the building constitutes the electrical service. This may be either an underground service, with special wiring buried in the ground, or overhead service, with the wiring run overhead from the pole to the building (Figure 26-1).

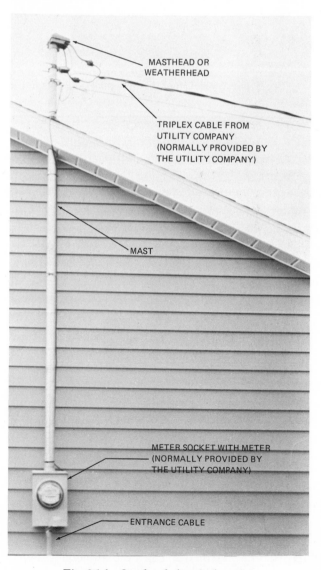

Fig. 26-1. Overhead electrical service.

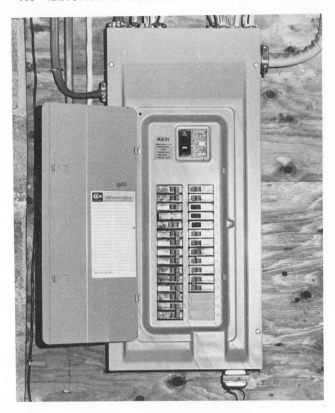

Fig. 26-2. Circuit breaker panel.

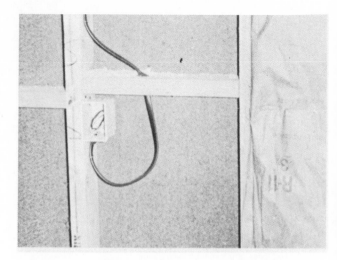

Fig. 26-3. Steel electrical box for a wall switch.

The *weatherhead* (see Figure 26-1) may be attached directly to the building or mounted on a mast made of 2-inch steel conduit (pipe). From the weatherhead, the entrance cable runs through the utility company's meter to the service panel. This service panel normally includes the main disconnect, which is required by the code, and a means of distributing the electricity to the various circuits within the building (Figure 26-2). The entrance cable from the service head to the service panel must be at least 4/0 gauge for a 200-ampere service.

BRANCH CIRCUITS

In the service (or distribution) panel, the entrance cable feeds several individual branch circuits. These are the circuits that carry the current to the various parts of the building. Each branch circuit must be protected against overloading by a circuit breaker or fuse. The size of the conductors in the circuit and the size of the overcurrent protection are governed by the code. In general, lighting circuits have 15-ampere circuit breakers and 14-gauge copper conductors; many outlet circuits are required to have 20-ampere circuit breakers and 12-gauge conductors. The estimator can base all general lighting and convenience outlet circuits on the 20-ampere,

12-gauge figures. Most houses are wired with at least 8–10 circuits. Of these, 2 are lighting circuits, 2 are kitchen outlet circuits (these are required), 2 are general outlet circuits, 1 is a laundry outlet circuit, 1 is an electric range circuit, and 1 is an electric clothes dryer circuit.

Some circuits are required to have additional protection provided by a ground-fault device. This is an electronic device that interrupts the flow of current in an extremely short amount of time in the event of excessively high current flow, such as occurs when the circuit is shorted. Ground-fault protection greatly reduces the hazard of electrical shock resulting from short circuits. Residences are required to have at least one waterproof outlet with ground-fault protection on the exterior of the house.

The material in a branch circuit includes: the cable, which runs from the distribution panel (circuit breaker) to the nearest outlet, and then to each of the other outlets in turn; a steel or plastic box (Figure 26-3) to house the receptacle, switch, or other device; and the device (duplex receptacle, switch, light fixture, etc.). All splices in the conductors must be made inside an approved box and must not be concealed in a wall or partition. Usually all splices not made at a device are made in junction boxes located in the cellar or attic.

In addition to switches, light fixtures, and convenience outlets, a variety of special electrical devices may be installed in a residence. These are described in the specifications and their location is indicated on the working drawings. Some sets of working drawings include the location of electrical equipment on the floor plan, others include a separate electrical plan. In our house, the location

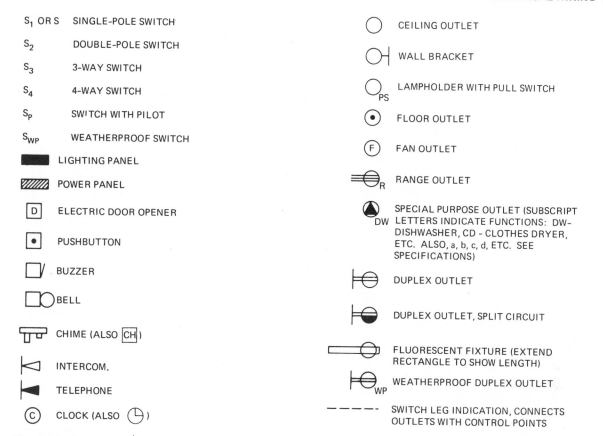

S₁ OR S	SINGLE-POLE SWITCH		CEILING OUTLET
S₂	DOUBLE-POLE SWITCH		WALL BRACKET
S₃	3-WAY SWITCH		LAMPHOLDER WITH PULL SWITCH
S₄	4-WAY SWITCH		FLOOR OUTLET
Sₚ	SWITCH WITH PILOT		FAN OUTLET
Sᵂᵖ	WEATHERPROOF SWITCH		RANGE OUTLET

S_1 OR S — SINGLE-POLE SWITCH

S_2 — DOUBLE-POLE SWITCH

S_3 — 3-WAY SWITCH

S_4 — 4-WAY SWITCH

S_P — SWITCH WITH PILOT

S_{WP} — WEATHERPROOF SWITCH

LIGHTING PANEL

POWER PANEL

D ELECTRIC DOOR OPENER

PUSHBUTTON

BUZZER

BELL

CHIME (ALSO CH)

INTERCOM.

TELEPHONE

C CLOCK (ALSO ⏰)

CEILING OUTLET

WALL BRACKET

LAMPHOLDER WITH PULL SWITCH $_{PS}$

FLOOR OUTLET

F FAN OUTLET

RANGE OUTLET $_R$

SPECIAL PURPOSE OUTLET (SUBSCRIPT $_{DW}$ LETTERS INDICATE FUNCTIONS: DW-DISHWASHER, CD - CLOTHES DRYER, ETC. ALSO, a, b, c, d, ETC. SEE SPECIFICATIONS)

DUPLEX OUTLET

DUPLEX OUTLET, SPLIT CIRCUIT

FLUORESCENT FIXTURE (EXTEND RECTANGLE TO SHOW LENGTH)

WEATHERPROOF DUPLEX OUTLET $_{WP}$

SWITCH LEG INDICATION, CONNECTS OUTLETS WITH CONTROL POINTS

Fig. 26-4. Common electrical symbols. There are many other symbols in use for commercial, industrial and special applications; those shown here are only the most common symbols for residential construction.

of the electrical equipment is given on the floor plan. Figure 26-4 shows some of the most common electrical symbols.

ESTIMATING ELECTRICAL WORK

The electrical subcontractor will probably prepare an itemized list of materials required to do the electrical work before commencing work on a job. However, because of the highly technical nature of the work, and the strict requirements imposed by electrical codes, the general estimator does not usually prepare such a list. The estimate is frequently based on the following:

- The type of electrical service to the building (underground or overhead).
- The ampacity (capacity in amperes) of the electrical service.
- Special equipment, such as garbage disposals and exhaust fans, that are included.
- The number of outlets for lighting, receptacles, and switches.
- The amount of money to be included as a fixture allowance.

The type of service to be provided is indicated in the specifications. The length of the entrance cable can be determined by reading the plot plan. Most estimators calculate the cost of burying underground cable on the basis of what it costs to bury one foot. If the service is to be overhead, the cost may be absorbed by the utility company.

The ampacity of the service should be given in the specifications. The estimator usually allows a fixed amount for the service equipment depending on the ampacity of the service. If a more accurate estimate is desired for this phase of the job, the material for the service can be itemized.

Any special electrical equipment which is to be included in the construction of the building is listed in the specifications. Some specifications do not indicate manufacturer's names, but sufficient information is always included to determine the quality and type of special equipment required.

The National Electrical Code specifies that outlets must be installed as follows:

- So that no point measured horizontally along a wall is more than 6 feet from an outlet.
- In any wall which is 2 feet or more in width.

- For each counterspace wider than 12 inches.
- At least one outlet in each bathroom.
- At least one outlet in the basement.
- At least one outlet in the garage.
- At least one outlet in laundry rooms.
- At least one outdoor outlet with ground-fault protection.
- Every room, hallway, stairway, garage, and outdoor entrance must have a lighting outlet, which is controlled by a wall switch.

The location of these outlets should be indicated on the floor plans or a separate electrical plan. If they are not shown, the estimator should include them in the estimate for electrical work. It is common practice to base the estimate on the cost of a typical outlet. When this is done the estimator determines the cost of installing one typical outlet, then multiplies this figure by the number of outlets of that type. In this manner the estimator does not need to determine the number of circuits or the quantity of cable required for the entire construction project.

The following is an example of what an electrical estimate might look like for a house.

Pieces	Feet	Material	Cost	Total
1		200-ampere service with 30-circuit panel		
1		exhaust hood with 300 cfm fan		
1		set of door chimes		
1		door chime transformer		
3		push buttons for door chimes		
29		lighting outlets		
19		switch outlets		
1		range outlet		
1		clothes dryer outlet		
2		weatherproof outlets with grounded faults		
33		convenience outlets		
2		telephone outlets		

27
ESTIMATING CHECKLIST

Specifications are written in the sequence in which the building would be constructed, with a few exceptions. (For example, the section that deals with moisture protection covers foundation damp-proofing or waterproofing and also roofing. These operations are completed at different times during the construction of a house.) Specifications are written to spell out in detail the information that cannot be clearly expressed on the working drawings (plans). For example, if the working drawings show wood floors, they might be oak, maple, or vertical grain fir. Tile flooring on the working drawings might be ceramic, asphalt tile, or vinyl tile. Where the working drawings show flashing, it could be galvanized iron, aluminum, or copper. Roof shingles could be asphalt or wood. The specifications describe the quality and type of materials, colors, finishes, and workmanship required. A typical set of specifications would include the following divisions for house construction:

Instruction to Bidders
General Conditions
Division I: General Requirements
Division II: Site Work
Division III: Concrete
Division IV: Masonry
Division V: Metals
Division VI: Carpentry and Millwork
Division VII: Moisture Protection
Division VIII: Doors, Windows, and Glass
Division IX: Finishes
Division X: Specialities
Division XI: Mechanical
Division XII: Electrical

You must constantly refer to the specifications as you study and make your material estimate from the working drawings. The following schedule will help you in making your material list and also in figuring the necessary labor to put that material in place according to the specifications and the working drawings.

We will start our estimate for the materials, starting at the site with the batter boards and stakes.

SITE WORK:

Figure 27-1 shows a typical right angle batter board. Depending upon the number of corners and the layout of the house, you should figure 3 2 x 4s approximately 6 feet long, 2 pieces of shiplap 1 x 8 about 5 feet long, and 1 stake for each corner of the house. You will also need a few nails and a roll of twine for the lines.

Figure the number of corners of the building. Multiply this by 3 for the number of 2 x 4s for the batter boards. Take the number of corners and multiply by 2 and this gives the number of pieces of 1 x 8 shiplap for the corner batter boards. Next count the number of corners and this will give you

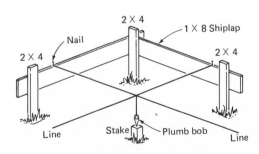

Fig. 27-1. Batter boards.

the number of stakes to lay out the house. We now start our material list as follows:

Pieces	Feet	Description	Cost	Total
		2 x 4 x 6' batter boards		
		1 x 8 x 5' shiplap — batter boards		
		stakes for batter boards		
		twine for lines on batter boards		

MASONRY ESTIMATE

Concrete Work — Footing Forms

Forms for the sides of the footings in residential construction are usually constructed using 1 x 8 low-grade lumber. To determine the amount of lumber needed, double the perimeter (distance around the building). Make sure that you have included any areas for support posts, fireplaces, and porches. Wood stakes are required to hold the forms in place until the concrete is poured. Figure 2 stakes for every 4 linear feet of perimeter and add 2 for each corner.

Pieces	Feet	Description	Cost	Total
		linear feet 1 x 8 shiplap — forms for footings		
		bundles wood stakes — for forms (12 to bundle)		

Concrete Work — Footings

To estimate the amount of concrete necessary for the footings, determine the linear feet around the perimeter of the building including the length of any other walls such as porches. Multiply the width of the footings in feet times the thickness in feet. This product is then multiplied by the total length of the footings in feet. This will give you the number of cubic feet in the footings. Divide this number of cubic feet by 27 to find the number of cubic yards. Transit mix is sold by the cubic yard. The amount of concrete to be used for the supporting posts and fireplace footings must be added to this amount. To figure this area, multiply the length in feet times the width in feet times the thickness in feet and divide by 27.

Pieces	Feet	Description	Cost	Total
		cubic yards concrete — footings		

Concrete Work — Floors and Porches

To estimate the amount of concrete needed for the floors and porches, determine the total square-foot area of all the places to be covered with concrete. Multiply this area by the thickness in feet and divide by 27 to get the number of cubic yards. *Example*: You have 908 square feet to be covered, including all floors and porches, and the thickness is 4 inches. You multiply 908 by $\frac{1}{3}$ (4 inches is $\frac{1}{3}$ foot) and you get 303. Divide 303 by 27 and you get $11\frac{6}{27}$ or rounded off $11\frac{1}{3}$ cubic yards of concrete is required.

Pieces	Feet	Description	Cost	Total
		cubic yards concrete — floors and porches		

Concrete Work — Sidewalks

Sidewalks are usually shown on the plot plan and are in the specifications. You estimate the concrete for the sidewalk the same as you did for the floors and porches. Length in feet times width in feet times thickness in feet divided by 27 gives you the number of cubic yards. *Example*: Your plot plan and specifications call for a sidewalk 4 feet wide and 3 inches thick. The plot plan shows the house sets back 30 feet so the walk will be 30-feet long. You multiply 30 (length) times 4 (width) = 120 square feet. Three inches thick means $\frac{1}{4}$ foot; $\frac{1}{4}$ × 120 = 30, and 30 ÷ 27 = $1\frac{3}{27}$ or $1\frac{1}{9}$ yards concrete required for the walk.

Pieces	Feet	Description	Cost	Total
		cubic yards concrete — sidewalk		

Concrete Work — Gravel Fill

The only place you will find this information on the working drawings will be on the section view, and not enough information will be given. You must refer to the specifications under Concrete

Work to find this. To determine the amount required, take the square-foot area where the concrete is to be placed and divide by 2, or the proportional part of a foot that the specifications call for in thickness. (In house construction, a 6-inch thick layer of gravel fill is usually specified.) This gives the amount in cubic feet. For cubic yards, divide by 27. Alternatively, length in feet times the width in feet times the thickness in feet divided by 27 gives the number of cubic yards needed.

Pieces	Feet	Description	Cost	Total
		cubic yards gravel fill — all floors		

Masonry Work — Cement Block

There are two methods of figuring cement block. In the first method, find the area of the wall. Each 8″ x 16″ block covers 0.888 square feet, so figure $112\frac{1}{2}$ blocks for every 100 square feet of wall. Add 5 to 10 percent for waste.

The second method is to figure 3 blocks for every 4 linear feet per course. Add the perimeter of the building in feet. Add the porches and fireplace base, if any. You multiply this by three and divide by four. This gives you the number of blocks for one course. You get 3 courses for every 2 feet in height. The standard height of a foundation in residential construction is 7 feet or 84 inches. This would be $10\frac{1}{2}$ courses high. You multiply the number of blocks required for one course by ten for the number of full-size blocks needed. The wall usually is capped with 4-inch solid cap blocks, so list the number of blocks required for one course for the cap. Corner blocks are listed separately, so count the number of corners in the foundation and multiply this by 10. You subtract this total from the total number of regular blocks you estimated.

You should figure two half sash blocks and two whole sash blocks for every 15 x 12 two-light cellar sash.

Example: Our total perimeter including porches and fireplace base is 139 linear feet. $139 \times 3 = 417$; $417 \div 4 = 106$ blocks for one course. You multiply this by 10 for a total of 1,060 regular blocks and 106 4-inch cap blocks. If we have 10 corners in the foundation, we need $10 \times 10 = 100$ corner blocks. We would subtract this 100 from our 1,060. You count the number of cellar sashes and then figure two whole and two half sash blocks for each window.

Pieces	Feet	Description	Cost	Total
		8 x 8 x 16 regular cement block		
		8 x 12 x 16 regular cement block		
		8 x 8 x 16 regular corner block		
		4 x 8 x 16 regular cap block		
		whole sash block		
		half sash block		

Masonry Work — Poured Foundation Walls

If the foundation is to be poured concrete instead of cement block, the number of cubic yards of concrete in the wall must be computed. To do this, multiply the length of the foundation times its thickness in feet. Multiply this times its height in feet. This is the number of cubic feet in the wall. Divide this by 27 to find the number of cubic yards needed.

Masonry Work — Basement Sash

To determine the number of cellar sashes needed, you count the number of windows shown on the foundation and basement plan. Check your specifications to make sure you are estimating what is called for. Also check to see if storm sash and screens are to be included.

Pieces	Feet	Description	Cost	Total
		15 x 12 steel cellar sash complete with storm and screen		

Masonry Work — Metal Areaways

Where the building sets close to the finished grade, an areaway has to be placed around the basement windows to allow for ventilation and to keep the earth away from the window. The areaways are shown on the foundation and basement plan and are also found in the specifications as to the type and material. List them as specified and the number should be the same as the basement cellar sash.

Pieces	Feet	Description	Cost	Total
		Galvanized iron areaways		

Masonry Work — Mortar for Block

In estimating mortar for cement block, allow 1 bag of cement for every 28 blocks. Check the specifications to find out if mortar is to be made from masonry cement or with portland cement and mason's lime.

Pieces	Feet	Description	Cost	Total
		bags masonry cement — mortar for block		
		bags regular cement — mortar for block		
		bags mason's lime — for mortar block-work		

Masonry Work — Washed Sand

A good grade washed sand is required to make a good mortar. You should figure 3 cubic feet of sand for every bag of cement. The easiest way is to figure 1 cubic yard of sand for every 10 bags of cement. This will allow for waste.

Pieces	Feet	Description	Cost	Total
		yards washed sand — mortar for blockwork		

Masonry Work — Wall Reinforcing

This information will be given on the section view and also in the specifications.

Determine the number of courses which require reinforcing. Take the linear feet where the reinforcing is required and multiply by the number of courses where it will be placed. This will give the linear feet required.

Pieces	Feet	Description	Cost	Total
		linear feet Dur-O-Wall reinforcing 8″		
		linear feet Dur-O-Wall reinforcing 12″		

Masonry Work — Steel Support Posts

These are found on the basement plan and are listed according to size and the number required. Check specifications to make sure both working drawings and specifications agree.

Pieces	Feet	Description	Cost	Total
		steel post w/ base and cap		

Masonry Work — Chimney

For a concrete block chimney, find the total height of the chimney in feet from the basement to the top of the chimney. (See the elevations.) To determine the number of flue liners needed, divide this height by 2. (Flue liners are 2 feet long.) A chimney is usually built of whole chimney block and flue liners up to the roof line. Brick is used where the chimney comes through the roof. From the top of the roof to the top of the chimney should be at least two feet above the ridge. To build an 8″ x 8″ furnace chimney, 3 whole chimney blocks are required for each 2 feet of height. Example: Consider a 24-foot high chimney which extends 6 feet through the roof with a single 8 x 8 flue liner. The total height of 24 feet requires 12 flue liners. Subtract the 6 feet from the total height and we have 18 linear feet of whole chimney block. Two divided into 18 gives us 9, which, multiplied by 3, gives 27 whole chimney blocks required. From the roof line to the top of the chimney is 6 feet. We multiply 6 by 27 (number of bricks needed for each foot of height for a single 8 x 8 flue chimney), for a total of 162 bricks. We would also list the cement and flashing for the chimney.

Pieces	Feet	Description	Cost	Total
		8 x 8 flue liners — chimney		
		whole chimney block — chimney		
		brick — chimney		
		bags masonry cement — chimney		
		bags regular cement — cap		
		linear feet flashing — chimney		

Masonry Work — Fireplace

In estimating a fireplace you must determine the location of the chimney to know what to figure. If the chimney goes up through the house, the only finished brick work would be the fireplace face and the chimney above the roof. If the chimney is built on the outside of the house, it requires face brick on the exposed face. The flue liners are figured the same, one for every two feet in height. You will need about four (4) fewer of the fireplace flues if the fireplace is located on the first floor. Check your plans to make sure how many fireplaces are on the chimney. In many plans, a fireplace is located in the recreation area in a basement and also in the living room or library on another floor.

Estimating a fireplace, with a prefabricated fireplace unit (which most fireplaces now use) you figure 30 firebricks for the firebox base. Your fireplace base is usually figured in with the foundation. You will need an ash dump, a 12″ x 15″ cleanout door for the ash pit, and an 8″ x 8″ cleanout for the flue liners of the furnace flue. List the prefabricated fireplace unit by size and manufacturer. Each unit will need an angle iron to hold the brick across the fireplace face opening. This length should be at least 8 inches wider than the opening. Estimate brick by the square foot, adding any necessary fills. You should also figure the hearth material in your list. This information is in the specifications and the floor plans.

Pieces	Feet	Description	Cost	Total
		regular firebrick – fireplace base		
		ash dump		
		cleanout doors – 12″ x 15″ – 8″ x 8″		
		No. prefabricated fireplace unit		
		8 x 8 flue liners		
		8 x 12 flue liners		
		$3\frac{1}{2}$ x $3\frac{1}{2}$ angle iron x feet		
		brick for fireplace face and exterior face		
		quarry tile fireplace hearth		
		bags regular cement – hearth and cap		

Masonry Work — Fireplace Mortar

The amount of mortar cement that is required to build a fireplace depends partly on the mix that is used. A typical mix would be 1 part cement, 1 part mason's lime, and 6 parts sand. If a $\frac{3}{8}$-inch mortar joint is used, than 21 cubic feet of mortar is required to lay approximately 1,000 bricks. One (1) bag of portland cement, one (1) bag of mason's lime, and 48 shovels of sand will lay approximately 400 bricks. Allow one-third for filling between the brick and flue liners and for fills.

Pieces	Feet	Description	Cost	Total
		bags regular cement – fireplace		
		bags mason's lime – fireplace		
		bags masonry cement – fireplace		
		cubic yards washed sand – mortar for fireplace		

Masonry Work — Brick Veneer

To estimate the quantity of common or standard face brick in veneer work, 675 bricks are needed for every 100 square feet of wall. Allow 5–10 percent for waste. You should also figure 70 wall ties for every 100 square feet of brick.

Pieces	Feet	Description	Cost	Total
		brick – veneer work		
		galvanized iron wall ties – brick veneer work		

Masonry Work — Brick Veneer Mortar

Figure 1 bag of regular cement and 1 bag of mason's lime for every 400 bricks. Add any angle irons for lintels, and flashing material.

Pieces	Feet	Description	Cost	Total
		bags regular cement – veneer		
		bags mason's lime – veneer		
		$3\frac{1}{2}″$ x $3\frac{1}{2}″$ x ″ angle iron over doors and windows – veneer		
		galvanized iron 30″ x 96″ – flashing, veneer and chimney		

Masonry Work — Foundation Plastering

One bag of portland cement and one bag of mason's lime with 48 shovels of sand will cover approximately 40 square feet. Determine the square-foot area of the foundation to be covered. Divide this by 40 to get the number of bags of cement and lime needed.

Pieces	Feet	Description	Cost	Total
		bags regular cement — parging		
		bags mason's lime — parging		

Masonry Work — Foundation Damp-proofing

Brush-type asphalt coating covers approximately 30–35 square feet of wall to the gallon. Figure the square-foot area to be covered and divide this by 30. If the specifications call for two coats, the second coat will require only about half as much. Asphalt coating is sold in 5-gallon cans, so divide the number of gallons required by 5.

Pieces	Feet	Description	Cost	Total
		5-gallon cans asphalt coating — foundation		

Masonry Work — Drain Tile

To figure the amount of plastic drain tile or pipe needed, find the distance around the perimeter of the house and divide by 10. This is the number of 10-foot pieces needed. A 90-degree elbow is required at every corner, and where the drain goes under the footings to a sump a tee is required. Crushed stone is sold by the ton and approximately one ton of stone is required for every 18 linear feet of drain pipe.

Pieces	Feet	Description	Cost	Total
		10-foot perforated plastic drain tile		
		90-degree elbows — drain tile		
		plastic tees		
		tons No. crushed stone — drainage		

CARPENTRY ESTIMATE: FRAMING AND EXTERIOR WORK

Carpentry Work — Girders

The girder and its supporting posts support the main bearing partitions as well as part of the weight of the floors and their contents. These are usually built up using three or more members. The lengths of the members should be determined so the ends of the wood members making up the girder should be joined only at the support columns. If possible use pieces that can run from wall to wall. Determine the length of the members and list them according to length and size.

Pieces	Feet	Description	Cost	Total
		2 x 8 (or 10 or 12) x length — built up girder		

Carpentry Work — Termite Shield

If the specifications and working drawings call for a termite shield, determine the size and length needed and list by either the linear foot or the number of pieces.

Pieces	Feet	Description	Cost	Total
		termite shield		

Carpentry Work — Sill Seal and Sill

Enough sill seal and sill must be figured to cover the perimeter of the building. If a steel supporting beam is used for the main bearing beam and the joists rest on top of the beam, then enough sill must be added to cover the length of the girder.

Pieces	Feet	Description	Cost	Total
		linear feet 6-inch sill seal		
		2 x 6 x (length) sill		

Carpentry Work — Box Sill

The box sill is the header joists placed at right angles to the ends of the floor joists. It is the same width as the floor joists and rests on the top of the sill. The box sill should be the same as the perimeter required for the sill.

Pieces	Feet	Description	Cost	Total
		2 x (8 or 10 or 12) x length box sill		

Carpentry Work — Floor Joists

To determine the number of floor joists needed, multiply the length of the foundation wall on which they rest by $\frac{3}{4}$ then add 1. This provides three joists for every four feet and one additional joist at the end of the wall. Also add one joist for every partition that runs in the same direction as the joists.

Pieces	Feet	Description	Cost	Total
		2 x (8 or 10 or 12) x length — floor joists		

Carpentry Work — Bridging

To determine the amount of bridging needed, multiply the number of floor joists by 3. This is the number of linear feet of bridging needed for one row. If the span of the joists is over 14 feet, figure two rows of bridging for that span.

Pieces	Feet	Description	Cost	Total
		linear feet 1 x 3 spruce — bridging		

Carpentry Work — Headers and Trimmers

Headers are used to support the cut ends of the joists at such places as stairs, chimneys, or fireplaces. The joists at the ends of the headers are called trimmers. Trimmers should be doubled to support the extra load. Headers and trimmers must be the same width as the floor joists and trimmers should be the same length as the floor joists. Measure the length of the headers from the working drawings and list the headers and trimmers by the piece.

Pieces	Feet	Description	Cost	Total
		2 x (8 or 10 or 12) x length — headers		
		2 x (8 or 10 or 12) x length — trimmers		

Carpentry Work — Subfloor

Estimating the subflooring is done by the square-foot area to be covered. If the subflooring is plywood, take the square-foot area to be covered and divide by 32 for the number of 4' x 8' pieces. If shiplap or T&G is used for the subflooring, take the square-foot area to be covered and add 25 percent to determine the amount needed.

Pieces	Feet	Description	Cost	Total
		(thickness) x 4' x 8' plywood — subfloor		

Carpentry Work — Plates and Shoe

This is estimated by the linear foot. Take the perimeter of the outside walls. Figure the linear feet of interior walls. Figure all walls solid. Do not deduct for door or window openings. Add the exterior linear feet of wall and the interior linear feet of interior partitions together and then multiply this total by 3 for the total linear feet of 2 x 4's needed for the plates and shoe.

Pieces	Feet	Description	Cost	Total
		linear feet 2 x 4 plates and shoe		

Carpentry Work — Studs

In estimating for studs, allow one stud for every linear foot of interior and exterior walls and partitions, and two extra for every corner.

Pieces	Feet	Description	Cost	Total
		2 x 4 x 8' studs		

Carpentry Work — Headers

Headers are required where openings for windows or doors are cut out of the walls to carry the vertical loads to other studs. The size of headers varies. They must be of sufficient size to carry the load. To find the length of headers, measure each opening for doors or windows. Allow 3 feet for single openings. Multiply the number of single openings by three and then double this (headers are double). Then determine the standard lengths that can be divided into this total.

Pieces	Feet	Description	Cost	Total
		2 x 4 x (length) door and window headers		

Carpentry Work — Second Floor Framing

If the working drawings show a second floor, you estimate the materials in the same manner in which you did the first floor. List your box sill first, then the second floor joists. The headers and trimmers should be figured, the bridging for the second floor joists, and the subflooring. You then list the plates and shoe for the exterior walls and the interior partitions on the second floor. Next list the studs and headers for all door and window openings.

Pieces	Feet	Description	Cost	Total
		linear feet 2 x size box sill – second floor		
		2 x (8-10-12) x length joists – second floor		
		2 x (8-10-12) x length headers and trimmers – second floor		
		linear feet 1 x 3 spruce – bridging, second floor		
		size x 4' x 8' plywood subfloor – second floor		
		linear feet 2 x 4 plate and shoe – second floor		
		2 x 4 x 8' studs – second floor		
		2 x size x length – headers for doors and windows, second floor		

Carpentry Work — Ceiling Joists

Ceiling joists support the ceiling of the top floor of the house. They also support the rafters and help secure them to the house. The ceiling joists must span from the outside wall of the house to an interior bearing wall or to the opposite exterior wall. To estimate ceiling joists you find the direction in which they are to run. This is usually on the second floor plan. Multiply that distance by 3, divide by 4, and add 1. The length is the span that they have to cover.

Pieces	Feet	Description	Cost	Total
		2 x size x length – ceiling joists		

Carpentry Work — Rafters

To determine the number of rafters 16 inches o.c. required on each side of a gable roof, multiply the length of the ridge by 3, divide by 4, and add 1. Assume jack rafters the same size and the same length as the common rafter. To find the number of jack rafters needed, take $1\frac{1}{2}$ times the width of the building and add 2. Allow 1 hip rafter for each hip and 1 valley rafter for each valley. Hip

and valley rafters should be 2 inches wider than the common rafters to allow for a solid nailing surface.

Pieces	Feet	Description	Cost	Total
		2 x size x length – rafters		
		2 x size x length – ridge board		
		2 x size x length – hip and valley rafters		

Carpentry Work — Trusses

If the plans and specifications call for preassembled roof trusses instead of rafters, these are usually spaced two feet on center. To determine the number required for a straight roof, divide the length of the building in feet by 2 and add 1.

Pieces	Feet	Description	Cost	Total
		Roof trusses – give pitch and tail (overhang)		

Carpentry Work — Collar Beams

Occasionally one collar beam is used for each pair of rafters, but most buildings have one for every third pair of rafters. If collar beams are spaced 16 inches on center, divide the number of rafters by 2 and add 1. If they are spaced 32 inches on center, divide the number of rafters by 4 and add 1.

Pieces	Feet	Description	Cost	Total
		1 x 8 x length collar beams		

Carpentry Work — Gable Studs

When a gable roof is framed, gable studs are necessary between the end rafters of the gable roof and the top plate of each end wall. On a straight roof with two gable ends, measure one end and figure one stud for each linear foot. The height of the gable from the top of the wall plate is the length of the studs.

Pieces	Feet	Description	Cost	Total
		2 x 4 x length gable studs		

Carpentry Work — Roof Sheathing

On a plain gable roof, multiply the length of the ridge times the length of a common rafter to find the area of one side. Multiply this by 2 to find the total roof area. If the roof area is to be shiplap or T&G, add 25 percent to this figure to find the number of board feet needed. If plywood is used then take the total area to be covered and divide by 32 to find the number of sheets of plywood.

Pieces	Feet	Description	Cost	Total
		size x 4' x 8' plywood — roof sheathing		

Carpentry Work — Roofing

Roofing material is estimated by the square. A square is enough material to cover 100 square feet. Divide the square foot area of the roof by 100 to find the number of squares required. For asphalt shingles, add $1\frac{1}{2}$ square feet for every linear foot of ridge, eaves, hip, and valley. When asphalt roofing is used, building felt is placed between the roof sheathing and the finished roof. Building felt is sold in 500 square foot rolls. Add 10 percent for overlapping the edges. Take the square-foot area of the roof, add 10 percent, and divide by 500 to determine the number of rolls of felt.

Pieces	Feet	Description	Cost	Total
		square asphalt roofing		
		rolls 15-pound asphalt felt — roof		

Carpentry Work — Drip Edge

When asphalt shingles are used for the finished roofing, a metal drip edge is installed along the eaves and the rake of the house. Drip edge is manufactured in 10-foot lengths. It is nailed over the top of the roofing felt and the asphalt shingles are installed over the top of the drip edge. To determine the amount of drip edge needed, measure the length of the eaves and the rake and add them together, then divide by 10 for the number of pieces required.

Pieces	Feet	Description	Cost	Total
		Pieces of galvanized drip edge — 10-foot lengths		

Carpentry Work — Cornices

All cornices are made up of finished millwork consisting of boards and moldings used in combination with each other to finish the ends of the rafters which extend beyond the face of the outside walls. Cornices are listed by the linear feet. The parts are named according to the position they are placed in. Usually a tight cornice the rake is made using 5/4 stock material. Start with your rakes and list the cornice material.

Pieces	Feet	Description	Cost	Total
		linear feet 5/4 x size rake		
		linear feet 1 x size fascia		
		linear feet 1 x size soffit and frieze		
		linear feet crown mold (size) — cornice		
		linear feet bed mold (size) — cornice		

Carpentry Work — Exterior Sheathing

To determine the amount of sheathing needed, take the distance (perimeter) around the house and multiply this by the height. If you have a gable roof, do not forget the gable ends. This will give you the total square-foot area to be covered. If shiplap is used, add 25 percent to this total. If plywood is used, take the total square-foot area to be covered and divide by 32 to give you the number of sheets of plywood required.

Pieces	Feet	Description	Cost	Total
		thickness x 4' x 8' plywood — sheathing		

Carpentry Work — Siding

Siding is estimated according to the material used. If it is shingles and sold by the square, then you figure the number of squares required. If it is sold by the board foot, then you find the number of square feet, add for cutting and waste and list it as board feet. All sheathing should be covered with a building paper before the finished siding is installed. Kraft building paper is sold in 500 square foot rolls. Take the square-foot area to be covered, add 10 percent for lapping, and divide by 500.

Pieces	Feet	Description	Cost	Total
		(thickness x size) bevel siding		
		bundles of wood shingles — siding		
		rolls kraft building paper — siding		

Carpentry Work — Exterior Finish Work

If you have any special work to be done to finish the outside of the building, you would list this work now. This would cover such items as corner boards, porches, special overhangs, entrance porch railings, etc.

Pieces	Feet	Description	Cost	Total
		linear feet 5/4 x size corner boards		
		thickness x 4' x 8' exterior plywood — overhang, etc.		
		wood louvers No. 600-L — gable ends		
		any framing necessary for entrance roofs, porches, etc.		
		finish work for any entrances, porches, etc.		
		wrought iron railings, etc.		

CARPENTRY ESTIMATE: INTERIOR WORK AND FINISHES

Carpentry Work — Insulation

In estimating ceiling insulation, figure the square-foot area to be covered. Fiberglass insulation is sold in batt form. The batts are 15″ x 48″ and come 10 pieces to the bag. Each bag covers 50 square feet. Determine the square-foot area to be covered and divide this by 50 to get the number of bags required.

Pieces	Feet	Description	Cost	Total
		bags (thickness) fiberglass insulation — ceilings		

In estimating insulation for the sidewalls, do not figure the gable ends. Multiply the perimeter of the house by the height of the walls to find the total area in feet; $3\frac{1}{2}$-inch thick fiberglass insula-

tion is sold in 70 square foot rolls so divide 70 into the total square-foot area to get the number of rolls required.

Pieces	Feet	Description	Cost	Total
		bags (thickness) insulation — sidewalls		

Carpentry Work — Wallboard

Estimating wallboard can be done in two ways. One is to lay out a sketch of each wall and ceiling, showing the dimensions. From this the number of lengths and pieces can be determined. The second system is to figure the square foot area to be covered. The amount of wallboard needed for the ceiling is the same as our floor area. If you have two floors, double this amount. In figuring wallboard, no deductions are made for openings, so take the length of the outside walls and multiply them by the height. The length of our interior partitions has been established in our framing. Add the total length of all interior partitions and multiply this by 8 (8-foot ceiling heights) for the square-foot area. Interior walls have wallboard on both sides, so double this figure. Add the ceiling areas, the outside wall areas, and the total interior wall areas for the total amount of wallboard required.

Pieces	Feet	Description	Cost	Total
		square feet (size) wallboard — ceiling and sidewalls		

Carpentry Work — Joint System

Joint compound is a premixed vinyl-based cement that is used directly from the container. It is used to apply the paper to the joints of the wallboard and for concealing the nail heads. The tape that goes over the seams comes in 250- and 500-foot rolls of tape. Joint compound is sold in 5- and 1-gallon cans. Estimate one (1) 250-foot roll of tape and 6 gallons of compound for every 1,000 square feet of wallboard.

Pieces	Feet	Description	Cost	Total
		rolls perfa-tape 250 feet — wallboard		
		5 gallon cans joint cement — wallboard		

Carpentry Work — Molding

Cove molding is estimated by the linear foot. The distance around the outside of the house is known. The total length of the interior partitions is already known. The cove mold goes on each side of the interior partitions. Double the length of the interior partitions and add this to the perimeter of the exterior walls.

Pieces	Feet	Description	Cost	Total
		linear feet (size) cove mold — ceiling		

Carpentry Work — Baseboard and Carpet Strip

Baseboard and carpet strip can be estimated in two ways. One is to find the perimeter of each room and deduct the width of the door openings. The other is to deduct twice the width of the interior door openings and the width of the exterior door openings and subtract from the total length of the ceiling moldings.

Pieces	Feet	Description	Cost	Total
		linear feet base (size) and carpet strip (size)		

Carpentry Work — Underlayment

To estimate the underlayment, take the square-foot area where the underlayment is required. Divide this area by 32 to get the number of pieces of 4′ x 8′ needed.

Pieces	Feet	Description	Cost	Total
		(thickness) x 4′ x 8′ underlayment — give location		

Interior Finish — Linoleum or Resilient Flooring

To estimate the quantity, find the square-foot area to be covered and then divide by 9 for square yards. If flooring is foot square, then list by the piece.

Pieces	Feet	Description	Cost	Total
		square yards inlaid linoleum — location		
		square feet resilient flooring — location		

Interior Finish — Slate Flooring

Slate is sold by the square foot. Find the square-foot area to be covered. You will also need 1 gallon of special adhesive for every 100 square feet of slate, and 5 pounds of grout for every 100 square feet.

Pieces	Feet	Description	Cost	Total
		square feet multicolored slate — location		
		quarts special slate adhesive		
		pounds grout — for slate joints		

Carpentry Work — Hardwood Flooring

Find the square-foot area where the hardwood flooring is required. Allow one-third for matching. Add the total square-foot area and the third for the total board feet. Deadening felt is laid between the subfloor and the underlayment and hardwood flooring. Felt is sold in 500 square-foot rolls. Divide 500 into the square-foot area where the deadening felt is required.

Pieces	Feet	Description	Cost	Total
		1 x 3 select (grade) oak (type) — flooring		
		rolls deadening felt		

Interior Finish — Ceramic Wall Tile

Find the square-foot area to be covered. Tile is sold by the square foot. Cap tile is sold by the linear foot. Inside and outside corners are sold by the piece. Ceramic fixtures such as soap and grab, grab bars, towel bars, etc., are sold by the piece. 1 gallon of ceramic adhesive will cover approximately 100 square feet; 5 pounds white cement grout will cover about 60 square feet of wall tile.

Pieces	Feet	Description	Cost	Total
		square feet $4\frac{1}{4}''$ x $4\frac{1}{4}''$ ceramic wall tile		
		linear feet ceramic tile cap		
		inside corners cap		

Pieces	Feet	Description	Cost	Total
		outside corners cap		
		ceramic fixtures (soap and grab, etc.)		
		quarts ceramic tile adhesive		
		pounds white ceramic tile grout		

Interior Finish — Ceramic Floor Tile

Floor tile is estimated by the square foot. One gallon of adhesive is required for every 100 square feet and 5 pounds of white grout is needed for every 100 square feet. A marble threshold is needed between doors where ceramic tile and other type floors meet.

Pieces	Feet	Description	Cost	Total
		square feet ceramic floor tile		
		quarts ceramic floor tile adhesive		
		pounds white grout for ceramic floor tile		
		6-inch marble threshold x length		

Carpentry Work — Exterior Doors

List exterior doors first. Door frames and sills are listed according to size. Doors are always listed first width and then height and then thickness. Any special entrance doors with sidelights or special heads should be listed first.

Pieces	Feet	Description	Cost	Total
		exterior door frame with sill 3'0" x 6'8" — front door		
		exterior door frame with sill 2'8" x 6'8" — side door		
		exterior door 3'0" x 6'8" x $1\frac{3}{4}''$ M-108 — front door		
		exterior door 2'8" x 6'8" x $1\frac{3}{4}''$ No. — side door		
		combination storm and screen door 3'0" x 6'8" No. — front		
		combination storm and screen door 2'8" x 6'8" No. — side		
		lineal feet $1\frac{3}{4}''$ door stop — exterior doors		
		lineal feet casing — exterior doors		

Carpentry Work — Interior Doors

Interior doors are listed the same as the exterior doors, according to size and style. You list any cased opening first.

Pieces	Feet	Description	Cost	Total
		door jambs 3'0" x 6'8"		
		door jambs 2'8" x 6'8"		
		door jambs 2'6" x 6'8"		
		door jambs 2'4" x 6'8"		
		door jambs 2'0" x 6'8"		
		louver doors (size x height x thickness)		
		doors (sample) 2'6" x 6'8" x $1\frac{3}{8}''$ birch flush		

Carpentry Work — Door Stop and Casing

This is listed by the linear foot. It requires approximately 17 linear feet of stop for each door and 34 linear feet of casing for both sides of a door opening. Count the number of doors and multiply by 17 for the stop and by 34 for the linear feet of casing.

Pieces	Feet	Description	Cost	Total
		linear feet door stop — interior doors		
		linear feet door casing — interior doors		

Carpentry Work — Closet Shelves and Rods

List this by the linear feet. Closet shelves are usually 1 x 12 pine boards and the supports are 1 x 3 pine. Closet rods are metal chrome plated and are usually adjustable.

Pieces	Feet	Description	Cost	Total
		linear feet 1 x 3 pine — closet shelves supports		
		linear feet 1 x 12 pine — closet shelves		
		adjustable closet rods		

Carpentry Work — Windows

Windows are listed according to size and style. Starting at the first floor, list each window on the first floor. Do the same for the second floor. Group them and list according to size.

Pieces	Feet	Description	Cost	Total
		single double hung 3'0" x 4'6" (style) location		
		single double hung 3'0" x 3'2" (style) location		
		mullion double hung 2'8" x 4'6" (style) location		
		awning window — size and style		
		picture window (sample) 2'0" x 4'6" x 5'0" x 4'6" x 2'0" x 4'6"		
		storm sash and screen 3'0" x 4'6" — location		
		storm sash and screen 3'0" x 3'2" — location		
		etc.		

Carpentry Work — Window Trim

Window trim is listed by the linear foot. Estimate trim required for each window separately and then group together. List according to part and style.

Pieces	Feet	Description	Cost	Total
		linear feet window casing (style)		
		linear feet window stop (style)		
		linear feet window stool (style)		
		linear feet window apron (style)		
		linear feet window mullion casing (style)		

Carpentry Work — Fireplace Mantel

List according to style and manufacturer. This is found on the working drawings and also in the specifications.

Pieces	Feet	Description	Cost	Total
		mantel M-1448 — fireplace		

Carpentry Work — Blinds or Shutters

List according to size and style.

Pieces	Feet	Description	Cost	Total
		pair blinds (size) and (style) — location		

Carpentry Work — Stairs

List by size and style. List each part separately.

Pieces	Feet	Description	Cost	Total
		set housed stringers with risers and treads		
		newel posts (style)		
		balusters (style and length)		
		linear feet handrail (style)		
		oak treads (size and length)		
		(material) risers (size and length)		
		(material) landing tread (size and length)		

Interior Finish — Hardware

This is usually made as an allowance and is found in the specifications. You list the hardware allowance on the list. Nails should have an allowance because this is not covered with the hardware.

Pieces	Feet	Description	Cost	Total
		finish hardware allowance		
		nails allowance		

Carpentry Work — Kitchen Cabinets and Bath Vanities

List by units. List the wall units first, the base units second, and then the lineal feet of counter tops. List giving size and manufacturer's finish. Width first then height.

Pieces	Feet	Description	Cost	Total
		W3018 wall cabinet — Oakmont		
		W1830 wall cabinet — Oakmont		

		Description		
		W3330 wall cabinet — Oakmont		
		W2430 wall with spin shelves — Oakmont		
		W3315 wall cabinet — Oakmont		
		B-18" base cabinet — Oakmont		
		SF-42" sink front — Oakmont		
		CB-45 square corner base — Oakmont		
		lineal feet laminated plastic counter top — molded edges		
		bath vanity RV-36" — location		
		bath vanity RV-30" — location		
		laminated plastic counter tops for above		
		$\frac{1}{4}$" silver-plated glass mirrors (size) — location		
		any medicine cabinets required (size and model — location)		

PAINTING

Exterior Painting — Siding

Figure the square-foot area to be painted. Allow 450 square feet to the gallon for one coat of primer. House paint should cover 600 square feet per gallon over the prime coat. Take the square-foot area to be covered and divide by 450 for the primer and by 600 for the house paint.

Pieces	Feet	Description	Cost	Total
		gallons exterior primer		
		gallons exterior house paint		

Exterior Painting — Cornices and Porches

Figure 2 square feet for every lineal foot of a box cornice and $\frac{1}{2}$ square foot for every lineal foot of a tight cornice such as the rake. The sash and frame of one side of an average window is equal to 25 square feet. Most doors are estimated at 25 square feet for one side.

Pieces	Feet	Description	Cost	Total
		gallons trim and shutter paint — exterior		
		lineal feet (size and material) cove molding		
		pints white glue		
		wedges		

Interior Painting — Walls and Ceilings

This is figured by the square-foot area. Allow 450 square feet to the gallon for wall and ceiling primer and 600 square feet to the gallon for the finish coat. Divide 450 into the square-foot area to be covered by the primer and 600 into the square-foot area for the second coat.

Pieces	Feet	Description	Cost	Total
		gallons interior wall primer		
		gallons interior flat wall paint		
		gallons interior satin wall paint		

Interior Painting — Trim and Doors

Paint for the interior trim is the same as that for the outside windows. Figure 25 square feet for each window and 25 square feet for each door. For each cased opening, figure 20 square feet. For the baseboard, figure 1 square foot for every 3 linear feet. Figure 500 square feet for each gallon of undercoat enamel and 500 feet for the finish coat.

Pieces	Feet	Description	Cost	Total
		gallons undercoat enamel		
		gallons enamel		
		gallons varnish for doors (natural)		
		gallons shellac — filler for doors		

Interior Painting — Finishing Hardwood Floors

Wood filler covers 600 square feet to the gallon. Varnish covers 600 square feet to the gallon. Take the square-foot area of the hardwood floors and divide by 600 to determine the number of gallons of filler and varnish. Varnish usually requires two-coat work.

Pieces	Feet	Description	Cost	Total
		gallons paste wood filler — floors		
		gallons varnish — floors		

Painting — Miscellaneous Supplies

You should make an allowance for miscellaneous paint supplies such as drop cloths, brushes, paint thinners, etc.

Pieces	Feet	Description	Cost	Total
		miscellaneous paint supplies allowance		

28
LABOR ESTIMATING GUIDELINE

In estimating labor, it is necessary to establish guidelines as a starting point. Conditions, labor market, and availability of competent workers will affect the amount of work performed. Just as material estimating has factors beyond the estimator's control, such as how the workmen will use certain materials on the job, also in estimating labor, you will find that there are factors beyond your control. This checklist is an estimate based upon the experience of many successful contractors. You will have to adjust your estimate, taking into consideration the ability of your workmen and their methods of getting a job done.

We will follow the same sequence in estimating labor, as we did in establishing the material list. In this manner, with experience, you will be able to estimate the materials required and the labor necessary to put those materials in place according to the working drawings and the specifications.

The first thing you should do in any estimating either for materials or labor, is to study and fully understand the working drawings and the specifications for that particular project. After you fully understand the size, scope, and requirements of the project by studying the working drawings and specifications, then you should start your estimate with *site work*.

The site should always be visited to ascertain the existing conditions. This is important in understanding what preparatory work must be done before excavation can be started. If there is a large amount of brush or any large trees to be removed, this must be figured in the cost. Large trees can be expensive to cut down and haul away. Also the site should be visited to ascertain if any excess earth has to be removed or additional fill brought in.

We will start our labor estimate with site work.

SITE WORK

1. Trees to be removed from premises _____ @ _____ = _____
2. Excess earth to be removed from premises _____ yards @ _____ = _____
3. Fill to be brought to premises _____ yards @ _____ = _____
4. Brush to be removed — hours unskilled _____ @ _____ = _____

MASONRY WORK

Foundation Work

Layout, stakes, batter boards, levels, etc. _____ hours unskilled @ _____ = _____
_____ hours skilled @ _____ = _____

Excavation — With a bulldozer, 100 cubic yards of earth can be removed in three (3) hours.
_____ hours skilled @ _____ = _____

Backfilling — With a bulldozer, 100 cubic yards of earth can be backfilled in one (1) hour.
_____ hours skilled @ _____ = _____

Digging footings — 100 linear feet based on 8″ x 15″ requires 5 manhours of unskilled labor.
_____ linear feet _____ hours unskilled @ _____ = _____

Footing formwork — 100 linear feet setting forms to level grade requires 2 hours skilled and 2 hours unskilled labor.

| | linear feet | | hours skilled | @ | = | |
| | linear feet | | hours unskilled | @ | = | |

Pouring concrete footings — 100 linear feet, placing ready mix concrete, average conditions and wheeling distance to forms, figure $\frac{1}{2}$ hour skilled and 4 hours unskilled.

| | linear feet | | hours skilled | @ | = | |
| | | | hours unskilled | @ | = | |

Concrete floors — base preparation — For every 100 square feet, placing, grading, tamping base material, figure $1\frac{1}{4}$ hours unskilled. Setting forms to level grade for screeding concrete, figure $\frac{1}{2}$ hour skilled.

| | square feet | | hours skilled | @ | = | |
| | | | hours unskilled | @ | = | |

Placing vapor barrier, polyethylene, sisal kraft, etc. requires $\frac{1}{2}$ hour unskilled for every 100 square feet.

| | square feet | | hours unskilled | @ | = | |

Placing reinforcing rods or wire mesh, for every 100 square feet figure $\frac{1}{2}$ hour unskilled labor.

| | square feet | | hours unskilled | @ | = | |

Pouring and finishing, for every 100 square feet of concrete 4-inch thick, figure $1\frac{1}{2}$ hours of skilled and $1\frac{1}{2}$ hours of unskilled

| | square feet | | hours skilled | @ | = | |
| | | | hours unskilled | @ | = | |

Porches and Sidewalks — For every 100 square feet of porches and walks for grading, leveling, and forming, figure $\frac{3}{4}$ hour skilled and $\frac{1}{2}$ hour unskilled. Pouring and finishing concrete, for every 100 square feet figure $1\frac{1}{2}$ hours skilled and $1\frac{1}{2}$ hours unskilled. Removing forms after the concrete has set, figure $\frac{1}{2}$ hour unskilled. For every 100 square feet of porches and sidewalks, figure $2\frac{1}{4}$ hours of skilled and $2\frac{1}{2}$ hours of unskilled.

| | square feet | | hours skilled | @ | = | |
| | | | hours unskilled | @ | = | |

Porch steps — For every 10 square feet of tread with an 8-inch rise and 12-inch tread to form requires 2 hours skilled and 1 hour unskilled. Pouring concrete and finishing requires 1 hour skilled and $\frac{1}{2}$ hour unskilled. Removing forms and finishing requires 1 hour skilled and $\frac{1}{2}$ hour unskilled. For every 10 square feet of tread figure 4 hours skilled and 2 hours unskilled in all for steps of poured concrete.

| | square feet | | hours skilled | @ | = | |
| | | | hours unskilled | @ | = | |

Brick treads and risers, tooled joints — For every 10 square feet of tread area, figure 5 hours skilled and $2\frac{1}{2}$ hours unskilled.

| | square feet | | hours skilled | @ | = | |
| | | | hours unskilled | @ | = | |

Rough slate or stone treads to 4-inches thick — For every 10 square feet of tread area, figure 2 hours skilled and 2 hours unskilled.

| | square feet | | hours skilled | @ | = | |
| | | | hours unskilled | @ | = | |

Concrete block — Based on 100 square feet of wall area with average conditions, struck or tooled joints, common bond, including openings:
8 x 8 x 16 cement block requires 6 hours skilled and 6 hours unskilled. 10 x 8 x 16 cement block requires $6\frac{1}{2}$ hours skilled and $6\frac{1}{2}$ hours unskilled. 12 x 8 x 16 cement block requires 7 hours skilled and 7 hours unskilled. Placing reinforcing rods requires $\frac{1}{2}$ hour unskilled.

Block Foundation

	square feet 8 x 8 x 16		hours skilled	@	=	
			hours unskilled	@	=	
	square feet 10 x 8 x 16		hours skilled	@	=	
			hours unskilled	@	=	
	square feet 12 x 8 x 16		hours skilled	@	=	
			hours unskilled	@	=	

Wall Reinforcing

_____ square feet _____ hours unskilled @ _____ = _____

Concrete foundation wall — Figuring a wall 8 inches thick with plywood forms, $\frac{1}{2}$-inch reinforcing, pouring concrete, removing forms and hand rubbing walls, for every 100 square feet of wall figure 3 hours skilled and $6\frac{1}{4}$ hours unskilled. Poured Concrete Wall with $\frac{1}{2}$-inch reinforcing rods 8 inches thick.

_____ square feet _____ hours skilled @ _____ = _____

_____ hours unskilled @ _____ = _____

Foundation windows — For each unit either wood or metal, figure $\frac{1}{4}$ hour skilled. Setting poured-in-place basement frames and sash figure $\frac{1}{2}$ hour skilled.

Basement windows _____ each _____ hours skilled @ _____ = _____

Fireplace — Estimating average fireplace for base, firebrick hearth, setting prefabricated unit, and brick facing, figure $13\frac{1}{2}$ hours of skilled and 14 hours of unskilled. For exterior face brick on chimney with one (1) 8 x 8 flue and one (1) 8 x 12 flue liner, figure 2 hours skilled and 2 hours unskilled for every foot in height. Scaffolding, figure 3 hours of unskilled labor for metal scaffolding erection and dismantling for an outside fireplace to 36-feet in height.

Fireplace — Interior work — brick, tile, and unit

_____ hours skilled @ _____ = _____

_____ hours unskilled @ _____ = _____

Exterior work _____ height _____ hours skilled @ _____ = _____

_____ hours unskilled @ _____ = _____

Scaffolding _____ height _____ hours unskilled @ _____ = _____

Brick Veneer — 100 square feet of facebrick based on $\frac{3}{8}$-inch flush joint, normal openings, sills and headers to 16-foot height, 4-inch brick veneer, stretcher bond, including wall ties, figure 12 hours skilled and 8 hours unskilled. Cleaning brick using muriatic acid and water, figure 1 hour skilled and $\frac{1}{2}$ hour unskilled.

Brick veneer _____ square feet

_____ hours skilled @ _____ = _____

_____ hours unskilled @ _____ = _____

Cleaning w/muriatic acid _____ square feet

_____ hours skilled @ _____ = _____

_____ hours unskilled @ _____ = _____

Masonry Dampproofing — 100 square feet of cement plaster on block, one coat work with two (2) coat brush coats asphalt or tar, figure 2 hours skilled and $2\frac{1}{2}$ hours unskilled.

Dampproofing _____ square feet

_____ hours skilled @ _____ = _____

_____ hours unskilled @ _____ = _____

Drain tile — 100 linear feet of digging and installing 6-inch perforated drain pipe, 10-foot lengths, and covered with No. 2 crushed stone, figure $14\frac{3}{4}$ hours unskilled labor.

Drain Tile _____ linear feet

_____ hours unskilled @ _____ = _____

CARPENTRY WORK

Girders — Built-up girders nailed together, set in place, leveled and support posts in place, for every 10 lineal feet, figure 1 hour skilled and $\frac{1}{2}$ hour unskilled.

Girder 3–2 x ___ x ___ lineal feet _____

_____ hours skilled @ _____ = _____

_____ hours unskilled @ _____ = _____

Basement stairs — Frame per stairway, average type straight flight to 4 feet wide and 12 feet long, rough cutting, framing, placing, figure 5 hours skilled and 2 hours unskilled.

Basement stairs 2 x 10s and 2 x 8s _____ hours skilled @ _____ = _____

_____ hours unskilled @ _____ = _____

Floor joists — first floor — 100 square feet including sill, plate, box sill, bridging and plywood subflooring, using 2 x 8 joists 16 inches o.c., up to 12-foot span, figure $5\frac{1}{2}$ hours skilled and $2\frac{1}{4}$ hours unskilled.

First Floor Platform with 2 x 8s–16 inch o.c. with $\frac{5}{8}$ x 4 x 8 plywood.

     _____ square feet   _____ hours skilled   @ _____ = _____

                                     _____ hours unskilled @ _____ = _____

First floor studding — exterior walls — For every 100 square feet, including normal openings, average type outside walls, frame or veneer construction with plates, headers, fillers, bracing, firestops, girts with 2 x 4s–16 inches o.c. from 8-foot to 12-foot heights, figure $2\frac{1}{2}$ hours skilled and $\frac{1}{2}$ hour unskilled labor. With 4′ x 8′ plywood sheathing to 25/32 inch thick figure 1 hour skilled and $\frac{1}{2}$ hour unskilled. For 100 square feet of exterior walls with 2 x 4 framing and $\frac{1}{2}''$ x 4′ x 8′ plywood sheathing, figure $3\frac{1}{2}$ hours skilled and 1 hour unskilled.

First floor exterior walls _____ square feet

                                     _____ hours skilled   @ _____ = _____

                                     _____ hours unskilled @ _____ = _____

First floor studding — interior walls — For every 100 square feet of wall including normal openings, plates, headers and studs 2 x 4s–16 inches o.c., figure 2 hours skilled and $\frac{1}{4}$ hour unskilled.

First floor interior walls _____ square feet

                                     _____ hours skilled   @ _____ = _____

                                     _____ hours unskilled @ _____ = _____

Second floor framing — For every 100 square feet including box sill, joists, bridging and $\frac{5}{8}$-inch plywood subfloor on 2 x 8s–16 inches o.c., figure $7\frac{1}{2}$ hours skilled and $2\frac{1}{2}$ hours unskilled.
Second floor platform 2 x 8s–16 inches o.c. with $\frac{5}{8}''$ x 4 x 8 plywood.

     _____ square feet   _____ hours skilled   @ _____ = _____

                                     _____ hours unskilled @ _____ = _____

Second floor exterior walls — For every 100 square feet 2 x 4 framing with $\frac{1}{2}$-inch plywood sheathing, figure $4\frac{1}{2}$ hours skilled and $1\frac{1}{2}$ hours unskilled.

     _____ square feet   _____ hours skilled   @ _____ = _____

                                     _____ hours unskilled @ _____ = _____

Second floor interior walls — For every 100 square feet 2 x 4 framing, figure $2\frac{1}{2}$ hours skilled and $\frac{1}{2}$ hour unskilled.

           _____ square feet   _____ hours skilled   @ _____ = _____

                                     _____ hours unskilled @ _____ = _____

Overlays or ceiling joists — For every 100 square feet, normal construction first and second floor levels, including bridging, trimmers to 16-foot spans using 2 x 6s–16 inches o.c., figure 5 hours skilled and 2 hours unskilled labor.

           _____ square feet   _____ hours skilled   @ _____ = _____

                                     _____ hours unskilled @ _____ = _____

Rafters — For every 100 square feet, average-type construction to 22 foot lengths, using 2 x 6s–16 inches o.c. and $\frac{1}{2}''$ x 4′ x 8′ plywood roof sheathing, figure $3\frac{3}{4}$ hours skilled and $1\frac{1}{2}$ hours unskilled.

           _____ square feet   _____ hours skilled   @ _____ = _____

                                     _____ hours unskilled @ _____ = _____

Gable studs — For every 100 square feet using 2 x 4s, $\frac{1}{2}''$ x 4′ x 8′ sheathing, figure $4\frac{1}{2}$ hours skilled and $1\frac{1}{4}$ hours unskilled labor.

           _____ square feet   _____ hours skilled   @ _____ = _____

                                     _____ hours unskilled @ _____ = _____

Roofing — Asphalt Shingles — For every 100 square feet of strip shingles 10″ x 36″ with 15-pound asphalt felt underlayment, figure 2 hours skilled and $1\frac{1}{4}$ hours unskilled.

Asphalt shingles _____ square feet

                                     _____ hours skilled   @ _____ = _____

                                     _____ hours unskilled @ _____ = _____

Wood shingle roof — For every 100 square feet using 16-inch wood shingle with 5-inch exposures, figure 4 hours skilled and $1\frac{1}{2}$ hours unskilled.

Wood shingles _____ square feet

                                     _____ hours skilled   @ _____ = _____

                                     _____ hours unskilled @ _____ = _____

Slate roofing — For every 100 square feet using 10″ x 20″, figure $3\frac{1}{2}$ hours skilled and 2 hours unskilled.

_____ square feet   _____ hours skilled   @ _____   = _____
_____ hours unskilled   @ _____   = _____

CARPENTRY FINISH WORK — EXTERIOR

Cornice — For every 100 linear feet, including fascia, crown mold, soffit, bed mold, and drip edge, figure 12 hours skilled and 3 hours unskilled labor.

_____ linear feet   _____ hours skilled   @ _____   = _____
_____ hours unskilled   @ _____   = _____

Wood siding — For every 100 square feet of $\frac{1}{2}$″ x 8″ bevel siding with kraft building paper applied underneath, figure $2\frac{3}{4}$ hours skilled and 1 hour unskilled.

_____ square feet   _____ hours skilled   @ _____   = _____
_____ hours unskilled   @ _____   = _____

Corner Boards — For every 100 linear feet, figure 3 hours skilled and 2 hours unskilled.

_____ linear feet   _____ hours skilled   @ _____   = _____
_____ hours unskilled   @ _____   = _____

Front overhang soffit — For every 100 linear feet exterior plywood, 1 x 8 pine fascia, wood drip cap 24 — 30 inches wide, figure 6 hours skilled and 1 hour unskilled labor.

_____ linear feet   _____ hours skilled   @ _____   = _____
_____ hours unskilled   @ _____   = _____

Louvers or vents — Each unit, figure $\frac{1}{2}$ hour skilled labor for each.
Louvers _____ each   _____ hours skilled   @ _____   = _____

Porch railings — Wrought iron set into masonry steps and porches and attached to the framing, figure 1 hour skilled and $\frac{1}{4}$ hour unskilled for each set.
Porch railing _____ sets   _____ hours skilled   @ _____   = _____
_____ hours unskilled   @ _____   = _____

CARPENTRY — INTERIOR WORK

Insulation — ceilings — For every 100 square feet of nonrigid type 15″ x 48″ batts, figure $\frac{1}{2}$ hour skilled and $\frac{1}{2}$ hour unskilled labor.
Ceiling insulation _____ square feet

_____ hours skilled   @ _____   = _____
_____ hours unskilled   @ _____   = _____

Insulation — sidewall — For every 100 square feet of nonrigid-type blanket 16-inch to 8-foot lengths, figure $\frac{3}{4}$ hour skilled and $\frac{1}{2}$ hour unskilled.
Sidewall insulation _____ square feet

_____ hours skilled   @ _____   = _____
_____ hours unskilled   @ _____   = _____

Wallboard — ceilings — Gypsum board ceilings, for every 100 square feet $\frac{3}{8}$″ x 4′-12′ lengths with joints finished ready for paint, figure $2\frac{1}{2}$ hours skilled and $1\frac{1}{2}$ hours unskilled.
Ceilings _____ square feet

_____ hours skilled   @ _____   = _____
_____ hours unskilled   @ _____   = _____

Wallboard — sidewalls — For gypsum board walls, for every 100 square feet $\frac{3}{8}$″ x 4′-10′ lengths with joints finished ready for paint, figure 2 hours skilled and 1 hour unskilled.
Sidewalls _____ square feet

_____ hours skilled   @ _____   = _____
_____ hours unskilled   @ _____   = _____

Cove molding $1\frac{3}{4}$-inch — For every 100 linear feet of ceiling cove mold, figure 3 hours skilled and 1 hour unskilled.

Cove mold _____ linear feet

_____ hours skilled   @ _____ = _____
_____ hours unskilled @ _____ = _____

Baseboard and carpet strip — For every 100 linear feet of baseboard and carpet strip, figure 4 hours skilled and 1 hour unskilled labor.

Base and carpet strip _____ linear feet

_____ hours skilled   @ _____ = _____
_____ hours unskilled @ _____ = _____

Underlayment — For every 100 square feet of $\frac{5}{8}''$ x 4' x 8' plywood, figure 1 hour skilled labor.

Underlayment _____ square feet

_____ hours skilled   @ _____ = _____

Resilient-type flooring — For every 100 square feet resilient flooring, figure 3 hours skilled and $\frac{1}{2}$ hour unskilled labor.

Resilient flooring _____ square feet

_____ hours skilled   @ _____ = _____
_____ hours unskilled @ _____ = _____

Slate flooring — For every 100 square feet of random cut sizes to 12" x 12" including grouting or pointing of joints, figure 2 hours skilled and 2 hours unskilled.

Slate flooring _____ square feet

_____ hours skilled   @ _____ = _____
_____ hours unskilled @ _____ = _____

Ceramic floor tile — For every 100 square feet $\frac{1}{2}$-inch to 2 inches with paper backing, figure 4 hours skilled and 4 hours unskilled labor.

Ceramic floor tile _____ square feet

_____ hours skilled   @ _____ = _____
_____ hours unskilled @ _____ = _____

Ceramic wall tile — For every 100 square feet $4\frac{1}{4}''$ x $4\frac{1}{4}''$ including installing waterproof gypsum wallboard backing, adhesive, grouting and installing ceramic fixtures, figure 8 hours skilled and 8 hours unskilled labor.

Ceramic wall tile _____ square feet

_____ hours skilled   @ _____ = _____
_____ hours unskilled @ _____ = _____

Hardwood flooring — For every 100 square feet 25/32" x $2\frac{1}{4}''$ face, including the placing of the deadening felt, figure $3\frac{1}{2}$ hours skilled and $\frac{1}{2}$ hour unskilled.

Hardwood flooring _____ square feet

_____ hours skilled   @ _____ = _____
_____ hours unskilled @ _____ = _____

CARPENTRY — MILLWORK AND TRIM

Exterior doors — Setting exterior door frames, wood residential type, setting wood threshold, hanging standard size door $1\frac{3}{4}$-inch with 3 butts, fancy outside lockset, and door trim, figure 5 hours skilled and $\frac{3}{4}$ hour unskilled.

Exterior doors _____ each

_____ hours skilled   @ _____ = _____
_____ hours unskilled @ _____ = _____

Exterior combination storm/screen wood door complete with hardware and closer, standard size, figure 2 hours skilled and $\frac{1}{4}$ hour unskilled.

Combination doors _____ each

_____ hours skilled   @ _____ = _____
_____ hours unskilled @ _____ = _____

Interior doors — Interior door jambs and heads, assembly from stock sections, setting interior door frames, standard sizes, hanging interior wood doors $1\frac{3}{8}''$, door trim both sides, installing hardware, figure $5\frac{1}{4}$ hours skilled and 1 hour unskilled labor.

Interior doors _____ each

| _____ hours skilled | @ _____ | = _____ |
| _____ hours unskilled | @ _____ | = _____ |

Closet shelving — For every 100 linear feet of shelving including 1 x 3 and 1 x 12 stock sizes and installing the adjustable closet rods, figure 3 hours skilled and 1 hour unskilled.

Closet shelving _____ linear feet

| _____ hours skilled | @ _____ | = _____ |
| _____ hours unskilled | @ _____ | = _____ |

Window units — Each unit, setting complete window units, all types glazed to prepared opening including casing, stool, stop, apron, and hardware for windows to 3' x 3' to 3' x 5'6'', figure $2\frac{1}{4}$ hours skilled and $\frac{1}{2}$ hour unskilled labor.

Window units _____ each

| _____ hours skilled | @ _____ | = _____ |
| _____ hours unskilled | @ _____ | = _____ |

Storm/sash and screen (prefitted wood combination), for each unit, figure $\frac{1}{2}$ hour skilled and $\frac{1}{2}$ hour unskilled.

Storm and screen _____ each

| _____ hours skilled | @ _____ | = _____ |
| _____ hours unskilled | @ _____ | = _____ |

Fireplace mantel — Each unit, setting average type factory-built mantel unit to prepared wall, figure $2\frac{1}{2}$ hours skilled and $1\frac{1}{2}$ hours unskilled labor.

Mantel unit _____ each

| _____ hours skilled | @ _____ | = _____ |
| _____ hours unskilled | @ _____ | = _____ |

Blinds or shutters — Each pair, wood, small, figure $\frac{1}{2}$ hour skilled labor. Medium, figure $\frac{3}{4}$ hour skilled labor. Large, figure 1 hour skilled labor.

Blinds	Small _____ pair	_____ hours skilled	@ _____	= _____
	Medium _____ pair	_____ hours skilled	@ _____	= _____
	Large _____ pair	_____ hours skilled	@ _____	= _____

Main stairs — Each flight, setting housed stringers, installing risers and treads, setting newel post, balusters, and handrail, figure 8 hours skilled and 3 hours unskilled.

Main stair, flight _____ each

| _____ hours skilled | @ _____ | = _____ |
| _____ hours unskilled | @ _____ | = _____ |

Kitchen cabinets — For each 100 square feet of face area, average-type work base cabinets, figure 4 hours skilled and 2 hours unskilled labor. For each 100 square feet of face area for wall cabinets, figure 3 hours skilled and 2 hours unskilled.

Kitchen cabinets (factory built)

Base units _____ square feet	_____ hours skilled	@ _____	= _____
	_____ hours unskilled	@ _____	= _____
Wall units _____ square feet	_____ hours skilled	@ _____	= _____
	_____ hours unskilled	@ _____	= _____

Counter tops — For every 10-square-foot area, placing factory-built top in place, figure $\frac{1}{2}$ hour skilled and $\frac{1}{4}$ hour unskilled labor.

Counter top _____ square feet

| _____ hours skilled | @ _____ | = _____ |
| _____ hours unskilled | @ _____ | = _____ |

Bathroom vanities — For each unit, factory built, placing, scribing and anchoring in place with laminated plastic top, figure 2 hours skilled labor.

Bathroom vanities _____ each

_____ hours skilled   @ _____ = _____

Medicine cabinets or mirrors over vanities – For each unit, figure $\frac{3}{4}$ hour skilled and $\frac{1}{2}$ hour unskilled.
   Mirrors over vanity _____ each

_____ hours skilled   @ _____ = _____
_____ hours unskilled   @ _____ = _____

PAINTING: EXTERIOR WORK

Exterior walls – For every 100 square feet with oil, stain, or water-based paint, doing 1 coat primer and 1 coat finish on wood siding, figure 2 hours skilled labor.
   Wood siding _____ square feet

_____ hours skilled   @ _____ = _____

Exterior cornice – For every 25 linear feet of standard box cornice, figure 1 hour skilled.
   Wood cornice _____ linear feet

_____ hours skilled   @ _____ = _____

Exterior window trim – For each window medium size 2 coats, figure $\frac{3}{4}$ hour skilled labor for each unit.
   Window trim _____ windows

_____ hours skilled   @ _____ = _____

Combination storm and screen windows, figure $\frac{3}{4}$ hour skilled for each.
   Combination storm sash _____ each

_____ hours skilled   @ _____ = _____

Exterior doors – For each exterior door finished two sides with 2 coat work, figure 1 hour skilled labor.
   Exterior doors _____ each

_____ hours skilled   @ _____ = _____

Combination storm doors – For each door finished 2 sides with 2 coat work, figure $1\frac{1}{2}$ hours skilled.
   Combination storm doors _____ each

_____ hours skilled   @ _____ = _____

Shutters or blinds – For each pair with 2 coat work, figure 1 hour skilled labor.
   blinds _____ pairs

_____ hours skilled   @ _____ = _____

PAINTING: INTERIOR WORK

Interior ceilings – For every 100 square feet of primer, flat, casein, or water-based paint, two-coat work, figure 2 hours skilled labor.
   Interior ceilings _____ square feet

_____ hours skilled   @ _____ = _____

Interior walls – For every 100 square feet of primer, flat, casein, or water-based paints for two-coat work, figure $1\frac{1}{2}$ hours skilled labor.
   Interior walls _____ square feet

_____ hours skilled   @ _____ = _____

Interior window trim – For each window, figure $\frac{1}{2}$ hour skilled labor.
   Window trim _____ each

_____ hours skilled   @ _____ = _____

Interior doors – For each door, casing, jambs, stops for two-coat work on two sides, figure $1\frac{1}{2}$ hours for each door.
   Interior doors _____ each

_____ hours skilled   @ _____ = _____

Baseboard and carpet strip – For every 25 linear feet for two coat work, figure 1 hour skilled.
   Base and carpet strip _____ linear feet

_____ hours skilled   @ _____ = _____

Hardwood floors — For every 100 square feet of hardwood for filling, wiping and two coats of varnish, figure $1\frac{3}{4}$ hours skilled labor.

Hardwood floors _____ square feet

_____ hours skilled   @ _____ = _____

Stairs — For each complete open type stairs for each flight, figure 4 hours skilled for either stain, varnish, or paint.

Main stairs _____ flights

_____ hours skilled   @ _____ = _____

Fireplace mantel — For each unit, figure $\frac{1}{2}$ hour skilled labor for finishing with either stain or paint with two-coat work.

Fireplace mantel _____ each

_____ hours skilled   @ _____ = _____

29
DESIGNING A HOUSE

In the design of a house, there are certain factors which affect the size and layout of the individual rooms and how they are grouped together. Almost any floor plan can be adapted to meet the desires of the owner in regards to the exterior appearance, from modern architecture to the classic colonial, by the choice of windows, material used for finished siding, type of cornice, style of roof, and choice of exterior doors.

Most houses have the same general requirements in relation to bedrooms, kitchens, baths, living rooms, dining rooms, family rooms, etc. It depends upon how these rooms are grouped together to make an efficient layout as well as an attractive house in appearance from the outside.

In designing a house, you should always be reminded of and consider *modular* planning. The *module* is a standard unit of measure and is used throughout the residential construction industry. Modular planning saves material and reduces costs, because construction materials are manufactured on a 4-foot standard modular unit.

Buildings are designed using 16 and 24 inches as minor modules and 48 inches as the major module. For example, plywood is manufactured in 4-foot widths and 8-foot lengths. The 4-foot width is three 16-inch modules or two 24-inch modules. This same principle applies to gypsum wallboard, insulated wall sheathing, underlayment, etc. Also, insulation is manufactured in widths to fit between either the 16-inch module or the 24-inch module.

Lumber is sold in lengths using two-foot increments such as 8'0", 10'0", 12'0", 14'0", 16'0", etc. This should also be considered in planning the sizes of the rooms.

The first thing you should do in designing your house is to make a word picture of your house by making a list of all the things that you want in the house such as the number of bedrooms, number of bathrooms, laundry, fireplace, family room, etc. This will give you a picture in your mind of what your house will be like when finished. You should try to keep in mind modular planning and determine the approximate room size in making your word picture.

After you have listed all the things that you desire in your house, you should then consider the grouping or layout. All residences have the same basic units. You have the sleeping and bathing area, the food preparation and dining areas, living and activities areas, and recreation and laundry areas. How this is grouped together will determine the floor plans of your house.

Things you should consider in designing your home is the flow of traffic between the kitchen and dining room, the bedrooms and bath, and the living room and the dining room. The dining room should be next to the kitchen and should be easily accessible. The kitchen should have an outside door. Each bedroom should have easy and quick access to the bathroom. Each bedroom should have an access to a hall. There should be ample closet space in each bedroom and a linen closet from the bath or the hall.

An entrance closet is desirable for coats when coming in from the out of doors. All rooms should have adequate windows and cross ventilation where possible. Also you should try to avoid having to cross through one room to get into another. This is not always possible, but is desirable.

We will start our design of our sample house by first making a word picture of what we would like in our house. Assume we are building in a community with houses that were built fifty to sixty years ago. We want modern appliances and today's conveniences built into our house, but we still want our house to blend with our neighbors and the surrounding community.

The following is a word picture of our house. A two story-colonial design with three bedrooms and a bath on the second floor. A living room at least 12'0" x 20'0" with a fireplace. A separate dining room, a recreation room in the basement, and a laundry. We would like a half bath on the first floor located near the kitchen, a porch off the living room, a full basement under the house, double hung windows, an eating space in the kitchen with a good kitchen cabinet layout. We would like hot air heat and air conditioning incorporated in the system.

This gives us our word picture of our house. Our first step in making our plans is to make a rough sketch of the first floor plan. Our areas are fairly well set. We have the area where the food is prepared and eaten. We have the area for reading and relaxation, and we have the area for sleeping and bathing. Our house is a two-story house and the bedrooms and bath will be on the second floor.

We wanted a living room approximately 12 x 20. We will need a front entrance and a rear entrance. We wanted a separate dining room and kitchen. The living room should have easy access to the dining room and the kitchen should be next to the dining room and also have an outside entrance of its own.

We have to be able to go from the first floor area to the second floor and from the first floor down to the basement. If possible we should plan the stairs to go one over the other. We wanted a fireplace in the living room and a porch off the living room.

We will make a rough sketch on $\frac{1}{4}$-inch graph paper. Our plans will be drawn at $\frac{1}{4}$-inch-to-the-foot scale, so using the $\frac{1}{4}$-inch graph paper with the $\frac{1}{4}$-inch spaced lines, we can make a fairly close rough sketch of the actual floor plan.

In making our rough sketch, we draw the living room 12 x 20. We need an entrance hall and a dining room. The dining room should be approximately 12 feet wide and our entrance hall should be a minimum of 5 feet wide. Adding our 12 foot width for the living room and the 12 foot width for our dining room gives us a total of 24 feet. We said our entrance should be at least 5 feet wide, so add this to our 24 feet and we get a total of 29 feet. We should plan on modular units of construction, so we will add one foot and figure the plan to be 30 feet across the front.

We have already decided that the living room would be 20 feet long. We need stairs to go to the second floor area, and they should be at least three feet wide. Our dining room should be fairly square. This leaves us about 12 feet in which to work an adequate kitchen and half bath.

We add our living room and stairs and also the dining room and kitchen and we get a minimum of 24 feet. This falls into our modular units, so our floor plan will be 30 feet across the front and 24 feet deep.

We wanted a porch off the living room and our living room is 20 feet long. If we step in one foot from each end, this will give us an 18 foot porch. We can make it as wide as we wish, but 8 to 10 feet is adequate, so we will make ours 8 feet wide.

We wanted a fireplace in our living room and not to take space away from our porch area, we can place it on the inside wall. We should also consider furniture layout while making our sketch.

We wanted our windows placed so we could have cross ventilation where possible. We need a door from our living room to our porch, a door at the front entrance, and a door that comes into the kitchen. We need storage closets as well as clothes closets near our entrance and our work areas. We can work these in by the basement stairs, the front entrance, and the kitchen.

Our first floor plan on our graph paper would be as shown in Figure 29-1.

After we have made a rough sketch of the first floor plan, we then make a rough sketch of our second floor. We have determined that the first floor plan is 30 feet wide and 24 feet deep. Our second floor must go directly over our first floor. On our graph paper, we will outline the second floor with the same perimeter as the first.

We have located the stairs from the first floor to the second floor. We have to place the stairs in the same location as the first floor, so we will know where they stop at the second floor level.

The fireplace is located in the living room on the first floor, so our chimney must be directly over the fireplace where it goes up through the second floor to the roof.

We decided in our word picture that we wanted three bedrooms and a bath on the second floor. The stairs should land in a hallway and all the bedrooms should have easy access to the bathroom.

We should have a master bedroom so we would rough in a bedroom over the living room for the master bedroom. We locate a bedroom over the dining room and one over the kitchen. We next locate the bathroom where it is easily accessible

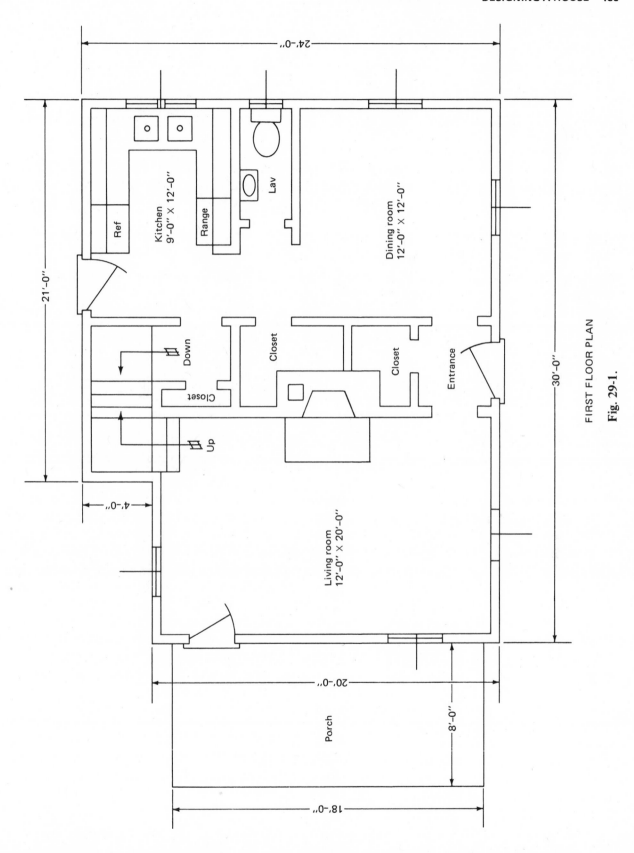

FIRST FLOOR PLAN

Fig. 29-1.

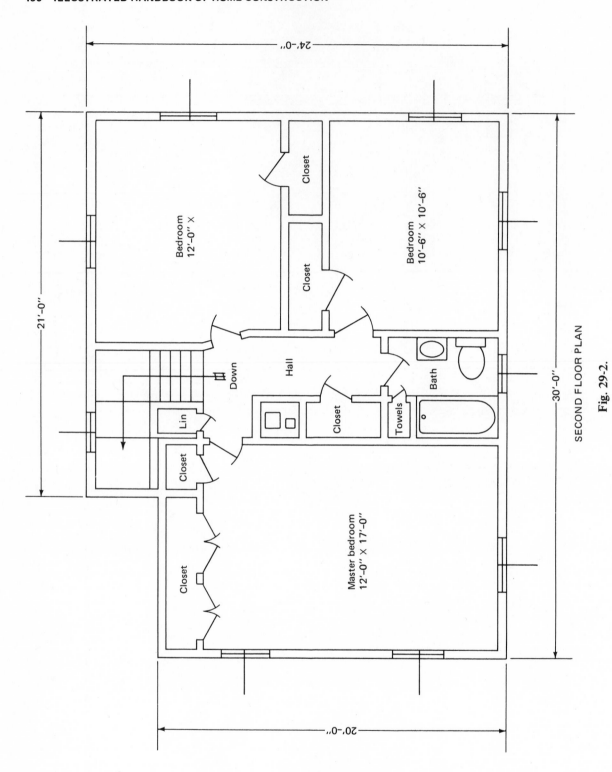

Fig. 29-2.

SECOND FLOOR PLAN

from all the bedrooms. We need a linen closet in the central hall and a closet for towels in the bath. We need ample closets for each bedroom and a large closet in the master bedroom.

In placing doors and windows, we keep in mind the placement of furniture: a double bed will need approximately six feet of wall space and a chest of drawers and dresser will require from three to six feet of space depending upon its size. We should locate our windows as close as possible over the ones on the first floor to give a balanced appearance from the outside. Notice that we made our walls, doors, and windows approximately the same size as we will draw them on the plan. You should figure six-inch width walls for all exterior and interior walls drawn at $\frac{1}{4}" = 1'0"$ scale. Figure 29-2 shows how our second floor plan will look on our graph paper.

30
DRAWING THE FIRST FLOOR PLAN

We have completed our rough sketch for the layout of the plan we are going to draw. The first step is to make our first floor plan of our working drawings. Place a 12″ x 18″ piece of tracing paper on your drawing board, the long way horizontal.

Locate your drawing near the center of your paper, making sure that you will have enough room for all dimensions and notes necessary for the completed drawings.

Referring to our rough sketch, draw the outside perimeter of the house according to the measurements at $\frac{1}{4}$″ = 1′0″ scale. This means that the front horizontal line will be 30′0″ long and 24′0″ deep. Across the back it will be 21′0″ and then turn

toward the front a distance of 4′0″ and then horizontal and parallel with the front wall a distance of 9′0″. Now draw the left side of the house from the starting point to the line you just drew. This will completely outline the perimeter of the house according to the rough sketch and at the dimensions given, drawn at $\frac{1}{4}$″ = 1′0″ scale. Your drawing should now be like Figure 30-1.

Next, draw the inside line of the outside wall after determining its thickness from the material used. In our house, we are using frame construction, so the wall should be drawn 6 inches wide. This is made up of the outside wall finish, the sheathing, the 2 x 4 studs, and the finished interior walls

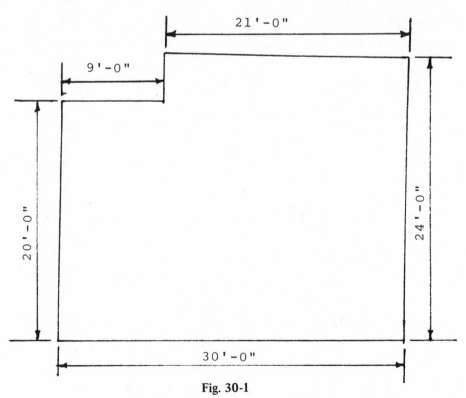

Fig. 30-1

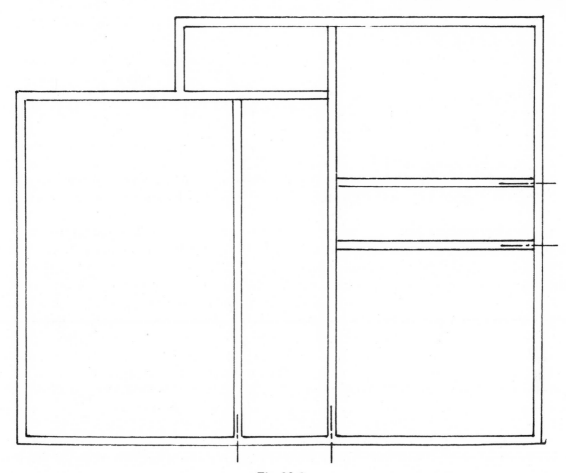

Fig. 30-2

of gypsum wallboard. In residential construction, all exterior and interior walls should be drawn six inches wide at $\frac{1}{4}'' = 1'0''$ scale.

Now locate by centerlines all the interior partitions. These should be drawn in lightly, and should be drawn in full regardless of the position of doors, etc. Your drawing should now be similar to Figure 30-2.

In a small house, the most economical place for the heating unit is near the center of the building, so the chimney should be located near the center of the house. We want a fireplace in the living room and we should center it on a wall if possible.

Place a light centerline on the center of the living-room inside partition. On plans, a fireplace may be shown with the fire box made of firebrick or with a metal fireplace form. Figure 30-3 shows how each would look. We will draw our fireplace 2'6'' thick and 5'6'' long. The opening of a standard fireplace firebox is 3'0'' in front and 2'0'' across the back and 1'6'' deep. This outlines the firebox. Next draw the ash dump in the center of the firebox next to the back wall. This should be 9 inches long and 6 inches wide.

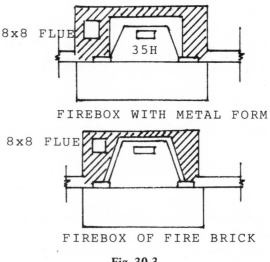

FIREBOX WITH METAL FORM

FIREBOX OF FIRE BRICK

Fig. 30-3

It is necessary to have a separate flue for the heating unit in the basement. Draw an 8 x 8-inch square to the left of the firebox. This represents the flue liner for the furnace. You should have at least 4'' inches of masonry between the firebox and the flue liner.

Our next step after outlining the fireplace is to locate the center of our windows and doors on the outside walls. We will use 3'0" windows, except the kitchen and the half bath. The kitchen window will be a 2'8" inch mullion (meaning two windows side by side each sharing the same 2-inch center mullion) and the half bath window we will make 2'4" wide.

The front door will be 3'0" wide; for the rear door from the kitchen and the side door from the living room to the porch we will use a 2'8"-wide door. We should now outline the side porch, making it 8'0" wide and 18'0" long. We should also outline the front and rear entrance porches. Our front entrance we will make 8'0" wide and 5'0" deep, and the kitchen entrance porch we will make 4'0" wide and 3'0" deep. Locate all windows and doors on the exterior walls with center lines.

Next measure and mark the width of all the window and door openings in the outside walls. Erase the lines between the openings for all the outside doors, then draw the line showing the door sill. The window and exterior door openings should now be similar to the ones shown in Figure 30-4.

We now locate our interior doors and erase the lines at the door openings so that our plan now shows the openings in the partitions where the doors are to be located. The doors from the front entrance going into the dining room and living room we will make 3'0" wide. The door between the kitchen and the dining room 2'6". The door from the kitchen to the basement stairs 2'8" wide, and all other doors to closets and the half bath we will make 2'0" wide.

Our plan should now be similar to that shown in Figure 30-5.

After locating the doorways and drawing the outline of the openings, we now show the way the various doors are to be hinged. The doors should be hinged so that they swing against a wall, so that when the door is open the room or closet is completely accessible and the person opening the door will not be pinned or pushed against a wall. Draw the placement of these doors showing which way they swing. The door itself is drawn from the jamb on a 30 degree angle to the swing line.

We now draw the placement and location of the kitchen sink and cabinets. Wall cabinets are drawn 12-inches deep and the base cabinets are drawn 24-inches deep. The sink is usually placed under the window. We are using a standard 21 x 32 double-bowl sink. We will make a detail of the kitchen cabinets on our detail section. Layout the range (which is 30 inches wide) and the refrigerator (33 inches wide) along the cabinet wall. You should plan your cabinet layout so that you do not have to take unnecessary steps while working in the kitchen. You take the food from the refrigerator, wash it at the sink, and cook on the range.

Now place your plumbing fixtures in the half bath next to the kitchen. Place the water closet (toilet) under the window facing the door. A small wall hung lavatory is placed on the kitchen wall. Try to layout your plumbing pipes in your mind when placing the plumbing fixtures so you save on piping and cutting of floor joists. All plumbing lines run into the basement area and are consolidated there into one system and run from there into a municipal sewer system or a septic system designed for the house.

Our next step is to finish our stair layout. We want to go from one area to the other with the least amount of effort. This works out in our plan to 15 risers and 14 treads. Our basement stairs will go directly underneath the main stairs. In drawing, make the landings 3'0" square and the treads 9-inches wide. Now draw your stairs with two landings and the treads. Our plan should now be similar to that shown in Figure 30-6.

We now place the location of our electrical wiring symbols. The meter is usually placed outside and toward the rear of the house, as in most cities electrical lines run from the rear of the lots. The meter is placed outside so that the man from the electrical company can read the meter without having to go into your home.

The lighting panel is shown by a black rectangle and is located in the stairway going down into the basement. In this panel are located the circuit breakers for the different circuits and the main cutoff.

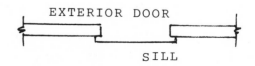

Fig. 30-4

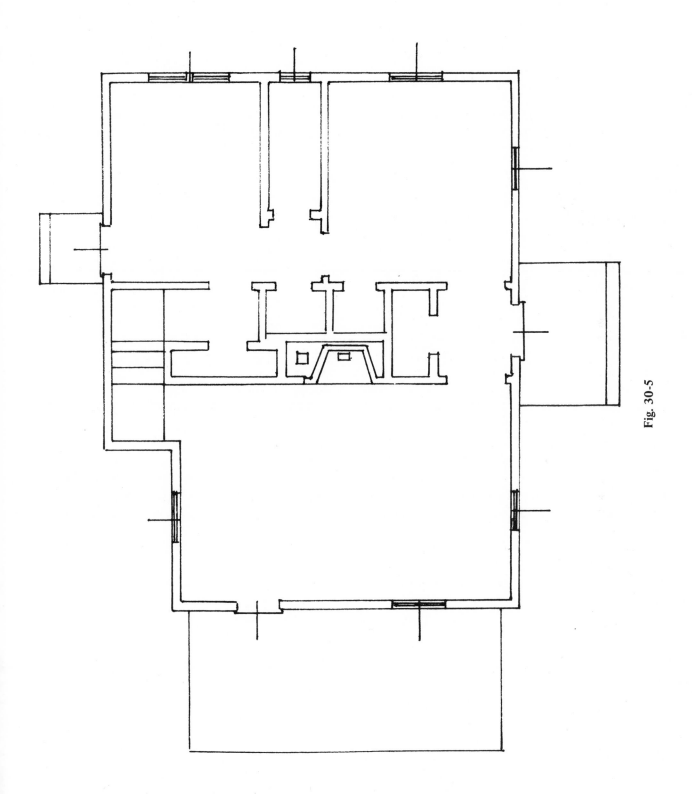

Fig. 30-5

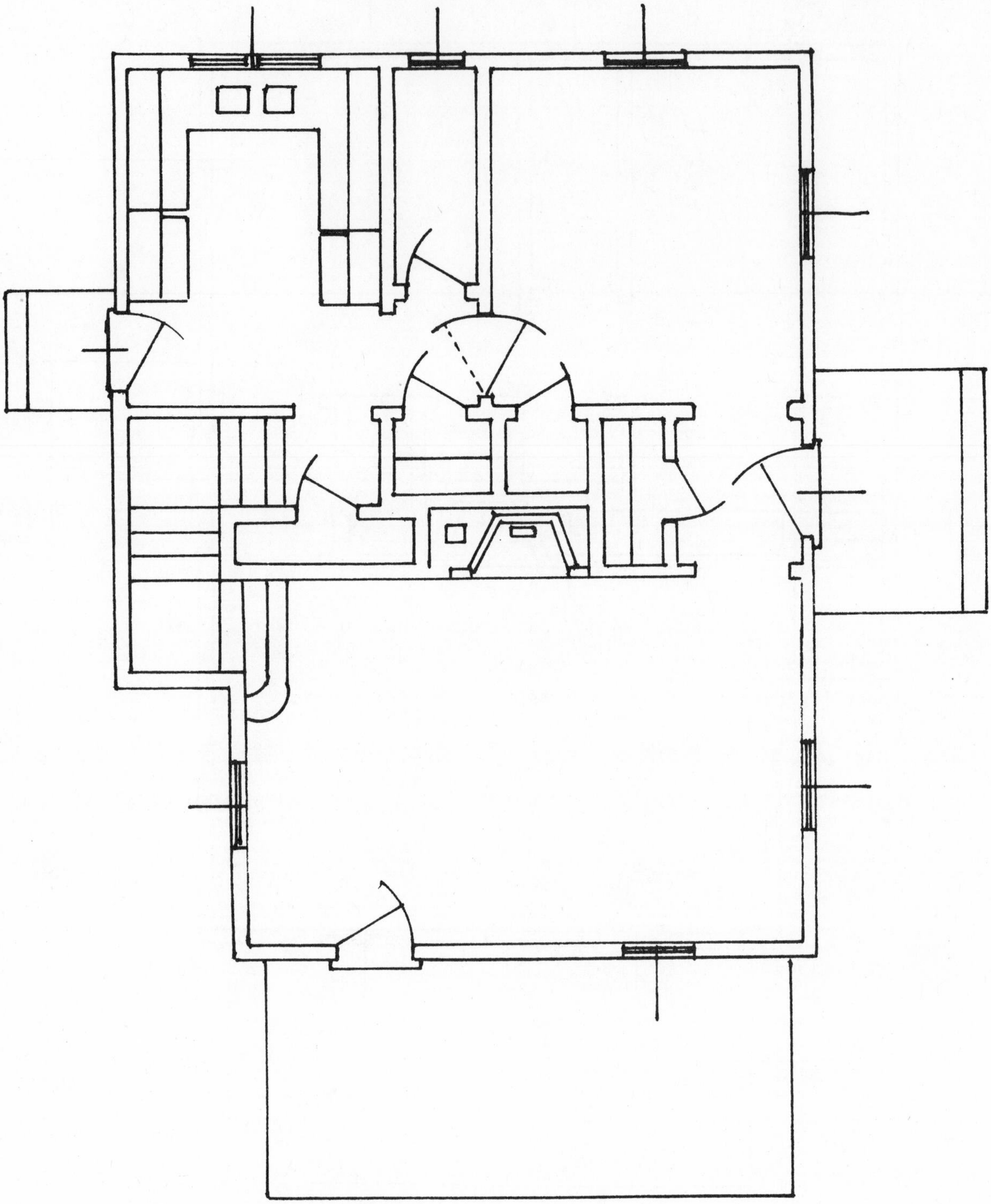

Fig. 30-6

In the living room, we should have a wall switch so that we can turn on a light when entering the room. This is usually done by having a wall plug work from a switch. Also, the ceiling light on the porch should work from a switch near the side door, and there should be a three-way switch from the foot of the stairs to the second floor hallway. Switches should never be placed behind a door so that the person entering a room must close the door first and then try to locate the switch.

The rest of the outlets in the living room should be located where they will be convenient for the placing of furniture and lamps. We will need a ceiling light in the entrance and a light over the front door. These are indicated and the wire from the light to the switch in drawn. A ceiling light from a wall switch in the dining room and also wall outlets along the walls. You should place one convenience outlet for every six feet of blank wall space you have.

The half bath has a ceiling light from a wall switch and a wall outlet near the lavatory and medicine cabinet. We will need a three-way switch from the basement stairway into the basement.

In the kitchen, we will have a ceiling light, a clock outlet, a buzzer and a bell. The range, if electric, will need a special 220-volt outlet. We will need an outlet for the refrigerator, the exhaust hood, and fan, and we should have convenience outlets in the wall between the countertops and the wall cabinets for small appliances such as mixers, toaster, coffee makers, etc.

Draw in your electrical outlets, using the correct symbols for each. You may choose to use a different plan from the one shown. Most closets have a drop cord pull chain fixture in the center of the ceiling. Instead, the closet light can be operated by a switch fastened to the door jamb that will turn it on when the door is opened and shut it off when the door is closed.

In accordance with long established custom, the location above the floor of standard electrical outlets is as follows:

- Wall switches are placed 4'0" off the floor.
- Switches between countertop and wall cabinets are placed 3'7" off the floor.
- Wall convenience outlets are placed 1'3" off the floor.
- Outlet for flatiron in laundry area is laced 4'0" off the floor.
- Outlet alongside a medicine cabinet is placed 5'2" off the floor.

After you have located all the electrical ou and placed them on your drawing, then make a key for the electrical wiring symbols. This can be placed anyplace on the drawings but is usually located where it can be easily found and yet not interfere with the plans or the reading of the plans. An example of a electrical key is shown in Figure 30-7.

Now that you have drawn all the electrical symbols and have them in the correct place, your drawing should be similar to that shown in Figure 30-8.

All sizes or space stipulations on plans are generally indicated by a system of lines, arrows, and figures, which are known as dimensions. In *all* cases dimensions are given in actual or *full* sizes or distances, regardless of the fact that the plans are actually drawn to a smaller scale. In addition to the overall dimensions, the exact location of windows, doors, and walls must be shown. Also the total of the dimensions used to locate or show sizes of items within the enclosure must equal the overall dimensions.

All partition dimensions are given from the face of the exterior sheathing to the centerline of the partition. This is done because variation of materials would make it impractical to dimension to the outside of partitions and the face of the finished exterior walls.

Now draw all the dimensions on your drawing. Make sure your dimensions are correct. After you have completed the dimensions, *print* in the names of the rooms, any notes such as hearth, ash dump, stair indicator, window sizes, door sizes, etc.

Notice that for all *window and door* sizes, the width is always *given first* followed by the height.

KEY TO ELECTRICAL WIRING SYMBOLS

SYMBOL	DESCRIPTION	NO. REQUIRED
	Lighting Panel	1
	Push Button	2
	Bell	1
	Buzzer	1
	Range Outlet	1
T	Transformer	1
	Duplex Convenience outlet	20
	Ceiling outlet	16
S	Single Switch	13
S_3	Three way switch	4
	Special purpose outlet	1
WP	Weatherproof outlet	1

Fig. 30-7.

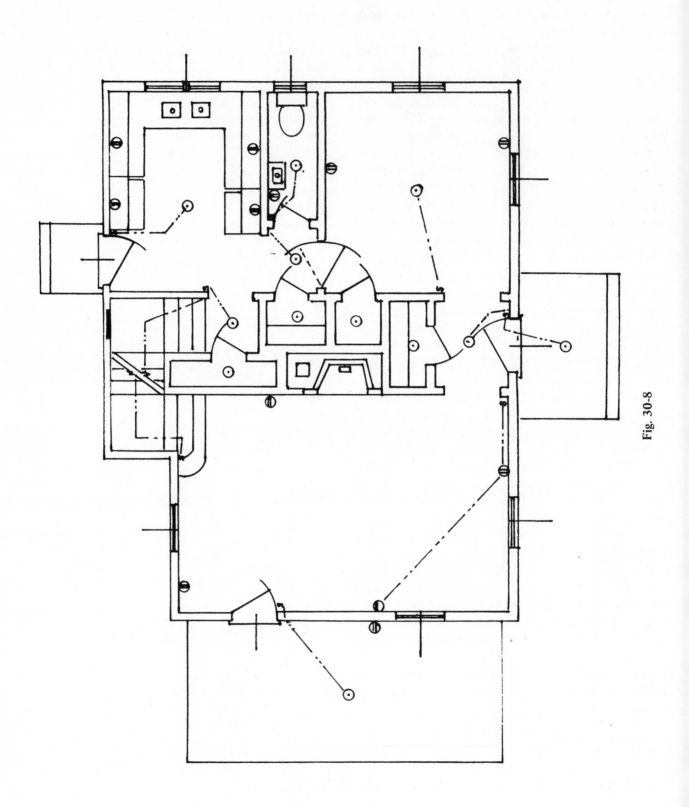

Fig. 30-8

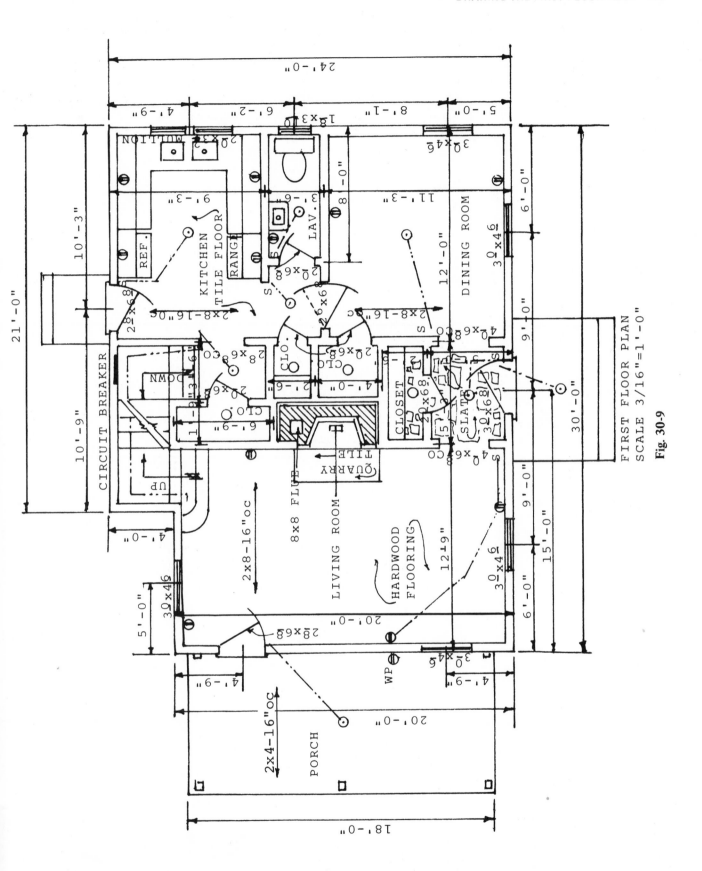

FIRST FLOOR PLAN
SCALE 3/16"=1'-0"

Fig. 30-9

Now note the size and direction of the ceiling joists. These are the framing members that will be used to support what goes above this floor plan. Notice that on the porch, they are 2 x 4s because nothing rests on top of them. They are used to support just the porch ceiling. The rest of the house shows 2 x 8s spaced 16 inches on center to support the second floor and its partitions.

You have now completed the first floor plan. Check to make sure that you have given all the dimensions, the various sizes of windows and doors. You have labeled each room and the equipment. Do not forget to indicate your masonry fireplace with the correct symbols. Your finished drawing of the first floor should be similar to that shown in Figure 30-9.

31
DRAWING THE SECOND FLOOR PLAN

The second floor plan must go directly over the top of the first floor. Lay out the outside walls the same as you did for the first floor. Next, draw the width of the exterior walls making, them six inches wide at $\frac{1''}{4} = 1'0''$ scale. Your drawing should now be similar to Figure 31-1.

Now locate the stairs coming from the first floor to the second floor and draw them. These must be directly over the ones drawn on the first floor plan. You will have to extend them into the second floor plan the distance necessary to show you have the correct number of risers and treads. This will locate where the stairs stop at the second floor level. When determining the stair layout, it helps to make a rough sketch which you will again use when you make your stair detail.

In house construction, the standard stair width is approximately three feet. You figure the approximate floor-to-floor height. In most residential construction this is almost standard. Most partition

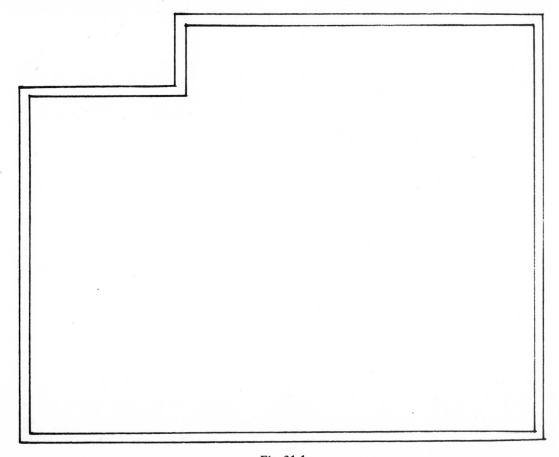

Fig. 31-1.

heights from the subfloor to the bottom of the ceiling joists or second floor joists are 8'1". The reason for this is that standard wallboard or paneling is manufactured in 8'0" lengths. The extra inch in height is to allow the workmen to use full sheets without having to scribe and fit them in place.

You know the size of the joist required by the span they have to cross. Most subflooring is either $\frac{1}{2}$"- or $\frac{5}{8}$"-inch plywood. You take the standard 8'1" and add the thickness of the floor joists, which in residential construction is mostly 2 x 8s or 2 x 10s, and then add the thickness of the subfloor. For example, a house using 2 x 8s would have a floor-to-floor height of 8'1" plus $7\frac{5}{8}$ inches (2 x 8) plus $\frac{5}{8}$ inch (plywood subfloor), which gives us a total of $105\frac{1}{4}$ inches. You should figure the riser (one step up to the other) between 7 and 8 inches. We divide 7 into $105\frac{1}{4}$ and we get 15+. This means we will need 15 risers and 14 treads. The width of the tread is from the face of one riser to the face of the other. This is referred to as the *run*. In house construction the run is between 8 and 10 inches. The actual tread width is $1\frac{1}{4}$ inch wider because each tread has $1\frac{1}{4}$-inch overhang beyond the face of the riser. If we figure a 9-inch run, this will give us a $10\frac{1}{4}$-inch tread, which will make a comfortable stair to walk up or down.

Our house requires that we use a platform to change directions in our stairs in going from one floor to the next. This is much better than installing winders and does not use any more room.

We must now locate the chimney over the fireplace on the first floor. The 8 x 8 flue liner should be directly over the one on the first floor plan. You can shift this one way or another if necessary to bring it alongside a partition. The fireplace must have a separate flue for the firebox. The size of this flue depends upon the size of the fireplace opening and the size of the damper as well as the total height of the chimney. If a prefabricated metal fireplace unit is used, the manufacturer will specify the flue liner size for the chimney height. We will use a 12 x 12 flue liner for the fireplace. You should separate the 8 x 8 and the 12 x 12 flue liners with 4 inches of masonry between them. Next draw a line representing 6 inches of masonry all around the outside of the flue liners. This will give us the chimney size. Your drawing should now be similar to Figure 31-2.

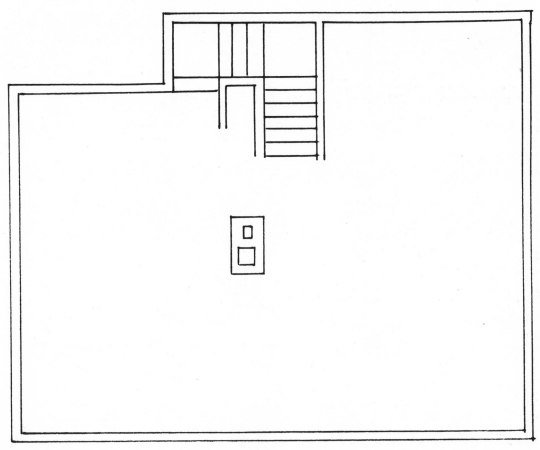

Fig. 31-2.

We made a rough sketch of our second floor layout giving us three bedrooms and a full bath. Our stair location and the chimney is fixed from the first floor plan. The stairs should land in a hallway and the bedrooms and bath should be off this hallway so that each room has full access to the hall and also to the bath without having to go through any other room.

If possible, do not have a chimney protrude into a room. Our stairs are three feet wide and this is also a good width for a hallway so draw light lines from where the stairs land on the second floor to the front of the house.

In our rough sketch, we had two bedrooms on one side and the master bedroom over the living room and the bath at the end of the hall. The main bathroom in a house should be adequate for the family and should be at least 6'6" to 7'0" wide. Our hallway should be widened to take care of this at the bath end of the hallway. This means we will make the front bedroom narrower than the back bedroom over the kitchen area.

Referring to our rough sketch layout for the second floor, locate and draw all the interior partitions. These should be six inches wide and should run full regardless of any doors. All closets except a walk-in closet are a standard 2 feet deep. It requires 20 inches to hang a garment on a coat hanger. The only exception to the 2-foot depth is a linen closet or a narrow storage closet that would be used for towels, etc. Your drawing should now be similar to that in Figure 31-3.

We next draw in the centerlines of our windows. The windows should be directly over the ones on the first floor if possible. All our second floor windows will be 2'8" wide and 4'6" high. The only two windows that will not fall directly over the ones on the first floor are the window centered on the stairway and the rear window on the back side of the bedroom located over the kitchen.

Draw in your windows to the correct size and location. Next locate and draw the door openings. All bedroom doors from the hallway to the bedroom should be 2'6" wide; the bathroom door, 2'4" wide; closet doors 2'0" wide, except the small linen closet door and the double doors on the large bedroom closets. The small linen closet door should be 1'0" wide and the closet doors in the master bedroom

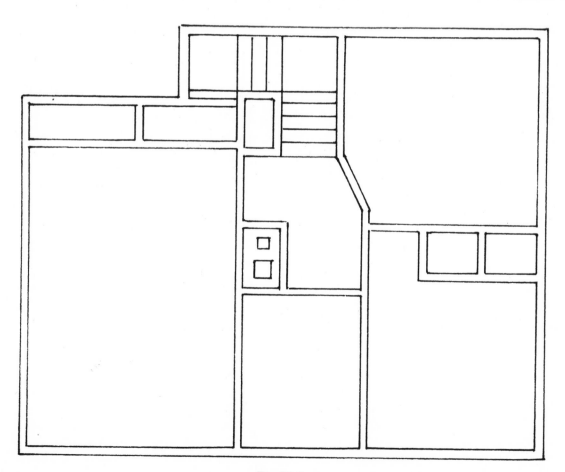

Fig. 31-3.

should be 1'6" each, making the opening 3'0" wide. Our next step is to draw a dotted line designating a scuttle opening 2'0" x 2'6" in the hall ceiling. This is an access to the attic area and should be located somewhere near the center of the house.

We next place the plumbing fixtures in the bathroom. The tub should be enclosed on three sides and have a combination shower and tub. The water closet should be located near a wall where it will be convenient to place a tissue holder. The built-in vanity should be located so it is easily accessible. Your drawing now should be similar to Figure 31-4.

We now determine the swing of the doors and draw them. Make sure that they open against a wall and will not be inconvenient to operate.

Next draw in the electrical layout for the second floor. It is standard practice to have ceiling outlets in the bedrooms, halls, and bathrooms, all of which are operated from a wall switch. The switch should be so located that you can operate it when entering or leaving the room. Closets have a drop cord which operates on a pull switch from the fixture. Your second floor plan should now be similar to Figure 31-5.

Now dimension the second floor plan. Make sure that your dimensions are correct. After you have completed the dimensions, *print* in the names of the rooms, any notes such as access scuttle to the attic, window and door sizes and the direction of the ceiling joists. Make sure that all interior dimensions total the same as the exterior dimensions. Your finished second floor plan should now be similar to Figure 31-6.

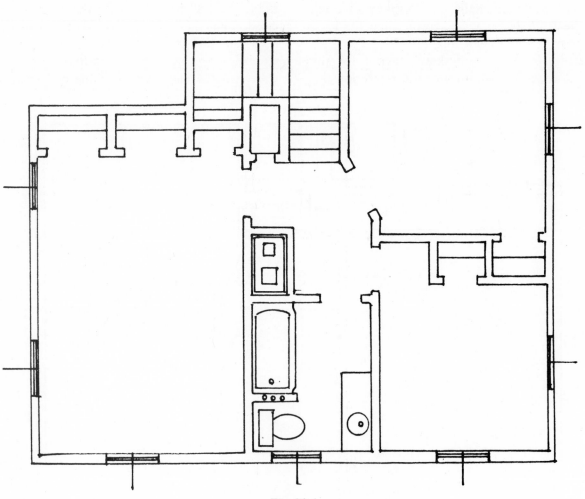

Fig. 31-4.

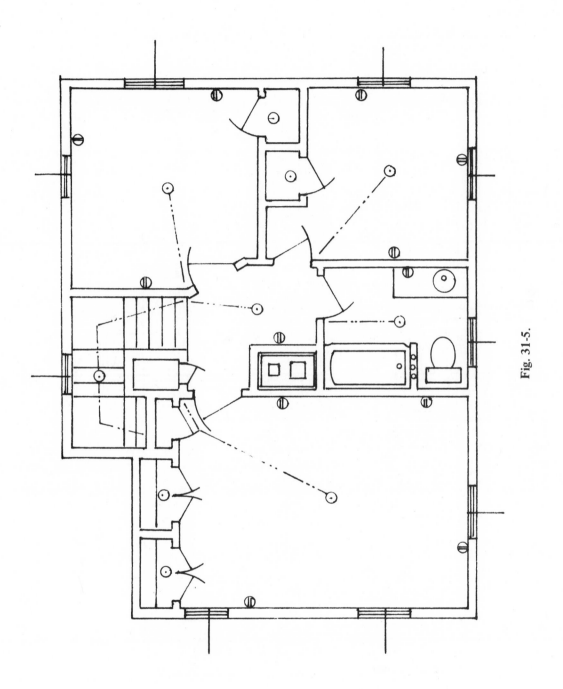

Fig. 31-5.

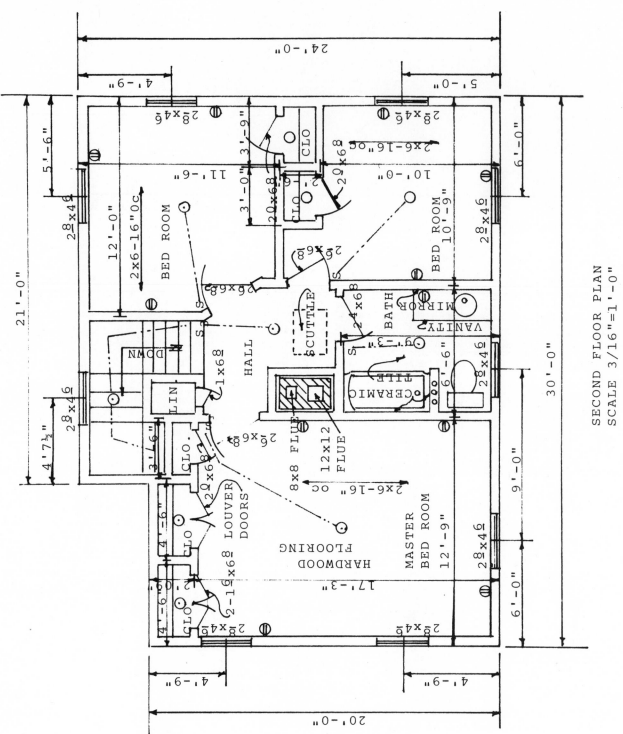

SECOND FLOOR PLAN
SCALE 3/16"=1'-0"

Fig. 31-6.

32
DRAWING THE BASEMENT AND FOUNDATION PLAN

After you have completed the first and second floor plans, the next step is to develop the basement and foundation plan from the first floor plan.

Lay out your outside walls the same size that you did for the first floor plan. This will be the outside of the foundation wall. After you have drawn the perimeter of the outside wall, determine the width of the wall and draw the interior wall line. You should refer to the local building code to find out what size of concrete block is required for residential construction in your area. We will use 8-inch cement block for our foundation wall, so we draw the inside line of the foundation making the wall 8 inches wide at $\frac{1}{4}'' = 1'0''$ scale. This would be the same for the foundation for the porch and the small entrance porches at the front and rear doors. The front entrance porch we will make 8'0'' wide and 5'0'' deep, and the rear entrance porch by the kitchen door we will make 4'0'' wide and 3'0'' deep. These should be centered on the door. Our side porch is 18'0'' long and 8'0'' deep. The porch wing of the house is 20'0'', so to balance the porch with the wing, we come in 1'0'' from each end.

Again our first step is to locate the stairway from the first floor area down into the basement. The stairway into the basement will be placed directly underneath the stairs going from the first to the second floor. On our basement plan, we run the stairs where they start in the kitchen and come down two landings and into our basement area.

Next we must locate the base for the fireplace in the living room. The fireplace base is also constructed using 8-inch cement block and must support the fireplace hearth in the living room. The fireplace base is 2'6'' wide and 5'6'' long. We place the furnace flue in its proper location and then we

make an 8-inch wall separating the flue from the ash pit. The fireplace base walls are all 8-inch cement block. Your drawing should now be similar to the one shown in Figure 32-1.

Remember in our word picture of the house, we wanted a recreation room in the basement and also a laundry room. The stairs leading into the basement should be in or near the recreation room.

If a basement is to be finished off into rooms, it is less expensive to use a 2 x 4 partition to support the floor joists than it is to use a built-up wood girder and supporting steel posts. When the rooms are finished, partitions have to be constructed anyway. The partition acts the same as a supporting partition either on the first or second floor.

Where the stairs come down into the basement, draw a 2 x 4 partition from the inside edge of the stairs across to the fireplace base and from the end of the fireplace base to the front foundation wall. This will enclose our recreation room area.

We also wanted a laundry area in the basement. A laundry will need plumbing for the sinks and washer, so it should be located underneath the kitchen area where the plumbing for the kitchen comes into the basement area. To close off the laundry into a room, we will run a 2 x 4 partition from the back side of the fireplace foundation wall to the outside foundation wall. Remember modular units and make the partition so it will take standard lengths for floor joists. This divides our basement into three general areas.

Your drawing should now be similar to that shown in Figure 32-2.

We now locate our basement windows. These should be placed under the windows on the first floor if possible, and we should locate them to provide cross ventilation. Basement windows are

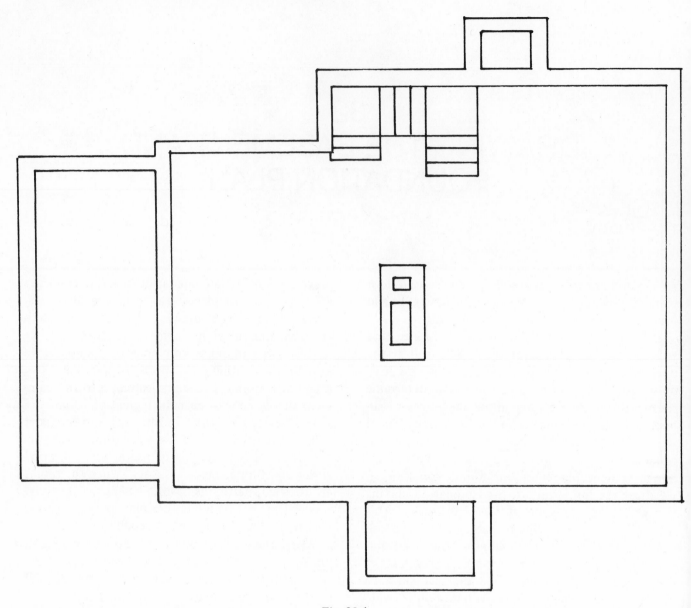

Fig. 32-1.

located by centerlines and are 32 inches wide. Do not place the cellar windows where the eaves of the house will drip into them if it can be avoided. This is not always possible. Basement windows are shown on the plan as in Figure 32-3.

A great many new residences being built today are so designed that they set low to the ground and become part of the landscape. These residences have anywhere from 9 to 12 inches of foundation wall showing above the finished grade. When a house is so designed, an areaway is required to fit around each window. An areaway is manufactured or constructed to hold back the earth and allow light and air to enter the window area. Areaways are made of metal or poured concrete. These are placed so that they are about two to four inches above the finished grade and are approximately 18 inches below the bottom of the cellar window. The earth is placed against the outside of the areaway up to finished grade. The inside of the areaway is usually filled with crushed stone up to within three inches of the bottom of the window. This allows any water to drain away from the windows.

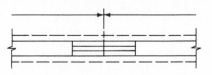

Note: All cellar sash to be 15 X 12 two light

Fig. 32-3.

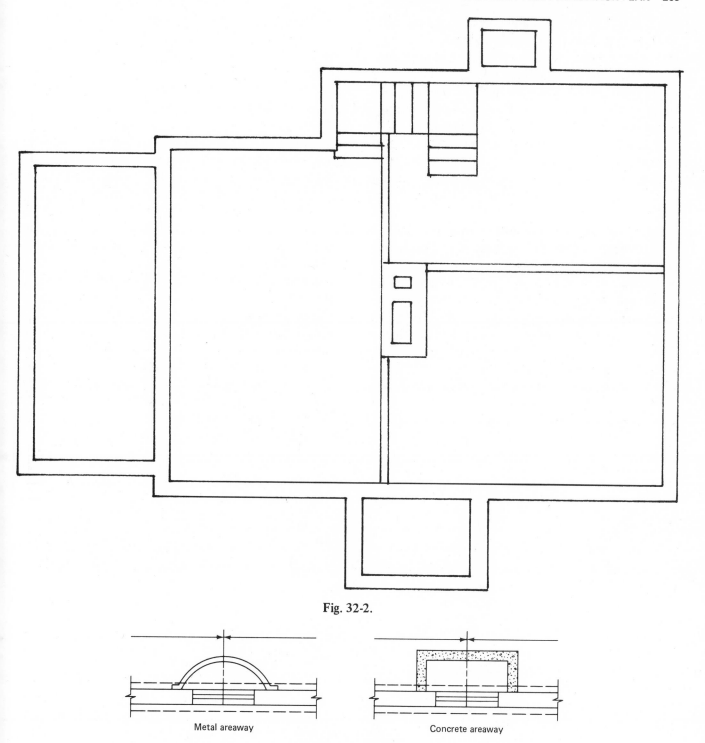

Fig. 32-2.

Metal areaway   Concrete areaway

Fig. 32-4. Areaways

Figure 32-4 shows how an areaway would look on the foundation plan.

We now draw the areaways around each basement window. We have the walls and partitions in place and should now layout the doorways into the room. We had to frame the basement stairs and platform with 2 x 4s, so where the stairs start down from the kitchen into the basement we can place a storage closet underneath and have the door into that closet from the laundry room.

We now locate and draw in the laundry set tubs and appliances. The set tubs should be located somewhere underneath the kitchen sink.

The heating unit is most effective if it is near the center of the house. We considered this in locating our chimney for the fireplace. We have already located the chimney flue, so we would now place the heating unit in its proper place. It should be near the chimney flue and out of the way in relation to the recreation room and laundry.

We now locate the electrical outlets. We will need a ceiling light in the laundry, one in the furnace room and two in the recreation room. We also need a three-way switch from the bottom of the stairway up to the kitchen. Our ceiling lights should work from switches located conveniently in the walls. We should have ample wall outlets in our recreation room and the laundry. We should locate a special purpose outlet near the furnace. Our drawing now should be similar to that in Figure 32-5.

We now draw our wall footings. These are shown as dotted lines and should be 4 inches wider on each side of the foundation wall. Our wall is 8-inch cement block, so our footings should be 16 inches wide. If we had used a 12-inch block, the footings would have been 20 inches wide.

We should now locate all extension and dimension lines. The foundation plan must have all dimensions and notes locating outside walls, interior partitions, fireplace location and size, basement windows, and areaways. Designate the rooms and stair indicator showing the direction of the stairs.

Areas to be filled and have concrete poured on top should be designated *fill*. This would apply to all our porches.

Notes should be added giving the size of the cellar windows, ash pit, cleanout doors, etc. If in place of 2 x 4 partitions, we had used a built-up girder and supporting posts, then these would be noted, giving size and exact location.

You should give the direction and the size and spacing of the first floor joists on the basement plan. This would complete the plan for the basement and foundation. You should check carefully to make sure that all dimensions are properly placed and are accurate. All the necessary information in terms of dimensions, notes, and symbols are placed on the plans. A plan is not complete unless the workman can take that plan and construct the building just the way it is drawn. Your finished basement and foundation plan should be similar to that shown in Figure 32-6.

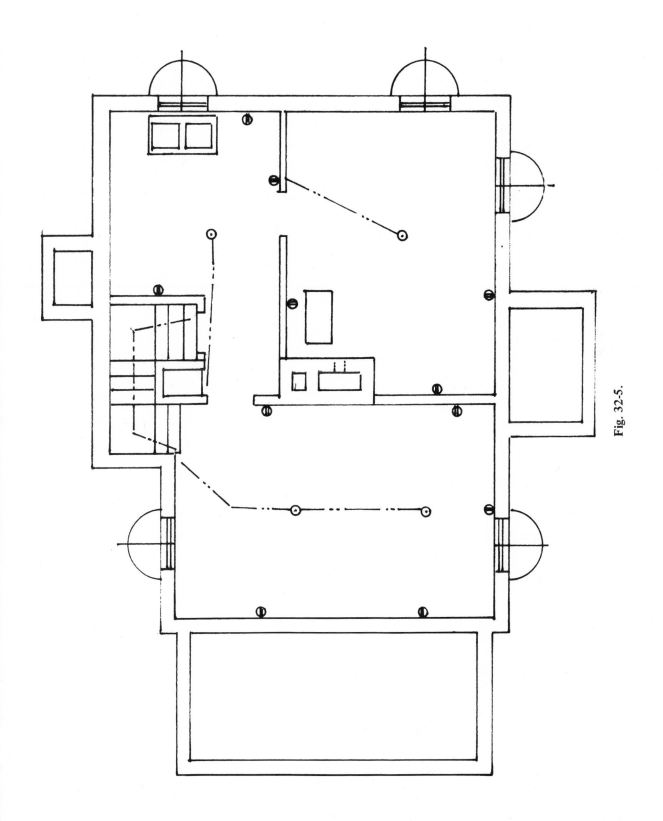

Fig. 32-5.

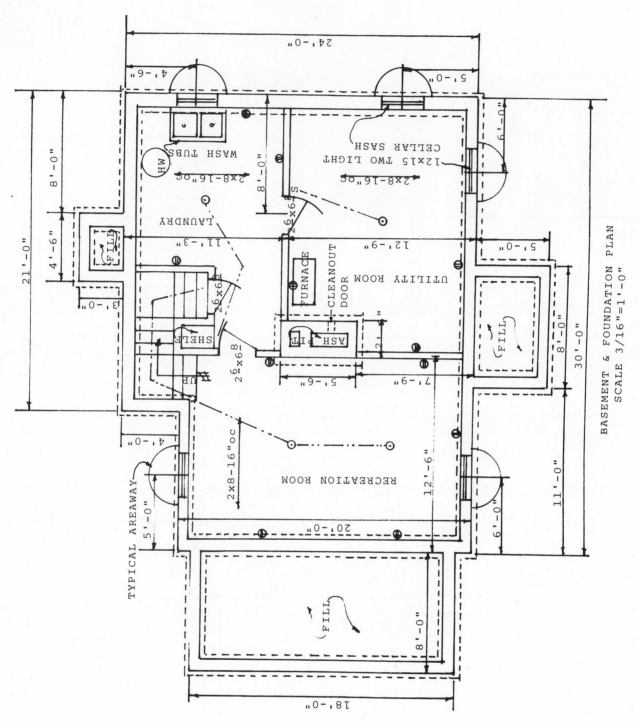

BASEMENT & FOUNDATION PLAN
SCALE 3/16"=1'-0"

Fig. 32-6.

33
DRAWING THE HALF SECTION VIEW
AND FRONT ELEVATION

After we have completed the plans for the first floor, the second floor, and the basement and foundation, the next step is to design the exterior and draw the elevations. The front elevation is the first elevation we draw.

The first step in developing the front elevation is to make a half view of the house at $\frac{1}{4}'' = 1'0''$ scale. Near the left border of your paper draw a centerline vertical from the bottom of your paper to the top. This will represent the center of our building. The left side of our house is 20'0'' wide, so half of that will be 10'0''. We measure from our vertical centerline 10'0'' at $\frac{1}{4}'' = 1'0''$ scale and draw a light, solid vertical line parallel with the vertical centerline.

Starting near the bottom of the paper, draw the footings and the cement block wall. The outside of the cement block should be on the 10'0'' light solid vertical line. The footings should be 4 inches wider than the wall and should be 8 inches deep. The blocks are 8 inches wide and start on the top of the footing. The foundation wall will be seven feet high from the top of the footing to the top of the wall. This is $10\frac{1}{2}$ courses of cement block. The half block is a solid cap block.

Starting at the top of the footing, mark off 10 courses, making them 8 inches high and the last course 4 inches high. Next draw a horizontal line from the top of the footing and from the inside edge of the block to the left where it meets the centerline. This is the bottom edge of the cement floor. Measure up 4 inches from this line and draw another line parallel to the first. This is the top of the cement floor in the basement. Note how this makes a key between the footing and the cement block and prevents any earth pressure from pushing the block into the cellar area.

Now draw the sill, box sill, floor joists and flooring for the first floor. These should be drawn to their exact size at $\frac{1}{4}'' = 1'0''$ scale. Draw the stud wall, plates, and the box sill and joists for the second floor. Figure 8'1'' for the ceiling height of the first floor.

Draw the second floor subflooring, and finished flooring. Next draw the second floor studs and plates. Make the second floor height 8'1''.

Draw the ceiling joists and the rafters. Each of these lines should extend from the solid vertical line to the centerline on the left. We use an 8/12 pitch for our rafters, which means that for every foot we go horizontally the rafter rises 8 inches. This means that in 10'0'' the bottom of the rafter should be 80 inches from the bottom of the ceiling joists on our centerline. Allow the rafter to extend 9 inches beyond the face of the exterior sheathing on the outside wall.

From the section view, extend a line designating the finished grade, the first floor line, the second floor line, and the second floor ceiling line. We should figure our finished grade about 9–12 inches below the top of the foundation wall.

Window and door heights are standard 6'8'' from the finished floor. This brings all the doors and windows in line. Locate the window height at 6'8'' above the finished floor on each floor. Designate each of the lines and measurements.

Your half section drawing should now be similar to that shown in Figure 33-1.

After we have completed the half section view, the next step is to develop the front elevation. To the right of our half section view, extend the lines horizontally to the right locating the finished grade, first floor line, the second floor line, second floor ceiling line, and the top of the ridge.

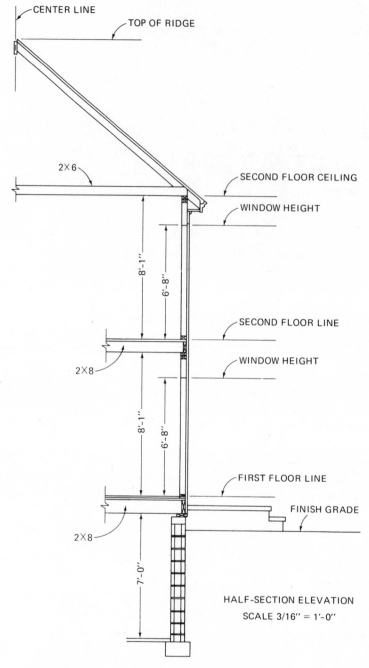

Fig. 33-1.

Locate the corners of the building from the floor plan. Our building is 30'0" long, so make sure you have enough room and then locate the right edge of the house. Measure over 30'0" at $\frac{1}{4}$" = 1'0" scale and draw the left corner of the building.

Referring to the first floor plan, locate the center of the windows and the front door and draw light vertical centerlines. Looking at the second floor plan, locate with centerlines the center of the windows on the second floor. They should be in line with the center of the windows and door on the

first floor. Locate the center of the chimney at the top of the roof line.

The high point of the half section view will give us the location of the ridge of our house. The same applies to the cornice. Extend the cornice lines from the half section view to the front elevation. Draw the openings for the windows and doors from the sizes we designated on our floor plans. The windows are centered over each other but the second floor windows are narrower in width.

Extend the window height line from the section view and then draw the correct height down from this line. This will locate the bottom of the window, which will also be the top of the window sill. The front door should be drawn down from the 6'8" line to within two inches of the first floor line. This is to allow us to show the exterior door sill.

We should also draw in the side porch. The beam that carries the rafters should be framed so that the bottom of the beam comes even with the top of the first floor windows. The cement floor on the porches is 4 inches thick and we will need posts to support the roof framing.

The porch roof is metal, so we need not make the pitch as high as the main roof. Also we have to consider the window height of the windows on the second floor so that the porch roof does not interfere with them.

Your drawing should now be similar to that shown in Figure 33-2.

The next step is to draw the cornice moldings. The rake (the line that extends from the eaves to the ridge) should extend about 4 inches beyond the corner of the house. We are using a tight cornice on our house, and this is made up by using a 5/4" x 6" pine board against the sheathing and a $3\frac{5}{8}''$ crown mold nailed over the 5/4 x 6 and the top edge of the crown mold comes flush with the top edge of the roof sheathing.

We now draw in our windows. We have already determined the size of our windows and we have the rectangle drawn outlining the windows and the front door. We now draw the outside window casing. This is 3 inches wide and goes up each side

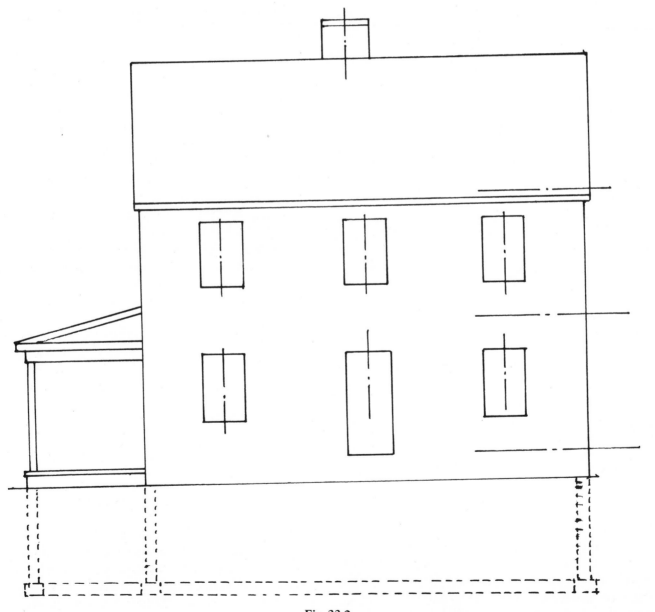

Fig. 33-2.

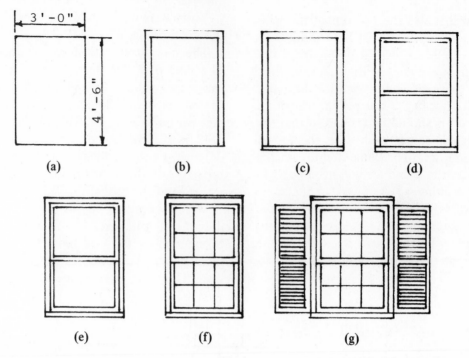

Fig. 33-3.

and across the top of the window. Figure 33-3 shows how to draw a window in elevation.

Draw the sill, making it 2 inches wide and extending it beyond the jambs to the edge of the outside casing. Your drawing should be similar to 33-3(c). Next locate the horizontal center of the opening and draw a line from one jamb horizontally to the other. This represents the lower edge of the meeting rail. Measure up 2 inches from this line and draw another parallel to it stopping approximately 2 inches from each jamb side. Measure down 2 inches from the head and draw a parallel line stopping it 2 inches from each side jamb. Measure up 3 inches from the top of the sill and draw a horizontal line stopping it 2 inches from each jamb. Your drawing should now be similar to 33-3(d).

Draw a vertical line 2 inches in from the side jamb on each side. The bottom sash line should go from the 3-inch line and touch the horizontal line that represents the bottom of the meeting rail. The top sash lines should go vertically from the line representing the top of the meeting rail to the 2-inch line representing the top of the upper sash. Your drawing should now be similar to 33-3(e).

This completes an elevation drawing of a window having one sash over one sash. If your design calls for a colonial style (divided lights), draw a single horizontal line across the center of the upper and lower sash. Next divide the sash into as many lights as the window calls for and draw the vertical

lines as in 33-3(f). If the design calls for blinds or shutters, you draw them at this time. Blinds should be half the width of the window and extend from the head of the window to the sill of the window.

All windows have the same basic parts which are jambs, casing, sill, and sash. If you were to draw a casement window in elevation, you would follow the same steps except you would eliminate the meeting rail. The outside casing would be the same and also the sill. You would draw the same side, head, and bottom heights for the sash. Figure 33-4 shows how a casement window with diamond lights would look in elevation.

The straight line over the top of the head casing represents the top of the drip cap. This will be on every window except where a wide overhang is used and the facia board comes even with the top of the window casing.

Figures 33-5–33-10 show how various windows look in plan and how they look in elevation.

Casement

Fig. 33-4.

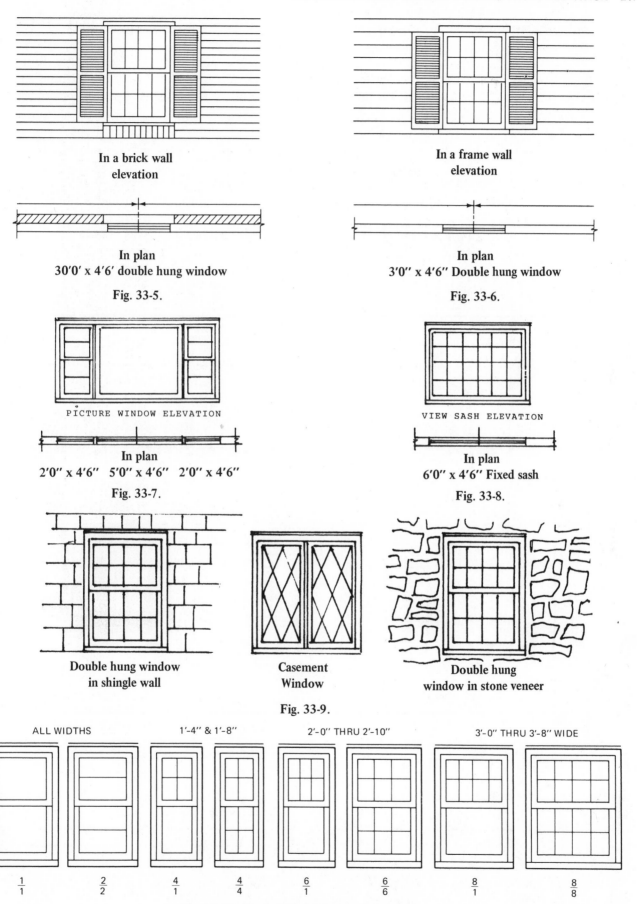

In a brick wall
elevation

In plan
30'0' x 4'6' double hung window

Fig. 33-5.

In a frame wall
elevation

In plan
3'0" x 4'6" Double hung window

Fig. 33-6.

PICTURE WINDOW ELEVATION

In plan
2'0" x 4'6"   5'0" x 4'6"   2'0" x 4'6"

Fig. 33-7.

VIEW SASH ELEVATION

In plan
6'0" x 4'6" Fixed sash

Fig. 33-8.

Double hung window
in shingle wall

Casement
Window

Double hung
window in stone veneer

Fig. 33-9.

ALL WIDTHS   1'-4" & 1'-8"   2'-0" THRU 2'-10"   3'-0" THRU 3'-8" WIDE

$\frac{1}{1}$ $\frac{2}{2}$ $\frac{4}{1}$ $\frac{4}{4}$ $\frac{6}{1}$ $\frac{6}{6}$ $\frac{8}{1}$ $\frac{8}{8}$

Fig. 33-10. Typical double hung window units.

We continue developing our front elevation by drawing the facia board, which will come even with the top of the second floor windows. Next draw a line at the top of the facia board and just underneath the main cornice. This represents the $1\frac{3}{4}$-inch bed mold. Now draw two lines across the 6-inch wide cornice that you roughed in from the section view. This represents the bottom of the $3\frac{5}{8}$-inch crown mold and its curved surface. Return each end with the crown mold profile from the rake to the main cornice. Locate and draw the side porch elevation showing the posts and cornice. The front entrance porch can be extended from the side porch elevation. The bottom of the beam should be the same height. Draw your front door. Your drawing should now be similar to that in Figure 33-11.

We now finish our front elevation. Draw in the blinds, the corner boards, the bevel siding, and the foundation windows, and locate the foundation depth by using hidden lines below the finished grade line. Locate the finished floor lines and designate them. Draw the finish on the front and side porches. Finish the chimney. Draw the brick cap and the flashing where the chimney comes through the roof. The chimney has a better appearance if you corbel out the last three courses of brick approximately $\frac{1}{2}$-inch all the way around.

Designate the type of roofing material, the wall finish on the walls. This should complete your front elevation. Place any necessary notes and print the view. Your drawing should now be similar to that shown in Figure 33-12.

Fig. 33-11.

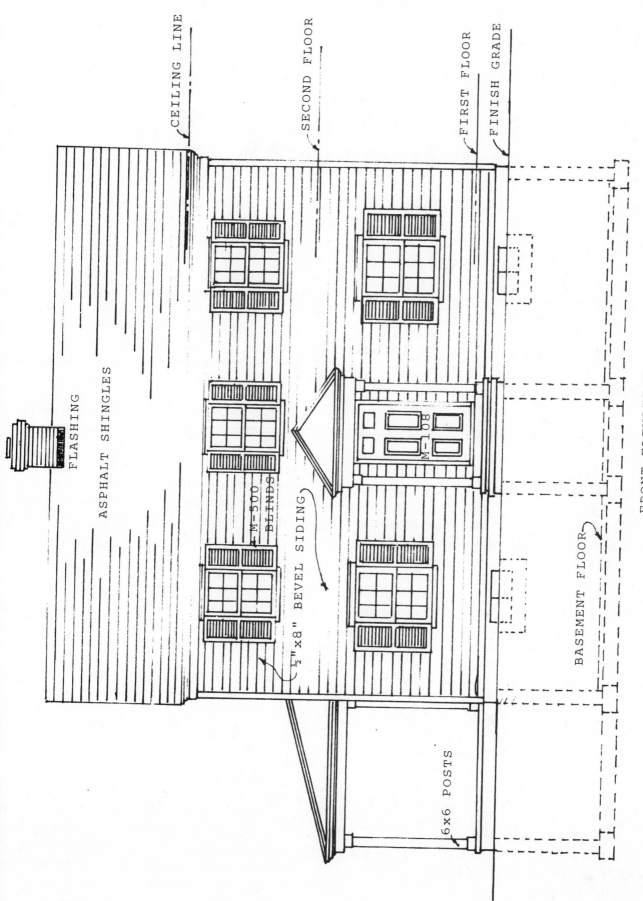

CEILING LINE

SECOND FLOOR

FIRST FLOOR

FINISH GRADE

FLASHING

ASPHALT SHINGLES

M-500 BLINDS

M-108

½"x8" BEVEL SIDING

6x6 POSTS

BASEMENT FLOOR

FRONT ELEVATION

Fig. 33-12.

34
DRAWING THE SIDE AND REAR ELEVATIONS

We already have the half section view drawn for use when we developed the front elevation. Using this half section view, extend the grade lines, floor lines, and cornice lines from the half section view to where you are going to draw the elevations.

Locate the end walls of the building and extend these up to the cornice line. We already have the ridge line established. The cornice should extend 9 inches in front of the sheathing. This is the face of the main cornice. Draw a line from the center of the side gable end down on a slant to where it meets the main cornice. Block in the cornice returns. These should be 2'6" on each side. The rake cornice comes down and rests on the top of the main cornice return.

Locate the center of all windows and doors that should be on this elevation by taking the measurements from the floor plans. Rough in the windows and doors according to the sizes on the floor plans. Draw the side porch showing the roof, cornice, and porch posts. The porch posts should be even with the top of the window and door casings. Your drawing should now be similar to that shown in Figure 34-1. Notice that we have located by center lines the center of the end gable, the center of the chimney, as well as the center of each window. Extend the front entrance porch, place your corner boards and the cornice returns.

You can get the height of the porch roof from the front elevation you just completed. Remember that in drawing an elevation, you draw that view as if you were standing away and looking directly at that side of the structure. This applies to all elevation views. The front as looking at the front, each side as looking directly at that side and the rear elevation as standing away and looking at the back side of the structure.

The next step is to draw in the cornice lines, draw the corner boards, porch roof lines, and the porch cornice. Draw the windows and blinds. Designate the roof material, exterior wall material, the louver vents for the attic area, and the finish on the gable end. Draw the foundation outline using dotted lines for that below the finished grade. The left elevation should now be similar to that shown in Figure 34-2.

The next elevation we will draw will be the rear elevation. This is looking at the floor plans from the back of the house. Again rough in the work the same as you did on the previous elevations. We rough in the work showing the ends of the building, the ridge, the chimney, gables, porches, finish grade line, and the location of the first and second floor lines.

Again locate the center of all window and door openings. Rough these in according to the sizes from the floor plans. Your drawing should now be similar to Figure 34-3.

Now finish the rear elevation drawing. Draw in the cornice lines, chimney and flashing. In this elevation we show the stairs leading from the first floor down into the basement and the stairs from the first up to the second floor. Draw the basement lines using hidden lines for that below the finished grade. Note that we did not use blinds on the rear elevation. The finished elevation should be similar to that in Figure 34-4.

The last elevation that we have to draw is the right side. Again we follow the steps we have taken in drawing the previous elevations. We again use our half section view to locate the finished grade, floor lines, cornice and window heights. Rough in the drawing by locating the corners of the building and draw from the cornice to the foundation. Lo-

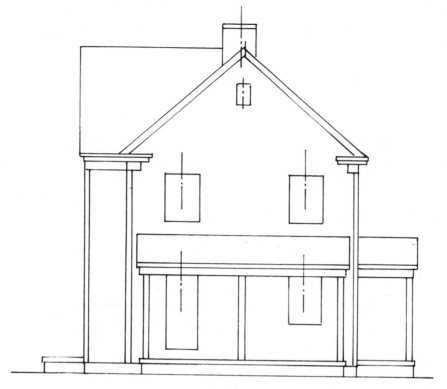

Fig. 34-1.

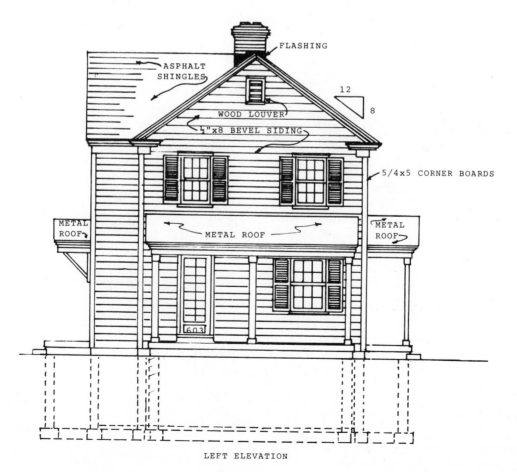

FLASHING

ASPHALT
SHINGLES

12
8

WOOD LOUVER

1"x8 BEVEL SIDING

5/4x5 CORNER BOARDS

METAL
ROOF

METAL ROOF

METAL
ROOF

603

LEFT ELEVATION

Fig. 34-2.

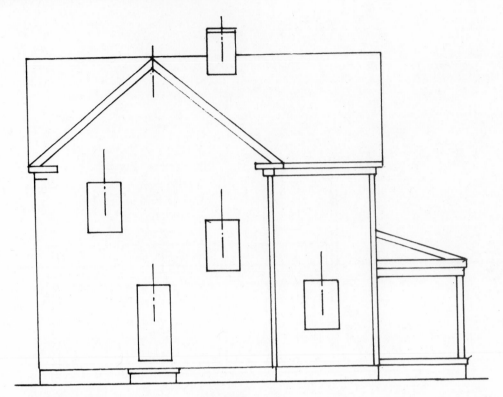

Fig. 34-3.

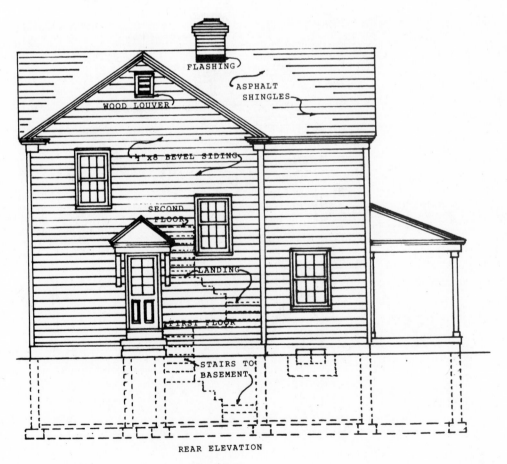

REAR ELEVATION

Fig. 34-4.

cate the center of the windows and rough in the openings. Your drawing should now be similar to Figure 34-5.

Now finish the right side elevation. Draw in the cornice lines and finish the blinds and windows. Show the roof pitch. Finish the lines for the chimney and flashing. Draw in the attic louver and wall finish. Designate the height for the window head. Your finished drawing should be similar to that in Figure 34-6.

Once you have developed the half section view and the front elevation of a house, the other elevations are easy to develop. You use the half section view to locate the height of the finished grade, the floor and ceiling lines, and the top of the ridge. We also locate the top of the windows from our half section view.

To develop the side and rear elevations, we draw them looking straight at that side, locating all door and window openings from the floor plans.

Any place that we have a gable end of a building with the cornice running up the rake and down the other side, the cornice running down the rake always cuts on top of the main cornice return. The main cornice runs along the eaves and usually returns against the side of the building. Usually the cornice will return against the house two-thirds and the overhang one-third. A 9-inch plumb cut on the rafter overhang will have a finished cornice approximately 12 inches wide. The return against the house would be 24 inches, giving a total length of level cornice against the gable end approximately 36 inches from the outside edge of the face of the cornice to where it returns against itself on the side of the gable end. This allows the facia board to return against the house the same distance as the overhang.

When designing a porch or entrance to a house, it is standard to have the bottom of the porch beams level with the top of the window and door casings. This allows the free swing of any storm doors that might be attached on the outside of the regular door.

Chimneys are always a minimum of two feet higher than any construction or roof line, to prevent a downdraft. This is why most chimneys either go up through the center of the ridge; if the house has an outside fireplace, the chimney runs up near the center of a gable.

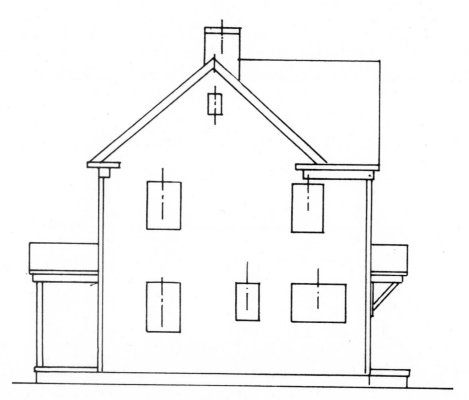

Fig. 34-5.

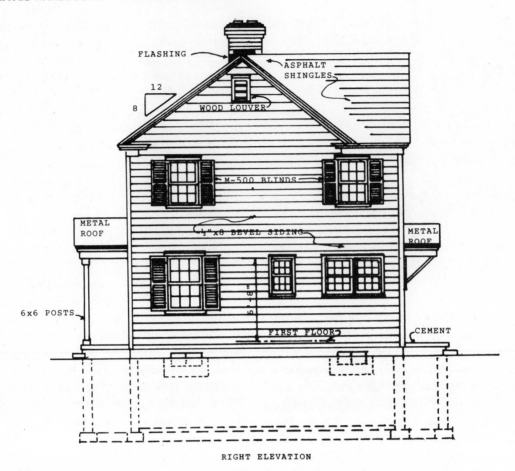

RIGHT ELEVATION

Fig. 34-6.

35
DRAWING THE WALL SECTION AND SPECIAL DETAILS

After we have completed the working drawings at $\frac{1}{4}'' = 1'0''$ scale, we now have to give the builder more detailed information on the construction of the building. We do this by drawing special details drawn to a larger scale.

The first detail we will draw will be a wall section detail showing the construction of the house from the footings to the ridge. This is different from our half section view in that it shows in detail the materials and how they are used. It is drawn at a larger scale, and because of this we will be required to use break lines in our drawing.

We start with a detail of the footing. We will draw this at $\frac{3}{4}'' = 1'0''$ scale. Our footing is 8 inches deep and 16 inches wide, so draw your rectangle at this size. On the top edge step in 4 inches on each side and draw a light vertical line. This is the outside and the inside edge of the 8-inch block wall. Measure from the top of the footing up 8 inches and draw light horizontal lines connecting the vertical ones. This is the cement block. Next draw a light line parallel to the one you just drew about $\frac{1}{2}$ inch at the correct scale. This is the joint of the block.

Next measure in $1\frac{1}{2}$ inches from each outside edge of the block and draw a vertical line on each side. This is the flange of the block. On the outside of the foundation wall, draw a line parallel to the outside edge making it approximately $\frac{1}{2}$ inch wide and cant the bottom to the edge of the footing. This is the cement plaster. On the inside measure up 4 inches from the top of the footing and draw a horizontal line to the right. Draw a horizontal line to the right from the top of the footing. This represents the cement floor. On the outside of the footing, draw the drain tile.

Now draw the correct symbols for each of the materials and place the names of the various parts with arrows identifying them. Your drawing should now be similar to Figure 35-1.

Our next section will be the top of the foundation wall and the first floor framing. We draw a break line between each to show that the construction between the break lines is the same. We continue our foundation lines directly over the ones below. We draw the cap block, making it 4 inches high. Our cement plaster comes even with the top of the foundation wall. From the top of the foun-

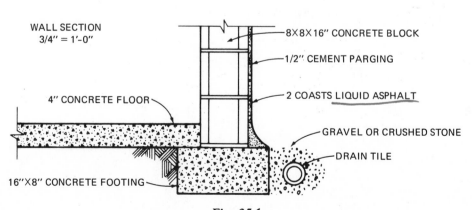

WALL SECTION
3/4'' = 1'-0''

8X8X16'' CONCRETE BLOCK

1/2'' CEMENT PARGING

2 COASTS LIQUID ASPHALT

4'' CONCRETE FLOOR

GRAVEL OR CRUSHED STONE

DRAIN TILE

16''X8'' CONCRETE FOOTING

Fig. 35-1.

dation wall we measure down 9 inches on the outside and draw a horizontal line to the left. This is our finished grade line.

On the top of the cap block, we measure in the thickness of our exterior sheathing, which is $\frac{1}{2}$ inch, and we draw our sill. This is $1\frac{1}{2}$ inches high and $5\frac{1}{2}$ inches wide. On the outside edge of this we draw our box sill which is a 2 x 8. On the top of the 2 x 8 we draw a horizontal line to the right. Measure up $\frac{5}{8}$ inch and we draw another line parallel to the first. This is the subflooring. From this line draw your floor joists, spacing them 16 inches on center. Next draw the diagonal bridging. On top of the subflooring, draw the 2 x 4 stud wall showing the 2 x 4 shoe plate. Extend the sheathing line and draw the interior wallboard line. Next draw the finished floor on top of the subfloor. Next to the stud wall and wallboard, draw the interior baseboard and carpet strip. Your drawing should now be similar to Figure 35-2.

We continue to show the construction of our house by developing the second floor area. We extend our lines directly over the ones below and show how the second floor is framed. On top of the 2 x 4 stud wall we have a double plate made up of two 2 x 4s. Our second floor box sill is next drawn and we extend our sheathing line. We show the 2 x 8s for the second floor, the subflooring and the finished floor. Our second floor studs are the same as on the first floor. We draw the wallboard, the interior base and carpet strip. On the outside of the sheathing we show our finished siding. Next identify each part. Your drawing should be similar to that in Figure 35-3.

We continue our wall section detail, using break lines between each section. We now show the ceiling, rafter, roof, and cornice framing. We continue to extend our lines directly over each other. At the top of the second floor stud wall we again have double 2 x 4 plates. Our ceiling joists rest on the top of these plates and our rafters are cut so they also rest on top of the 2 x 4 plates and are nailed to the side of the ceiling joists. We are using a box cornice, so we extend beyond the face of the exterior sheathing 9 inches and make a plumb cut on the rafter. We then draw our 1 x 6 facia, the 1 x 8 soffit, and the 1 x 8 facia against the sheathing. Draw the $1\frac{3}{4}$-inch bed mold and the $3\frac{5}{8}$-inch crown mold. The top edge of the crown mold should be even with the top edge of the roof sheathing. Next draw the roof shingles, wallboard, insulation and interior cove mold. Label each part. Your drawing should now be similar to Figure 35-4.

This completes the wall section detail. Note that we have shown in detail how each floor and the roof is framed. The various materials and how they are installed. Make sure that you have correctly labeled each part. Your finished wall section detail should now be similar to that shown in Figure 35-5.

The next detail we draw is the detail of the porch cornice and the main cornice of the house. This is again drawn at $\frac{3}{4}'' = 1'0''$ scale. When we constructed our porch, we said we would use 2 x 4s for the overlay because they only had to carry the finished porch ceiling. Also in doing this we decided we would bearfoot the rafters on the top of the 2 x 4s. We should show this in our details. We show the

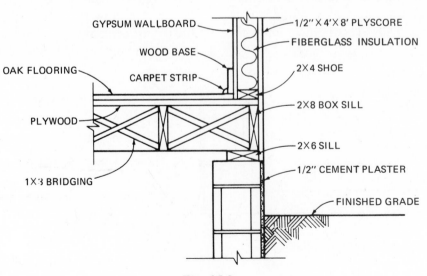

Fig. 35-2.

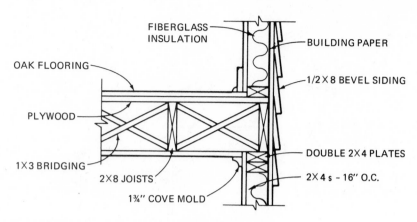

Fig. 35-3.

construction at the porch floor and the base of the posts and also the cornice and roof detail. This is shown in Figure 35-6.

In our foundation plan, we used a 2 x 4 stud partition in place of supporting posts and a girder (built-up-beam). If we had used a post and girder construction, we would also make a detail of the post, the footing, and how the post and girder are fastened together. Figure 35-7 is an example of how this detail would be drawn.

Our next detail is of the fireplace. Here we draw a section view showing details of the footing, wall, ash pit, cleanout door, hearth, firebox, ash dump, damper, and flue liners. On this section we also show the outline of the wood mantel and its height. To the right of the section view we extend our lines over to the right and draw a half section elevation, looking at the front. Figure 35-8 shows how the fireplace detail should be drawn.

The next detail we will draw will be the layout of the kitchen cabinets and the elevation drawings.

We will use the scale of $\frac{1}{4}'' = 1'0''$ for both the plan layout and the elevations. We first draw the walls and the window of the area where the cabinets are to be installed. Wall and base cabinets are manufactured in multiples of 3-inch increments, and most manufacturers make various size fillers. In our kitchen we will place the sink under the window. All base cabinets are 24 inches deep. We want to be able to go from the refrigerator to the sink and then to the range with a minimum of steps, so we will place the refrigerator near the back door and the range on the opposite wall. The range is 30 inches wide and the refrigerator will require a 36-inch space. We then figure the remaining space for wall and base cabinets. Figure 35-9 shows our cabinet layout.

We now make an elevation drawing of each wall showing the cabinets. All wall cabinets are installed so that the top of the cabinets are 84 inches off the floor. The space between the top of the wall cabinets and the ceiling is framed and finished the

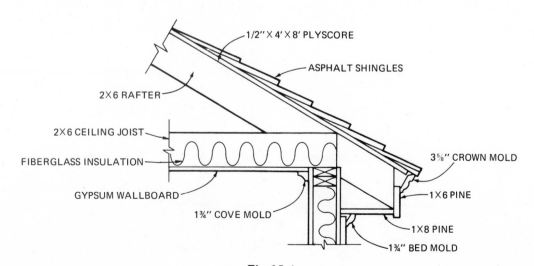

Fig. 35-4.

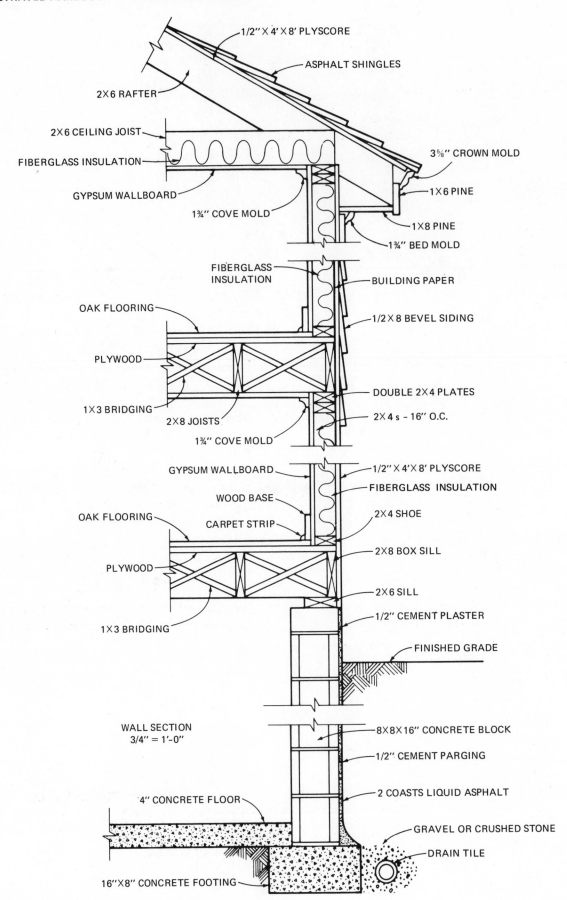

1/2″ × 4′ × 8′ PLYSCORE

ASPHALT SHINGLES

2×6 RAFTER

2×6 CEILING JOIST

3⅝″ CROWN MOLD

FIBERGLASS INSULATION

GYPSUM WALLBOARD

1×6 PINE

1¾″ COVE MOLD

1×8 PINE

1¾″ BED MOLD

FIBERGLASS INSULATION

BUILDING PAPER

OAK FLOORING

1/2 × 8 BEVEL SIDING

PLYWOOD

1×3 BRIDGING

2×8 JOISTS

DOUBLE 2×4 PLATES

2×4 s – 16″ O.C.

1¾″ COVE MOLD

GYPSUM WALLBOARD

1/2″ × 4′ × 8′ PLYSCORE

WOOD BASE

FIBERGLASS INSULATION

OAK FLOORING

CARPET STRIP

2×4 SHOE

2×8 BOX SILL

PLYWOOD

2×6 SILL

1/2″ CEMENT PLASTER

1×3 BRIDGING

FINISHED GRADE

WALL SECTION
3/4″ = 1′-0″

8×8×16″ CONCRETE BLOCK

1/2″ CEMENT PARGING

2 COASTS LIQUID ASPHALT

4″ CONCRETE FLOOR

GRAVEL OR CRUSHED STONE

DRAIN TILE

16″×8″ CONCRETE FOOTING

Fig. 35-5.

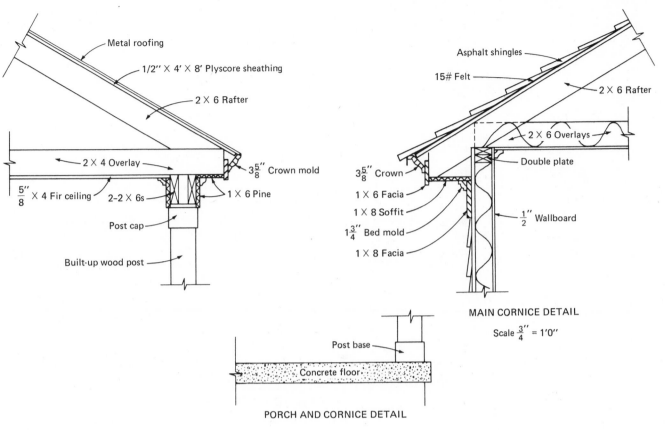

Metal roofing

1/2″ × 4′ × 8′ Plyscore sheathing

2 × 6 Rafter

2 × 4 Overlay

$3\frac{5}{8}$″ Crown mold

$\frac{5}{8}$″ × 4 Fir ceiling

2–2 × 6s

1 × 6 Pine

Post cap

Built-up wood post

Asphalt shingles

15# Felt

2 × 6 Rafter

2 × 6 Overlays

Double plate

$3\frac{5}{8}$″ Crown

1 × 6 Facia

1 × 8 Soffit

$1\frac{3}{4}$″ Bed mold

1 × 8 Facia

$\frac{1}{2}$″ Wallboard

MAIN CORNICE DETAIL

Scale $\frac{3}{4}$″ = 1′0″

Post base

Concrete floor

PORCH AND CORNICE DETAIL

Fig. 35-6.

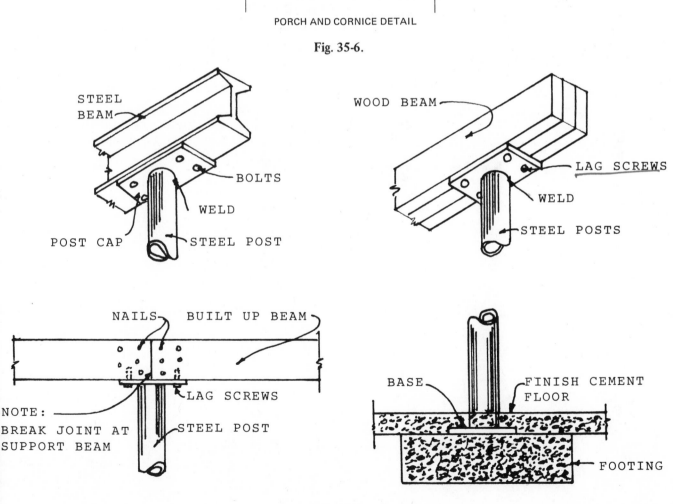

STEEL BEAM

BOLTS

WELD

POST CAP

STEEL POST

WOOD BEAM

LAG SCREWS

WELD

STEEL POSTS

NAILS   BUILT UP BEAM

LAG SCREWS

NOTE:
BREAK JOINT AT
SUPPORT BEAM

STEEL POST

BASE

FINISH CEMENT FLOOR

FOOTING

Fig. 35-7.

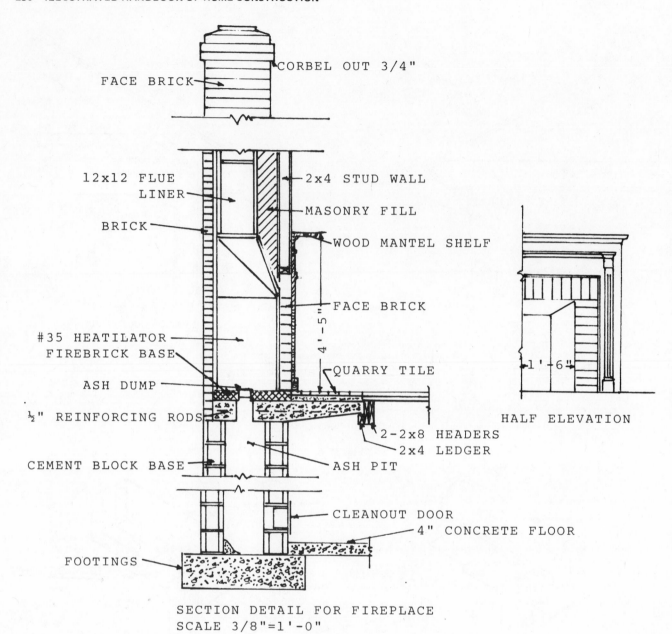

FACE BRICK
CORBEL OUT 3/4"

12x12 FLUE LINER
2x4 STUD WALL

MASONRY FILL

BRICK
WOOD MANTEL SHELF

FACE BRICK

4'-5"

#35 HEATILATOR FIREBRICK BASE
QUARRY TILE

ASH DUMP

½" REINFORCING RODS
2-2x8 HEADERS
2x4 LEDGER

CEMENT BLOCK BASE
ASH PIT

CLEANOUT DOOR
4" CONCRETE FLOOR

FOOTINGS

SECTION DETAIL FOR FIREPLACE
SCALE 3/8"=1'-0"

HALF ELEVATION

1'-6"

Fig. 35-8.

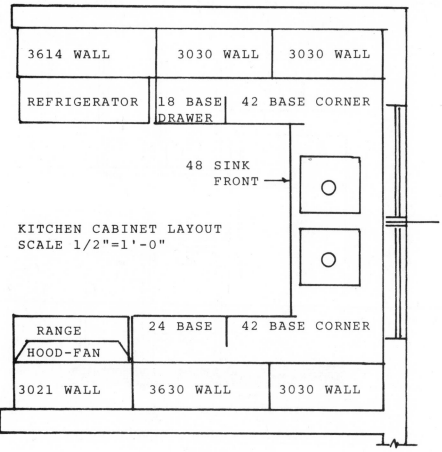

Fig. 35-9.

same as the wall. Wall cabinets vary in height according to where they are placed. A standard wall cabinet is 30 inches high; wall cabinets placed over refrigerators range from 12 to 15 inches high; and cabinets over ranges vary from 18 to 21 inches high. Base cabinets are 34 inches high and a standard counter top is $1\frac{1}{2}$ inch thick and has a 4-inch backsplash. We will draw the refrigerator wall first. Your finished drawing should be similar to Figure 35-10.

We now develop our range wall and the sink wall. On the range wall we show the cabinets and the range and also the exhaust hood over the range, Figure 35-11. On our sink wall, we show the base cabinets, the windows and the faucets. Notice that the valance extends across over the window to tie all the cabinets together and also to provide a place for a kitchen clock (Figure 35-12).

The last detail we will make will be of our stairs. This we will also draw at $\frac{3}{4}'' = 1'0''$ scale. Our stairs have two platforms and are a U-turn set of stairs. We will draw them starting at the living room floor and show a detail up to the second floor landing. The handrail should be continuous from floor

to floor. We use a 9-inch run, which means we will have a $10\frac{1}{2}$ inch tread and a $7\frac{1}{2}$-inch rise. The stair detail would look like Figure 35-13.

We have now completed our working drawings showing the plans of all the floors, basement and foundation, elevations, section views, and construction details. On some plans we might find it necessary to show details of the bathroom tub area and vanities. If the closets are large and are walk-in type with built-in trays and storage cabinets, we would have to make a detail drawing of these. For any construction that is special, or for a particular section that requires more information for the builder so that he can construct it to the specifications and the desires of the owner, we will need a detail drawing to show exactly how it is to be built.

We will now review the frame construction of our house. The mason has completed the foundation wall and the basement floor. We are finishing the basement area and dividing it into rooms, and we are using 2 x 4 stud partitions instead of supporting posts and girders.

The top of the partition plates must be level with the top of the 2 x 6 sill we place on the top of

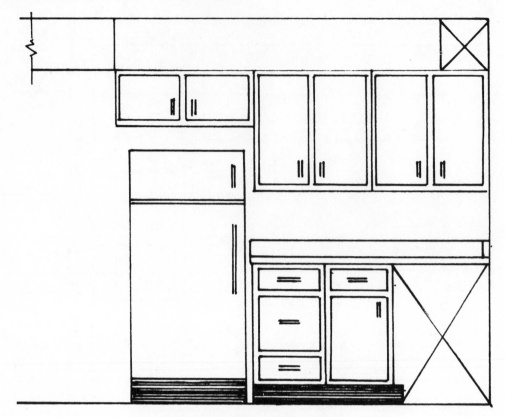

Fig. 35-10. Refrigerator wall.

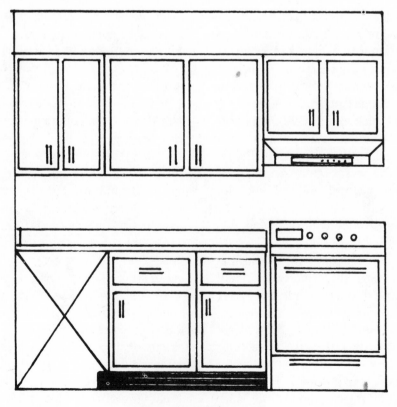

Fig. 35-11. Range wall.

Fig. 35-12. Sink wall.

the foundation wall. Under the sill is our sill seal.

The 2 x 8 box sill is nailed to the top of the 2 x 6 sill at a 90-degree angle. The floor joists are spaced 16 inches on center and go from the box sill to the partition plates on top of the partitions in the basement. These 2 x 8 joists are nailed to the box sill, the 2 x 6 sill, and are toe-nailed to the plate on the top of the 2 x 4 partition. The 2 x 8s are doubled wherever a partition is to go over the top of them. The doubled joists are spread $3\frac{1}{2}$ inches apart. This passageway is for hot air ducts through the joists and partitions to the rooms above. After the joists are in place, the bridging is nailed to the joists at the top edge only.

When the joists are all placed in position and the framing for the various openings such as the stairs and fireplace are finished, the entire floor area is covered with the subfloor. This is usually 4' x 8' sheets of plywood; if shiplap is used, it is installed diagonally.

After the subfloor is installed a 2 x 4 bottom plate is installed around the outside edge on which are placed the 2 x 4 studs. On top of these studs are placed two 2 x 4s called wall plates.

All framing for window and door openings is done at this time. Many carpenters will frame the wall on the subfloor and then tip it up in place and

brace the wall until it can be nailed in place with the top plate installed.

The inside partitions are now framed in place, using a 2 x 4 bottom plate and 2 x 4 studs spaced 16 inches on center with two 2 x 4 wall plates on top and the same level as the outside walls. The top plate is lapped at each corner and where the interior walls meet the outside walls. This ties the outside and inside walls into one unit.

The 2 x 8 box sill is again placed around the outside walls of the house flush with the outside face of the 2 x 4 studs. The 2 x 8 second floor joists are placed 16 inches on center and are nailed into the box sill, the plates of the outside wall and the plates of the bearing partition on the inside walls.

Again the 1 x 3 bridging is nailed to the top of each floor joists running through the center of the span. When all the second floor joists and framing for openings has been completed, the subfloor for the second floor is installed.

The outside walls and the interior partitions are now framed in place using 2 x 4s 16 inches on center. Again all window and door openings are framed at this time.

The next operation is covering the outside walls with the exterior sheathing. The framing is braced and checked to make sure it is plumb and true be-

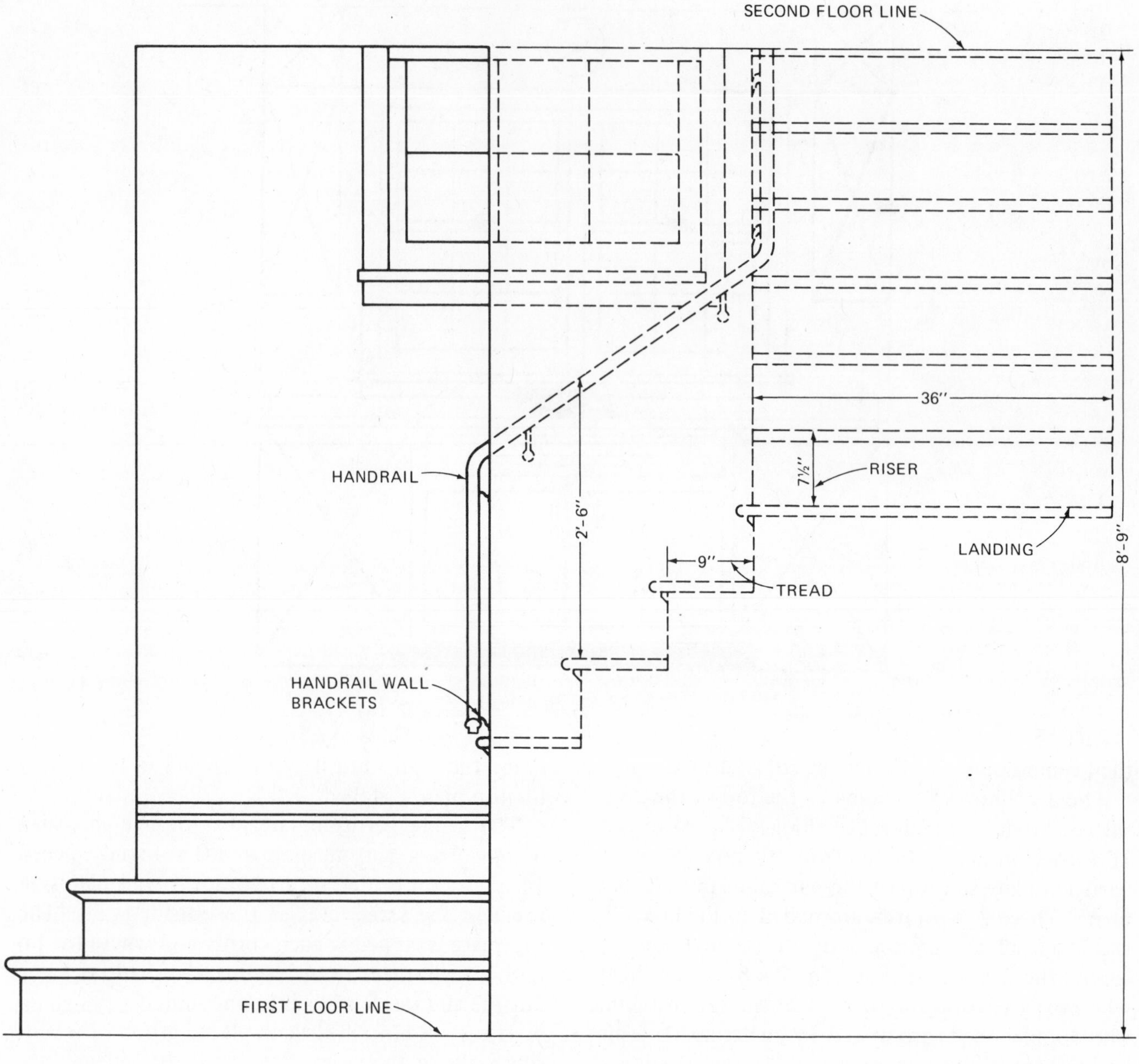

SCALE 3/4" = 1'-0"

Fig. 35-13. Stair detail.

fore nailing the exterior sheathing in place.

Once the sheathing is nailed in place, then the ceiling joists are placed 16 inches o.c. and are fastened to the wall plates. The rafters are then cut and placed on top. These are nailed to the top of the wall plate, the ceiling joists, and the ridge board. Collar beams are then nailed from rafter to rafter as a cross tie. These are usually placed on every other set of rafters or 32 inches o.c.

The ends of the rafters protrude beyond the face of the exterior sheathing according to the style of cornice designed for the house. Usually 2 x 4 lookouts 32 inches o.c. are attached to the ends of the

rafters and come back against the face of the sheathing. These are used to fasten the *plancer board* or soffit. (The plancer board is the finished trim covering the underside of the roof projection that sticks out beyond the face of the exterior sheathing. This is also called a soffit.) The outside edge of the plancer board is flush with the plumb cut at the end of the rafter. To the ends of the rafters is nailed a 1 x 6 facia board. The facia board is the facing trim covering the ends of the rafters facing the front of the roof projection. The $3\frac{5}{8}$-inch crown mold is nailed to the face of the facia board, and the top of the crown mold comes even with the top edge of the sheathing used to cover the roof rafters to make a base for the roofing material.

The roof sheathing is usually 4' x 8' sheets of plywood. The roof sheathing is then covered with a 15-pound asphalt felt building paper, and then the roof is covered with asphalt strip shingles. Two thicknesses of shingles are laid as the first row; thereafter they are exposed 5 inches up to the ridge.

After the roof, conrnice, gable ends, and gable end sheathing is in place, the window units and the exterior door frames are installed. Building paper is used to cover the outside wall sheathing, and then the exterior finished material is installed. On our house we are using $\frac{1}{2}'' \times 8''$ bevel siding with a 6-inch exposure. This means that 6 inches of the bevel siding is left exposed to the weather.

The outside of the house has been closed in. The rough plumbing and electrical wiring are now installed. The heating contractor also installs the rough heating ductwork. Insulation is attached to all the outside walls and the second floor ceiling. This is stapled to the studs and the ceiling joists, usually 3 inches of insulation in the sidewalls and 6 inches in the ceiling.

The electrical work and plumbing must be inspected before the wallboard can be installed. After the electrical and plumbing work has been approved, then the builder can install the interior wall finish. This may be either gypsum wallboard, wood panels, plaster, or whatever material is designated for the room. If wallboard is used, then the joints have to be filled, sanded, and made ready for paint or paper. When the interior walls are ready to be finished, the interior ceiling cove mold, and door and window trim are installed. The subflooring is covered with

building paper and the finished floor is nailed in place. The base and carpet strip is then installed. The interior doors are hung, kitchen cabinets and countertops are installed. Resilient flooring is then installed. Electrical fixtures, plumbing fixtures, paint, paper, heating system completed, wall to wall carpeting installed, etc. The windows are washed and the house is given a general cleaning. The house is now ready to move into.

Window shades, drapes, shower curtains, landscaping, etc. are the responsibility of the owner.

If we are going to go to a lending institution to get a mortgage on our house, then we have one more project to do. The lending institution will require that we show proof of ownership of the land on which we wish to build our house and they will also require that we have a plot plan showing the location of the house on the lot. We will have to get a surveyor to give us the information we need. The surveyor will make a plot plan showing the boundaries, land elevations, location of all sewers, curbs, sidewalks, compass direction of the plot of ground, and any other information such as the location of trees, etc.

When we have this information, then we can draw a plot plan showing the location of the house on the lot. The plot plan is usually drawn at $\frac{1}{16}''=$ 1'0'' scale. Figure 35-14 shows what our plot plan would look like. Notice we have selected some trees to be left and we have shown the elevation of the first floor. Our house sets back 30 feet from the property line and has a $1\frac{1}{2}$-foot slope in that 30 feet. We placed our house on the lot so that at a future time we can have a driveway and a garage on the right side of the house.

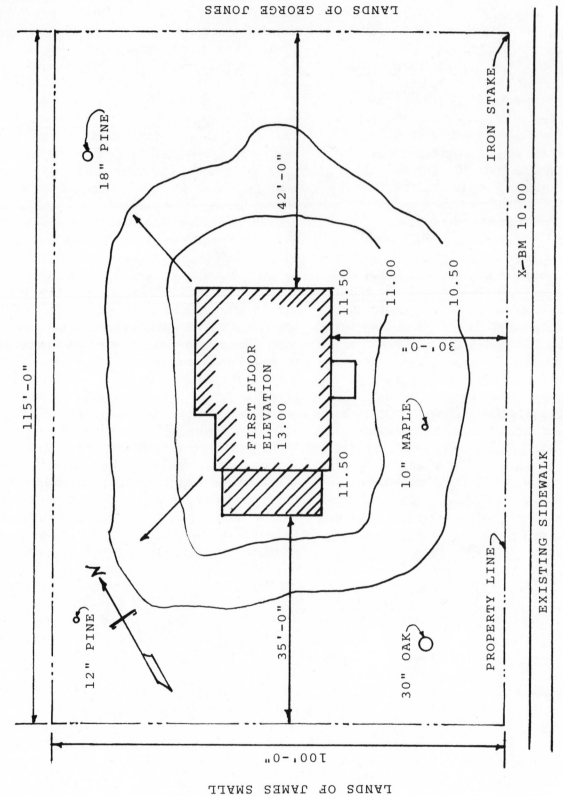

Fig. 35-14. Plot Plan.

36
DRAWING A RANCH HOUSE WITH A BRICK VENEER FRONT

We have completed the drawing of a two-story house with all the details. Now we will draw a ranch house with a brick veneer front. Our first step, the same as in the first house, is to make a word picture of what we would like in the house.

We would like a one-story ranch house with brick veneer on the front, three bedrooms, a living room–dining room combination, a fireplace in the living room. We would like an entrance foyer and a hall-way so we can go from one room to another without having to go through any other room, a bath with a tub and shower combination and a vanity in the bathroom, and a kitchen with adequate cabinets and an eating area. We would like an attached garage and a porch facing the backyard. We need ample closets, double hung windows with cross ventilation, wood shingles on the exterior walls except where the bricks are attached, removable grilles on the windows, asphalt shingle roof, and Early American design.

This gives us a word picture of what we would like in our house. We should try to keep our sketch within the modular system. In making our rough sketch, we should consider the flow of traffic between the rooms. We should group the rooms into areas. The bedroom and bath area should be in one group, the kitchen and dining area should be in a group, and the living room and activity room should be in another group. This gives us a general location in which to arrange our room layout.

Using our word picture, we can start to make our rough sketch, using a piece of graph paper with $\frac{1}{4}$-inch squares. We want an attached garage and a porch facing the backyard. This we can place on one side. We want three bedrooms and a bath, and these we can place on the other side. We can now place our living room–dining room, the kitchen,

entrance, and stairs to the basement between the bedroom area and the garage area.

The garage should be accessible from the house, and we wanted an entrance hall so we can go from one room to another without having to pass through another room. We have to be able to go down into the basement from the first floor, and our basement stairs should be near the kitchen or the central hall.

To save on plumbing materials, we should have the bathroom near the kitchen if possible; the bathroom should also be readily accessible from all the rooms. We want double hung windows and we want cross ventilation wherever possible.

This house does not have to be large to adequately fit this word picture. When designing a house you should always keep in mind the cost factor. The larger the house in square-foot area, the greater the cost for materials and workmanship to build that house.

Figure 36-1 gives us a layout covering our word picture and all of this is in a compact design.

After we have made the rough sketch for our house from the word picture, we now start to draw the working drawings. When using brick veneer on a house, we still consider modular planning and use standard lengths of lumber to our best advantage. Brick or stone veneer will only add 6 inches to the size of the building, and this is taken care of when we draw the foundation and basement plan.

When veneer is used on a residence and the foundation is to be constructed with cement block, where the veneer work is to be installed the foundation under the veneer work is usually 12-inch cement block. These blocks are used for 9 courses and then $1\frac{1}{2}$ courses are 8-inch cement block laid on top, with the inside face of the 8-inch block flush with the inside face of the 12-inch block. This

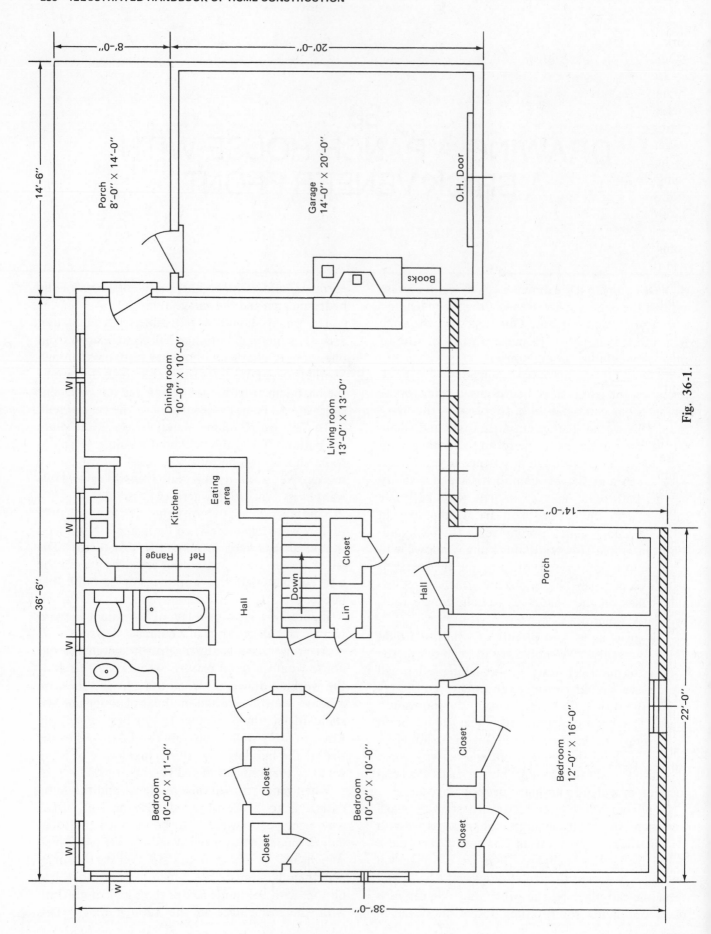

Fig. 36-1.

gives a 4-inch shelf on the outside to start the veneer work.

Our first step is to draw the floor plan, using $\frac{1}{4}'' = 1'0''$ scale. We add the room sizes and closets we have rough-sketched. We must then add the thickness of each interior and exterior wall and partition. Working with the modular unit in mind, we get a total of 38'0'' for the bedroom wing of the house. The back bedroom, bath, kitchen, and dining room areas plus the partitions and walls total 36'6''. The garage, which is on the right side of our plan, we will make 14'0'' wide. This is a one-car garage but with the fireplace in the living room protruding into the garage area, we will need this extra width. The length of the garage and back porch totals 28'0''. We have our living room–dining room area a total of 24'0''. This gives us a 2'0'' setback from the front of the garage front wall. We carry this line across the front of our living room to the front bedroom wing which protrudes in front of this wall. When we extend this front wall straight through to the left side of the house, we have a supporting partition to allow the roof framing to rest on this wall. If this is done, the front gable for the front bedroom wing of the house can then bearfoot back on the main roof, eliminating the need of cutting and fitting valley and jack rafters.

We wanted a porch to protect the front entrance door, so we make this 6 feet wide. We now determine the thickness of our outside walls and draw the inside line representing the inside of the exterior walls. Locate by centerlines all the interior partitions and draw them in with light, solid lines. Where the veneer work is to be installed, draw a light solid line 6 inches in front of the line you drew to represent the face of the exterior wall. Rough in the general location of the fireplace. Your drawing should now be similar to that in Figure 36-2.

Our next step after we have drawn the outside walls and the interior partitions is to locate by centerlines all the windows and exterior doors. Our first window will be over the kitchen sink. This we will center on the kitchen wall and we will use a 3'0'' x 3'2'' double hung window. Next, we locate the center of the dining room wall; here we will use a 2'8'' x 4'6'' mullion window. In the bathroom we will use a 2'4'' x 3'10'' single, double hung window. In the back bedroom we will use 2'8'' x 3'10'' single units. We will place them in the corner to give us better wall space for furniture in this room.

In the center bedroom we will use a 2'4'' x 3'10'' mullion, and we should center it on the wall.

The front bedroom we will use two single units 3'0'' x 4'6''. On the front wall, we will center this window in the center of the whole front wall, which includes the front porch area. This places the window off center in the bedroom, but will give a balanced appearance from the outside. You should keep in mind what the exterior of the house will look like and try to keep it in good proportion.

The side window of the front bedroom, we will place near the closet partition to give us better wall space for a dresser or chest of drawers. For the front windows in the living room we will use 3'0'' x 4'6'' double hung; we should space them equally between the edge of the garage wall and the wall of the front porch, making sure we have room for blinds on each side. This locates all our windows.

Locate the front and side doors by centerlines. Locate the center of the overhead garage door. The front door will be 3'0'' x 6'8'', the side porch door 2'8'' x 6'8'', and the garage door a 9'0'' x 7'0'' four-section.

Next locate and show the swing of all the interior and exterior doors, erase the wall lines where the door openings are. Swing all doors so that they swing against a wall or are convenient to open and have full access to closets and rooms. For the two doors from the kitchen into the dining room we will use louver doors and they should swing into the dining area so they can be opened against the dining room wall.

We now lay out the kitchen cabinets. This house is considered a small house and we are limited to the size of the kitchen. Place the sink under the kitchen window. The longest solid wall is along the bathroom, so we will place our appliances on this wall. We will place the refrigerator on the end and we will use a 33-inch cabinet. Place a cabinet between the refrigerator and the range. The range is 30 inches wide. We now fill in the remaining spaces with cabinets. The wall cabinets are placed on each side of the kitchen window, a corner cabinet and then wall cabinets against the appliance wall. Make wall cabinets 12 inches deep, and base cabinets 24 inches deep.

Next we lay out the bathroom. Locate the toilet so that it is hidden from the door when the door is open. Place the tub on the right of the door. A standard 5-foot tub with shower will leave us enough room for a linen closet between the end of the tub

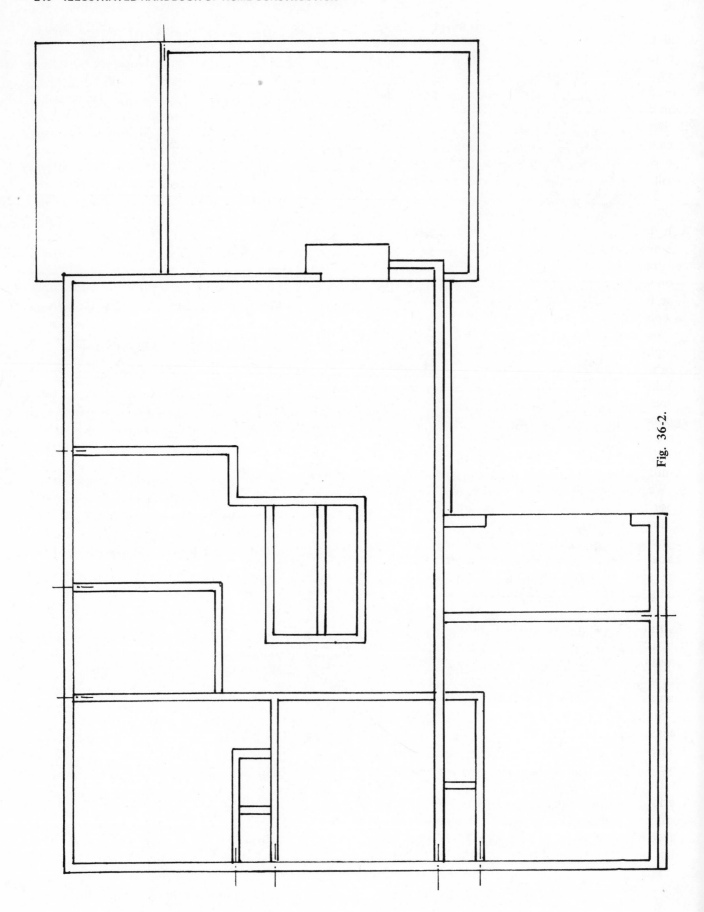

Fig. 36-2.

and the toilet. On the left wall entering the bath, we have room to place a vanity with a built-in lavatory. Notice we have placed all our plumbing in the same general area, which means lower installation cost.

Draw the stairs going down into the basement area. Use a 9-inch run. Now design the fireplace. We draw this so that the face of the fireplace is flush with the inside wall. The front opening will be 3'0" wide. The back of the firebox will be 2'0" wide and will have a depth of 1'6". Place the ash dump, centering it on the rear of the firebox. We need a separate flue liner for the heating system, so draw an 8" x 8" flue to the left of the firebox. You should have 4 inches of masonry between the flue liner and the firebox and 6 inches between the fireplace opening and any wood construction.

We have room to place a bookcase between the fireplace and the corner, and this we can also protrude into the garage area so that the face of the bookcase is flush with the inside wall.

Next designate the brick veneer with the correct symbol. The brick stops on each side of a window or door where the outside casing of the door or window is located. Your drawing should now be similar to Figure 36-3.

We now complete the first floor plan. Give the door size on all doors. Locate the center of all interior partitions where you are going to place the dimensions. Place all the dimensions and notes. Label the name of each room and closet. Draw the stair indicator showing the direction of the stairs leading to the basement area.

Give the correct size for all windows, list the size of the kitchen cabinets. Draw in the slate flooring on the front entrance foyer and have it go into the closet as well. We do this so that wet rubbers or umbrellas can be placed in the closet without doing damage to the floor.

Show the hearth finish on the fireplace. In dimensioning your drawing, remember that all the dimensions must add up to be the same as the overall dimensions.

Dimensions are given from the face of the exterior sheathing to the centerline of the interior partitions. Also remember that the dimensions are given in actual feet and inches regardless of the fact that the drawing is drawn to a smaller scale.

This should finish the first floor plan. Your drawing should now be similar to Figure 36-4.

We have now completed the first floor plan. The next step is to develop the basement and foundation plan. Remember that the foundation must be directly underneath the first floor. Our first step is to draw the outside perimeter of the building exactly the same as we did for the first floor plan. Now determine the thickness of the foundation wall. All walls will be 8-inch cement block, except where the 12-inch block for the veneer work is to be placed.

You should also rough in the location of the fireplace base. This is part of the foundation wall and protrudes into the garage area. Note that the foundation wall for the rear porch and the front of the garage where it meets the main house foundation is offset.

The reason for this is the framing along the dining room–living room–fireplace wall runs out to the front of the frame wall on the front of the garage. This wall runs in a straight line, so we have to offset the masonry foundation so that the face of the main foundation wall and the outside face of the foundation of the garage which returns against the main foundation is the thickness of the frame wall. This same condition exists where the rear porch comes back against the main foundation wall.

Your drawing should now be similar to that in Figure 36-5.

Our next step in developing our foundation and basement plan is to locate the stairs going down into the basement area. These must be in the exact location as the one drawn on the first floor plan.

Now locate and draw the correct position of the girders. We will need a main girder running from the garage wall to the bedroom foundation wall. We will also need a supporting girder across the front wing of the house.

Remember, lumber is manufactured in standard lengths in increments of 2 feet. To follow modular construction, we should run the main girder so it will accommodate standard lengths. The width of the main foundation wall is 24'0" wide, so we measure in 12'0" from the face of the rear foundation wall and draw the girder.

The girder for the front wing should be so placed that it also will accommodate standard lengths of lumber. This should be 12'0" from the center of the main girder. This will allow all joists to be 12'0" long, except the joists under the front bedroom wing, which will be 14'0" long.

Next we locate the supporting posts and the footings. We will need a supporting posts in the center of the girder for the front wing of the house. The center of the main girder should have a supporting

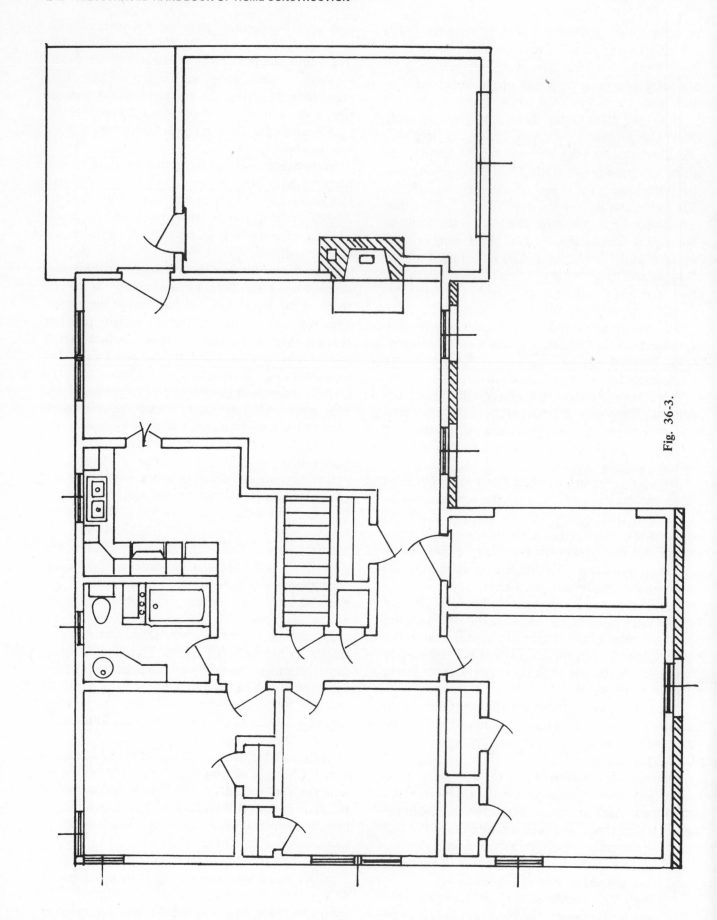

Fig. 36-3.

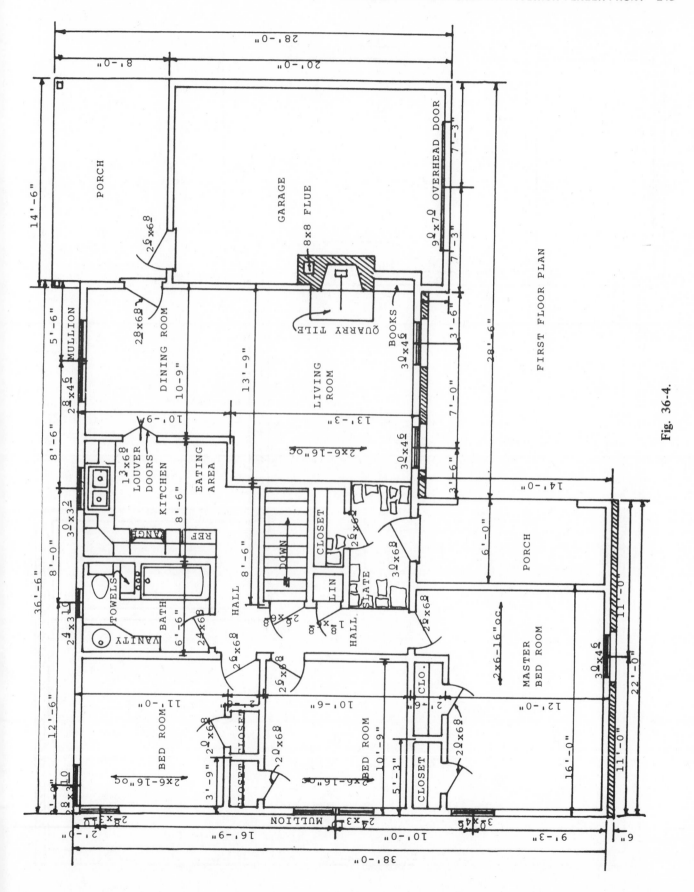

Fig. 36-4.

FIRST FLOOR PLAN

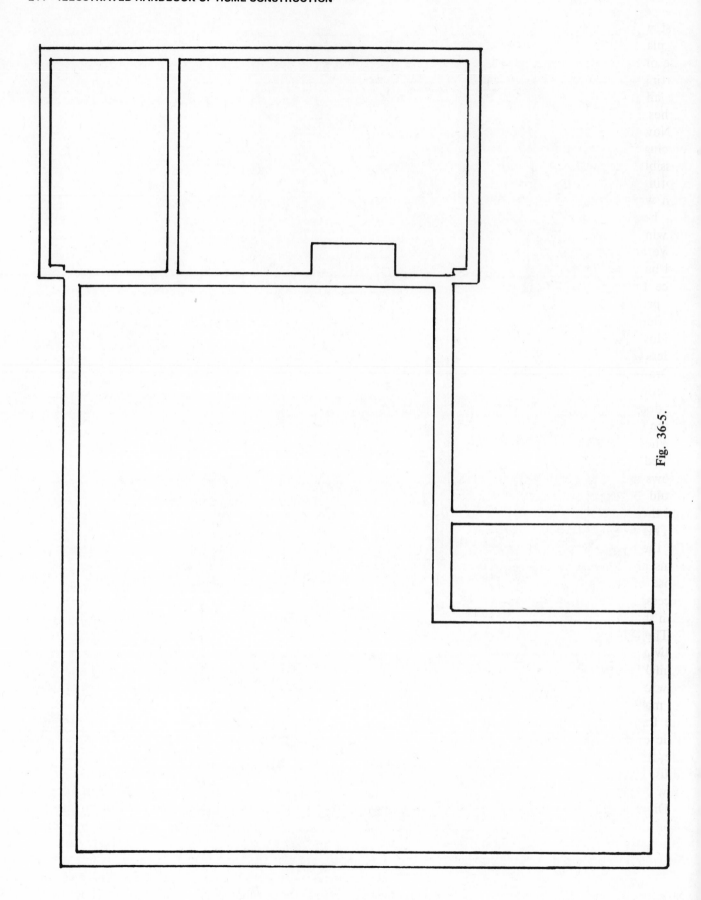

Fig. 36-5.

post which would be 18'3" from each end. Then we place another post a distance of 9'0" on either side of the center post. This will give us three supporting posts under the main girder. The footings for all posts should be 30 inches square and 12 inches deep.

Now locate and draw the basement windows, placing them under the ones on the first floor if possible. We will need five of them. Draw your footing lines 4 inches on each side of the foundation wall. Draw the ash pit and the flue liner for the heating system in the fireplace base. Your drawing should now be similar to Figure 36-6.

We are now ready to complete the foundation and basement plan. Draw your extension lines and place the dimension lines and the dimensions in the proper place. Show the direction and size of the floor joists. Give notes showing the sizes of the windows, areaways, steel posts, and supporting girders (beams).

Draw and show the stair indicator, showing the direction of the stairs. Label the ash pit and cleanout doors on the fireplace base. Show the thickness of the foundation walls.

Where concrete floors and porches are to be poured, these spaces should be marked *Fill* and arrows should be drawn indicating the areas. These would be the front porch, the garage floor area, and the rear porch.

This completes the foundation and basement plan for the house. Check to make sure that the dimensions are correct and that they will conform with those of the first floor plan. Place notes that are necessary. Your finished basement and foundation plan should now be similar to Figure 36-7.

The next procedure in making our set of working drawings is to develop our elevations. To assist us in making the elevations, we first develop a section view of the house. In the first house we designed, we made a partial view to locate the basement floor, the footings, the first floor, the second floor, the attic ceiling, and the ridge of the roof, as well as the finished grade. We will do the same for this house, except we will develop a full size section view drawn at $\frac{1}{4}$" = 1'0" scale.

Draw a light line across the paper horizontally near the bottom of your paper to represent the finished grade line. From this finished grade line, we will develop our cross section view.

The top of the foundation wall should be approximately 12 inches above the finished grade line. The basement wall is 7 feet high. The floor joists are 2 x 8s. The subfloor and finished floor are about 2 inches. The ceiling height for the first floor is 8'1". The ceiling joists are above this point. The rafters run down each gable and rest on the top of the double plate and are nailed into the plate and the ceiling joists.

Use a box cornice and come out 9 inches from the face of the wall sheathing and make the plumb cut for the face of the rafters. You can now develop the cornice.

In drawing the full section view, use a 6/12 pitch for the rafters. Our building is 26'0" from the front of the garage to the back wall of the house. Multiply half of this, or 13, by 6 to get the height of the ridge: 13 × 6 = 78 inches or 6'6", which will be the ridge height from the top of the double plate. Your drawing should now be similar to Figure 36-8.

Next finish your full section view. Place your floor heights. Draw in the courses of cement block and the brick on the front. Your finished full section view should now be similar to that in Figure 36-9.

With our full section drawn, we now draw our left elevation. We already have the outline of this drawn in our full section view. From the full section view, rough in the left elevation as shown in Figure 36-10. This is the same as the full section view except we block in the cornice, locate and rough in the windows, and locate the chimney.

After roughing in the cornice, windows, and chimney, now complete the left side view. Place the triangular louvers in the gable end. These are wood, are 6 feet across the bottom, and have a wire screen backing. The sidewalls are wood shingles, as is the roof. Draw the blinds on each side of the windows after you have drawn the windows to scale. Show the foundation below grade with dotted lines. Your finished drawing should now be similar to Figure 36-11.

We next rough in the front view. We have the heights of the finished grade, the first floor, and the cornice line. We get the location of the windows and doors from our floor plan. We can locate the chimney and block in the windows and doors. In the front gable we will use a small rectangular wood louver for ventilation. Draw the foundation lines below grade with hidden lines. Note that you do not see the top of the foundation where the veneer work goes because we start the brickwork just below the finished grade. Figure 36-12 is our rough front view.

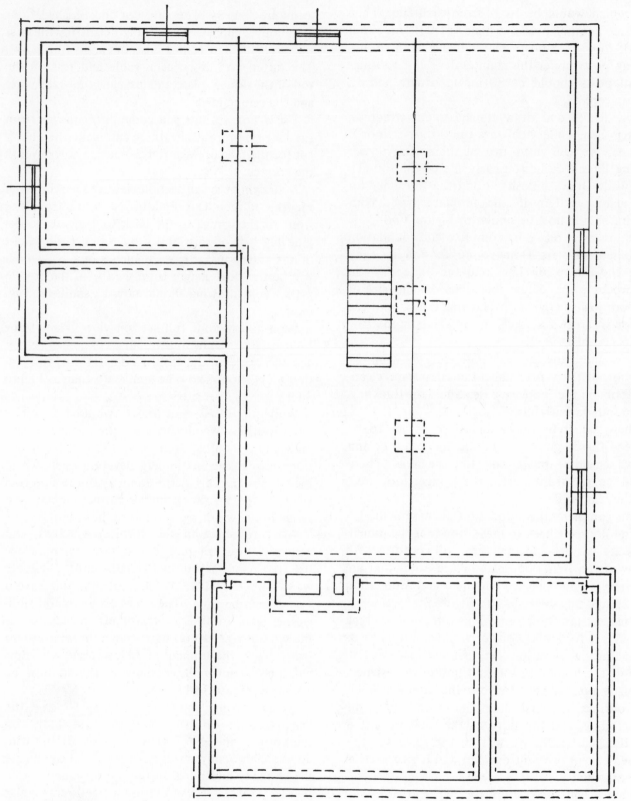

Fig. 36-6.

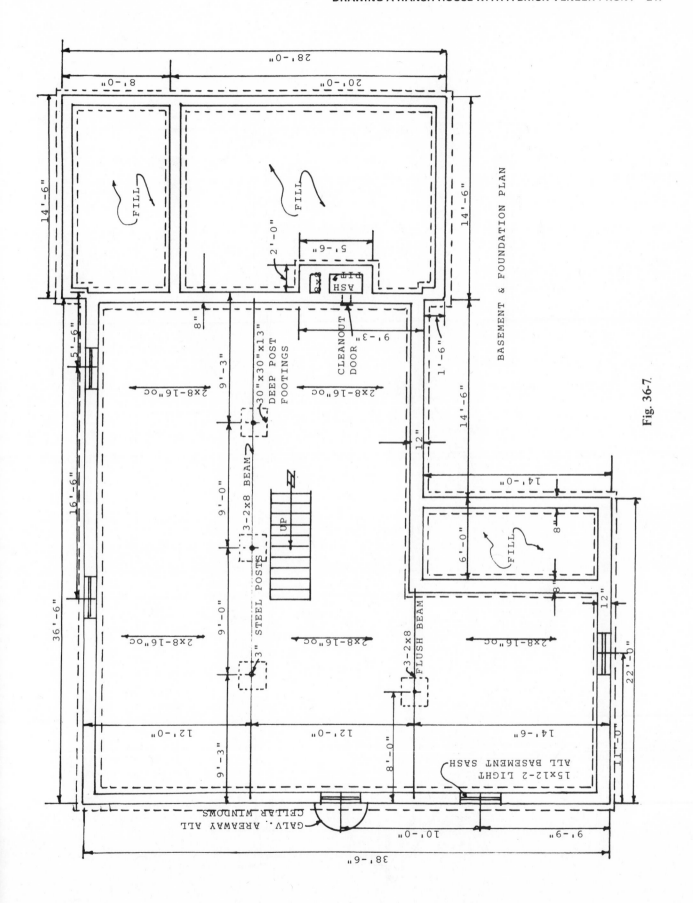

Fig. 36-7.

BASEMENT & FOUNDATION PLAN

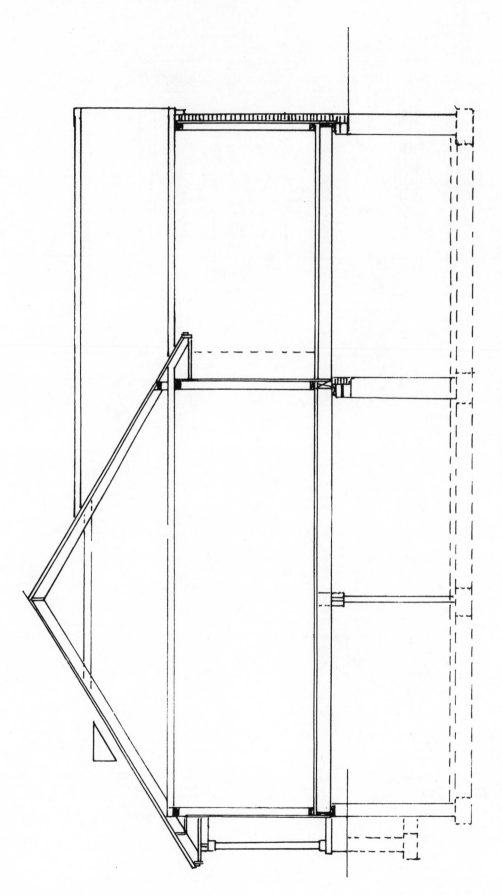

Fig. 36-8.

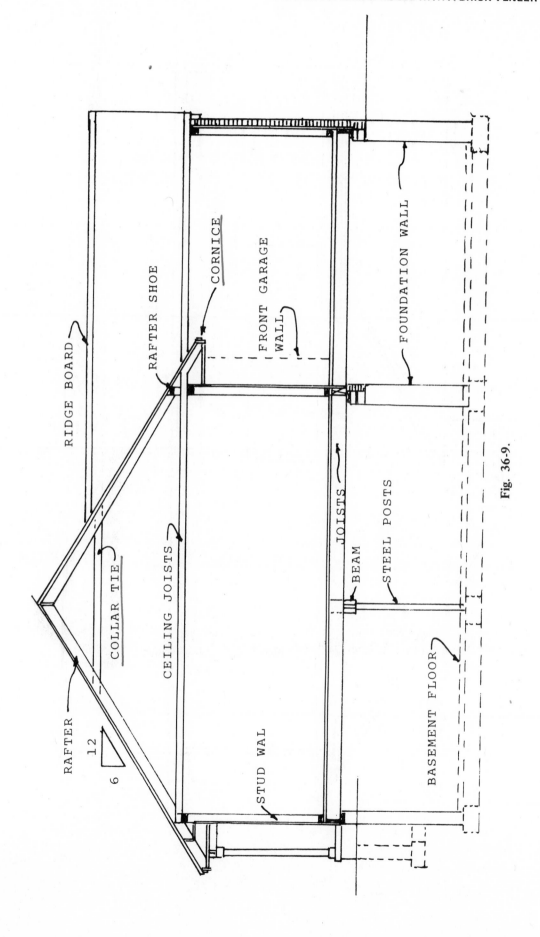

RIDGE BOARD

RAFTER SHOE

CORNICE

FRONT GARAGE WALL

FOUNDATION WALL

COLLAR TIE

CEILING JOISTS

JOISTS

BEAM

STEEL POSTS

RAFTER

12

6

STUD WAL

BASEMENT FLOOR

Fig. 36-9.

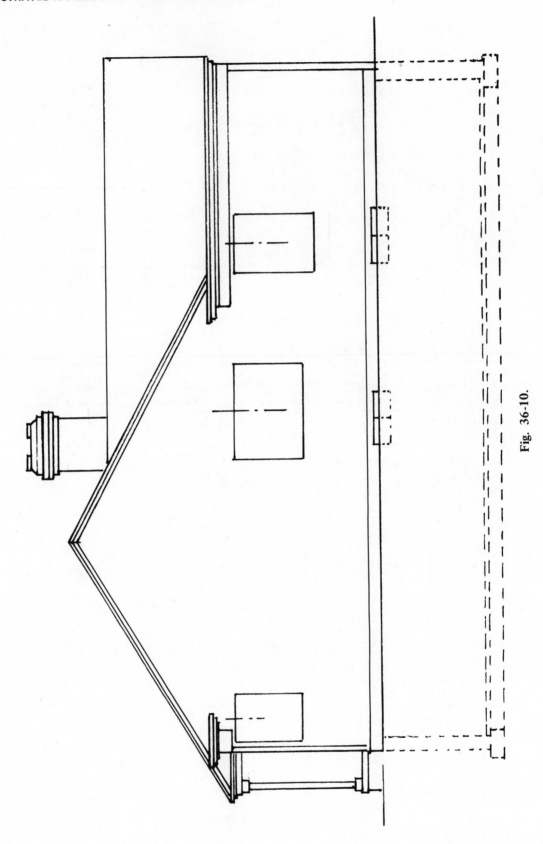

Fig. 36-10.

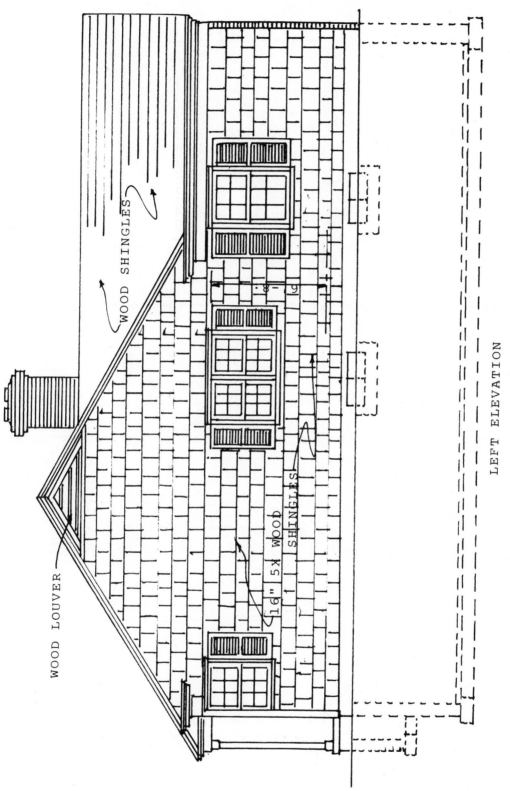

LEFT ELEVATION

Fig. 36-11.

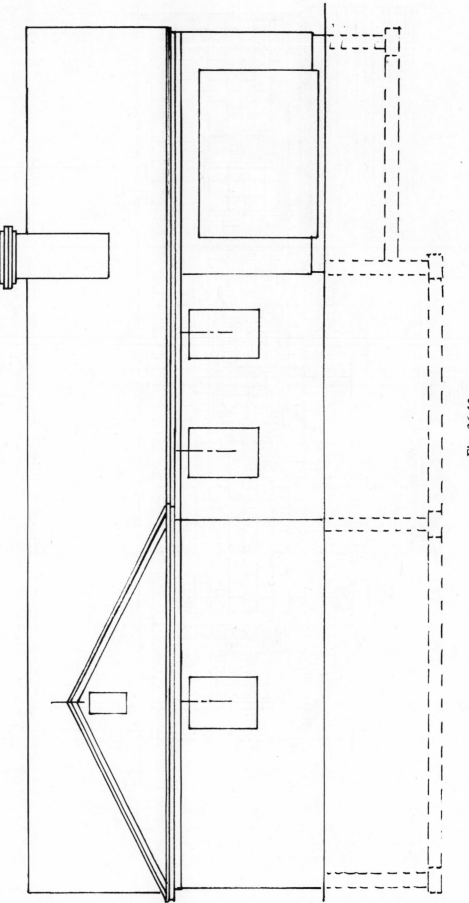

Fig. 36-12.

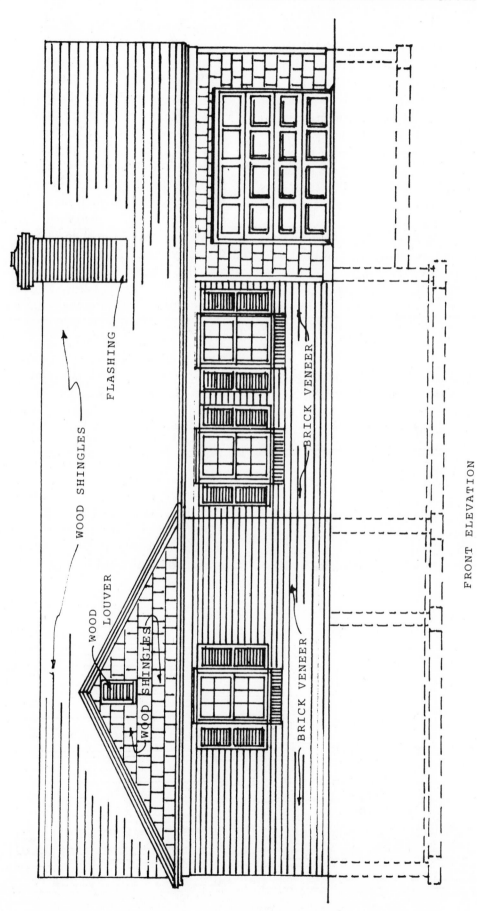

FRONT ELEVATION

Fig. 36-13.

We now complete the front view of our house. Draw the brick and the cornice, and finish the windows. Draw the overhead garage door, the corner boards, and the wood shingles. We framed the front wing gable out over the face of the brick and carried our cornice level. If we had run brick up to the top of the gable, then we would have to have a cornice return on each end of this gable. Draw the chimney, showing the flashing. Label the materials and draw the blinds on the windows. This completes the front elevation, and your drawing should now be similar to Figure 36-13.

Notice that in the front elevation we carried the cornice the same level across the front of the house and garage. On our first floor plan, we have the garage extending two feet out in front of the living room wall. In order to have the cornice at the same level across the front we have to construct a knee wall over the top of the living room wall to carry the rafters.

If we had the rafters of the main house rest on the double plate of the living room wall, then the cornice along the front of the garage would be 12 inches lower. The reason for this is we are using a 6/12 pitch and the front of our garage is two feet in front of the main part of our house. If for every foot horizontally we raise our rafter six inches, then in 2 feet we would raise it 12 inches.

Either way of framing would be correct. Without the knee wall and resting the rafters on the double plate of the living room wall, the appearance of the front elevation would change and the cornice would be run differently. The drawing in Figure 36-14 shows how the appearance would change where the garage and house join.

Our next step is to develop the rear elevation. Again rough in the elevation from our full section view. We have established the foundation line, the grade line, the cornice line, and the ridge line. Locate the windows and rough them in. We have the same condition at the back porch as we had on the front of the garage. Our roof is 12 inches lower at the porch. Our rough drawing of the rear elevation should now be similar to Figure 36-15.

We now finish our rear elevation drawing. Draw in the cornice and moldings. Finish the windows and the porch posts. Again notice that the beam carrying the rafters across the back porch is just above the top of the door casing going into the garage. Designate the material used for the finish. On many plans the draftsman will place notes telling what the finish is instead of drawing it in. This is also true in the foundation below grade. Depending upon the space available, some draftsmen will eliminate the dotted lines showing the foundation below grade once they have established this on other elevation views and the section views.

Your finished rear elevation should now be similar to Figure 36-16.

The last elevation we have to develop is the right elevation. Again rough in the view from the full section view. Locate the cornice lines, the chimney, and the back porch. Your drawing should now be similar to Figure 36-17.

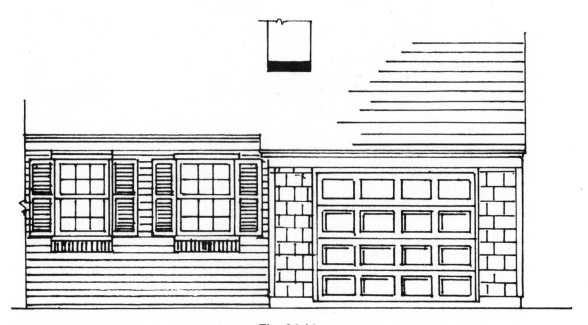

Fig. 36-14.

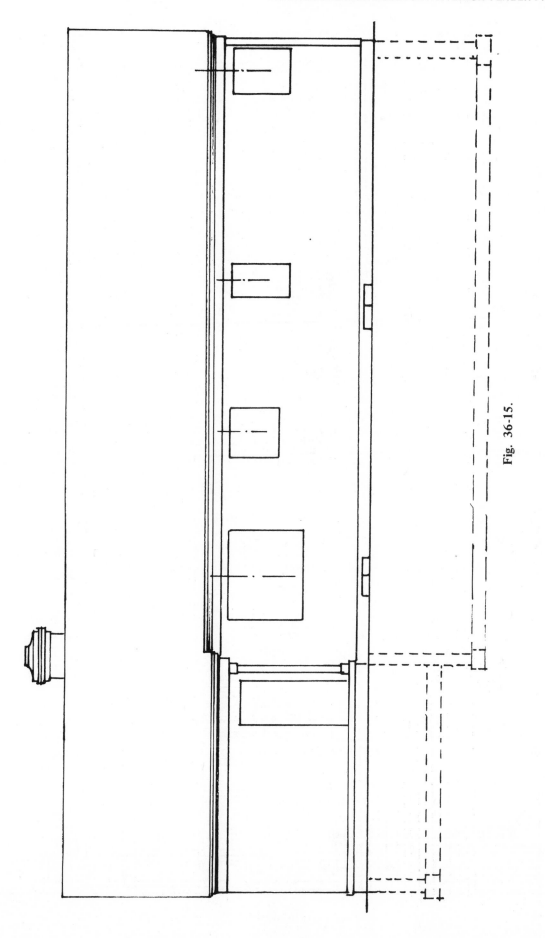

Fig. 36-15.

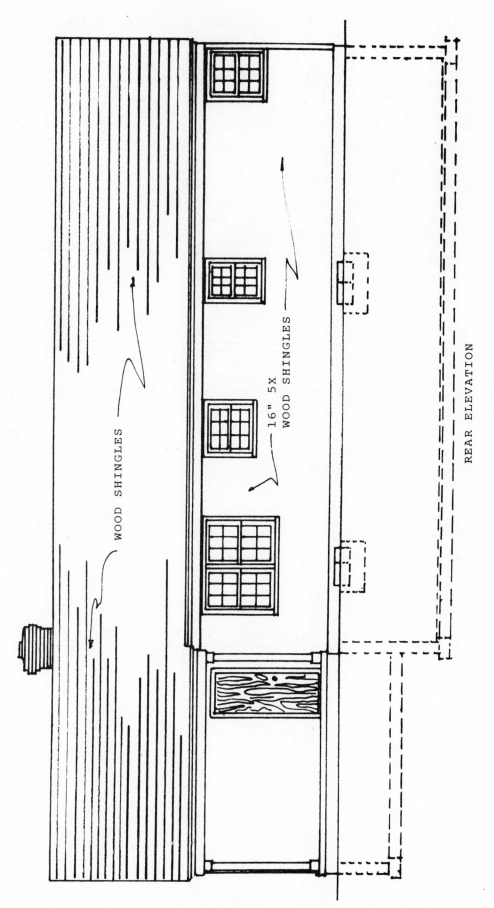

WOOD SHINGLES

16" 5X WOOD SHINGLES

REAR ELEVATION

Fig. 36-16.

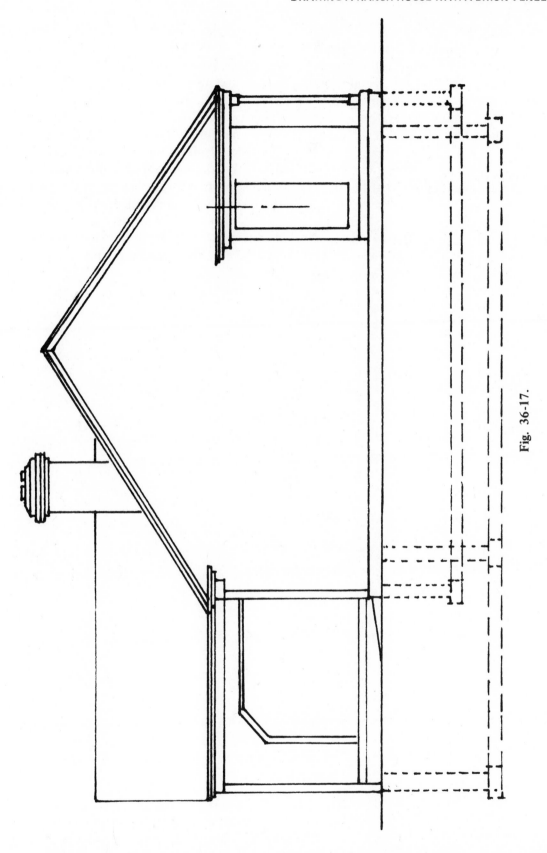

Fig. 36-17.

Now complete the right side elevation. Finish the cornice, the chimney, and the flashing. Draw the louver in the gable end the same as the one on the left elevation. Draw the front porch and show the brick on the front wall. Place any notes necessary. This completes the right side elevation. Your right side elevation should now be similar to Figure 36-18.

This completes our working drawings, except for the details. You should have realized that the floor plan which we called completed lacked the electrical layout. We did this to show how some architects and draftsmen will make a special floor plan and show just the electrical layout. This takes more time on the part of the draftsman but makes the electrical plan much more complete and much easier to read and understand. This is very helpful for the estimator and the electrical contractor. Figure 36-19 gives us the floor plan and the electrical layout for our house.

You have now completed the working drawings for a one-story house except for the special details. You should make a detail of the cornice, the fireplace, the kitchen cabinets, and the girder and supporting posts, as well as a detail of the wall where the veneer work is installed. You have made a detail of each of these on the first drawing and the details on this drawing would be similar. Figures 36-20–36-22 show similar details.

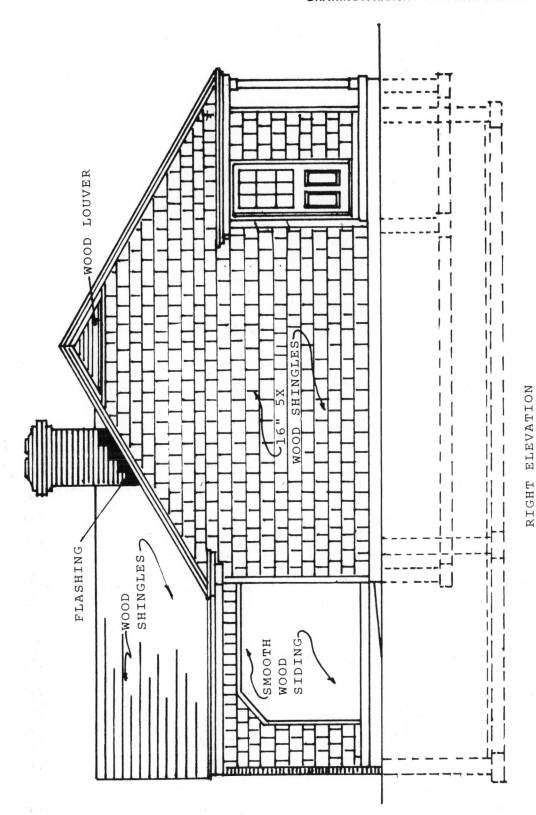

WOOD LOUVER

FLASHING

WOOD SHINGLES

SMOOTH WOOD SIDING

16" 5X WOOD SHINGLES

RIGHT ELEVATION

Fig. 36-18.

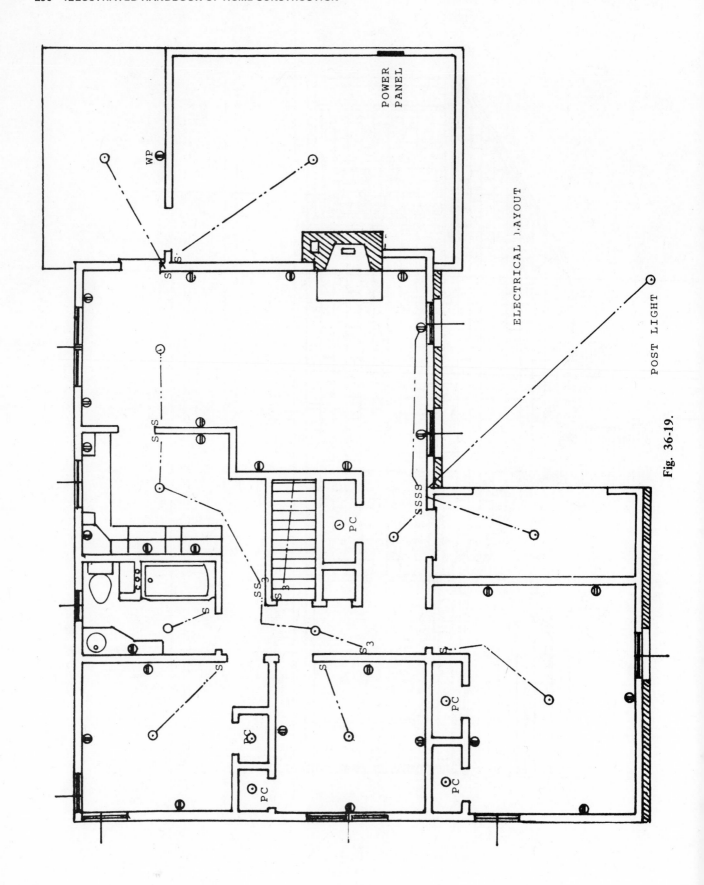

POWER PANEL

WP

ELECTRICAL LAYOUT

POST LIGHT

Fig. 36-19.

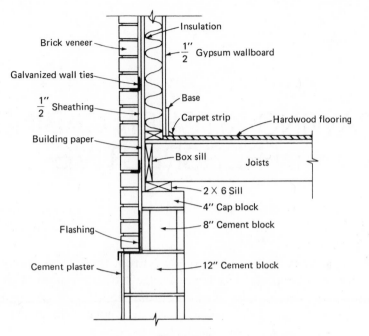

Fig. 36-20.

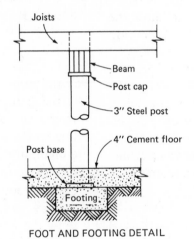

FOOT AND FOOTING DETAIL

Fig. 36-21.

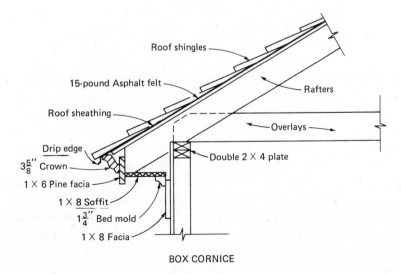

BOX CORNICE

Fig. 36-22.

37
DRAWING KITCHEN CABINETS
IN PERSPECTIVE

In drawing a perspective of a detail such as a living room wall with a bookcase and fireplace, kitchen cabinets, bathroom tub areas, etc. a perspective chart will be of great assistance. Usually such details are drawn at $\frac{1}{2}$" = 1'0" scale. You can use any scale, depending upon the size paper you use and the space available; $\frac{1}{2}$" = 1'0" is adequate to show the details and is an easy scale to work with. To develop a chart, follow the procedures outlined below.

Use a piece of 12" x 18" paper and place it horizontally on your drawing board. Locate the center of your paper and draw a light vertical line. Approximately two thirds of the way up this vertical line from the bottom, draw a horizontal line straight across your paper. This is the eye level line. Mark the point where this eye level line cross the vertical line with the number 6.

Using your $\frac{1}{2}$" = 1'0" scale, measure down from this line 6 feet on the vertical line and mark each foot 5, 4, 3, 2, 1, and 0, respectively. Next measure and mark two feet above the eye level line on the vertical line, and number them 7 and 8. This vertical line which measures from 0 to 8 is our wall measuring line and represents the corner with an 8 foot ceiling height.

Now equal distances on each side of the wall measuring line on the eye level line (line 6) mark your vanishing points, left and right. The vanishing points can be placed anywhere on the eye level line. You will find through practice that between 9 and 10 inches on each side of the centerline will give you a good proportion for the drawing.

Place a straightedge from the right vanishing point through the 0 at the bottom of the wall measuring line. Starting at the 0, draw a line *to the left* along the edge of the straightedge. This will give us the left floor measuring line.

From the left vanishing point, place the straight edge from this point through the 0. Draw a line from point 0 *to the right* along the edge of the straight edge. This gives us the right floor measuring line.

From point 0 to the left on the floor measuring line mark every foot for as long as this wall is to be at $\frac{1}{2}$" = 1'0" scale. Repeat the same procedure starting from the 0 and measuring every foot to the right. Now number each of these points starting at the zero.

Using the vanishing point as a pivot, draw lines from the foot marks on the left floor measuring line to the right. Using the vanishing point right as a pivot, draw lines from the foot marks on the floor measuring line. This will divide the floor area into one foot squares.

Draw ceiling lines in much the same way as you drew the floor measuring lines, using the fanishing points and the 8-foot point. Your chart should now be similar to Figure 37-1. Actually the chart will be much broader because of the width of your paper. Do not try to make the chart too narrow. If you do, the perspective drawing will be at too sharp an angle and the picture will appear distorted.

Now that we have our chart completed, we will make a perspective drawing of a kitchen corner showing the cabinets and the sink. Figure 37-2 is the plan layout of the kitchen cabinets we will draw. The cabinet layout plan is drawn at $\frac{1}{4}$" = 1'0" scale.

Place your drawing paper over the top of the chart and tape the corners, making sure the drawing paper is smooth and flat against the chart. Starting at the 0 at the bottom of the wall measuring line, measure out three feet along each wall. Come out two feet from the floor measuring line and draw a

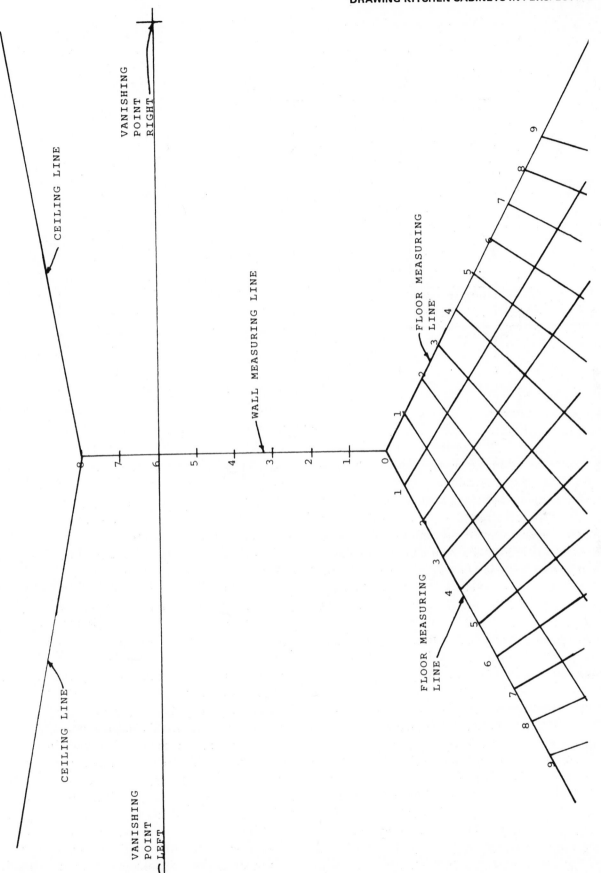

Fig. 37-1. Perspective drawing chart.

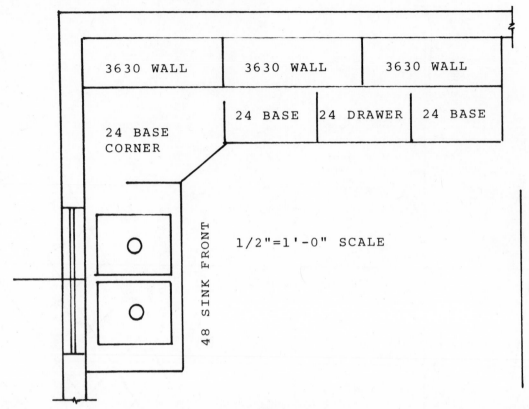

Fig. 37-2.

horizontal line 12 inches long. This is the face width of the base corner cabinet.

From the 0 on the right floor measuring line, measure 9 feet and draw a vertical line from the floor measuring line up to the ceiling line. This is the back wall of both the base and wall cabinets.

From the 0 on the left floor measuring line measure 7 feet and draw a vertical line from the floor measuring line up three feet. This is the end of the sink cabinet at the back wall of the cabinet. Our base cabinets are two feet deep, so from the left end of our sink cabinet, which we measured three feet high, we measure out two feet on our floor measuring line to find the front corner of this unit. We do the same from the 9-foot wall line on our right. Your drawing should now be similar to Figure 37-3.

From our right and left vanishing points we can now block in our base cabinet outline. After you block in the base cabinets with light lines, measure back 3 inches from the face of the base cabinets and measure up $3\frac{1}{2}$ inches. This is the toe space. From the 3-foot point on the sink cabinet measuring line project this line right and left. This is the top of the base cabinets at the wall. From the wall at the back of the cabinets project the top out to the front of the base cabinets, using the left and right vanishing points. Measure the top thickness

$1\frac{1}{2}$ inches and draw the counter top, having it project in front of the base cabinets. Your drawing should now be similar to Figure 37-4.

Next block in the wall cabinets, using the right and left vanishing points. The wall cabinets are 12 inches deep and 30 inches high. Start your measurement on your wall measuring line and project them from these points. Measure up 4 inches on the top of the base cabinets and project from these lines right and left. This is the countertop backsplash. Your drawing should now be similar to Figure 37-5.

Our next step is to mark off the cabinets to their correct sizes. These are measured from the wall measuring line and then projected out to the front of the cabinets, then the vertical lines are drawn.

In perspective drawing all vertical lines are parallel. After you have the cabinet sizes blocked in, then start with the base corner cabinet and draw the drawer and the door. This is a front view, so all lines are vertical and horizontal.

From the width and height of the drawer and door front on this cabinet, draw the door and drawer fronts on the other cabinets. The units on the right are drawn from the left vanishing point through the edge points of the door and drawer on the corner cabinet. The sink front is drawn in the same way, except that the line is drawn from the right vanishing point through the left edge of the corner cabinet drawer and door front.

The wall cabinets are each 36 inches wide, which means they will have two doors on each cabinet. Along the floor measuring line, locate the center of each of the wall cabinets and project the vertical lines up.

Starting at the wall measuring line, locate the top and bottom of the wall cabinet doors. This height is then projected from the left vanishing point to the right. Block in all your doors and drawers showing, the thickness on the side you are looking at.

Locate and draw in the door hardware. Draw the sink using both the right and left vanishing points. Sketch in the faucets.

Draw the kitchen window by using the vanishing points. The window is centered over the sink. Draw the interior window trim.

Make your perspective drawing as realistic as you can. By measuring along the wall measuring line and the floor measuring line and then projecting the lines from the right and left vanishing points, you can develop good perspective drawings. Your finished perspective drawing should now be similar to Figure 37-6. You can use this chart to show almost any detail along any wall and corner.

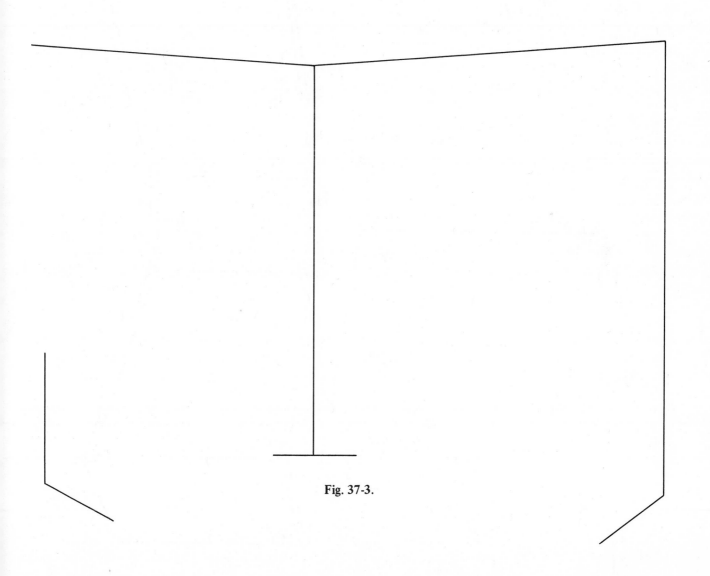

Fig. 37-3.

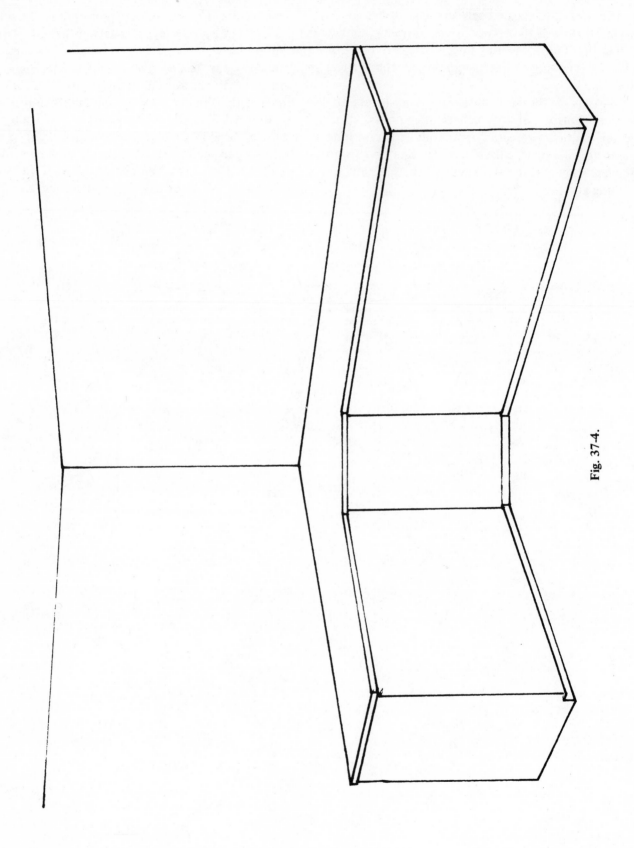

Fig. 37-4.

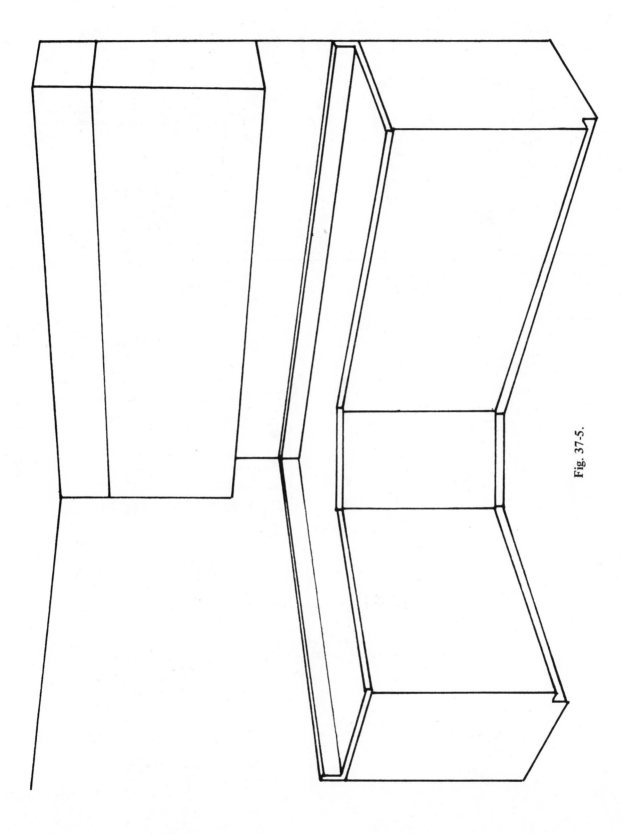

Fig. 37-5.

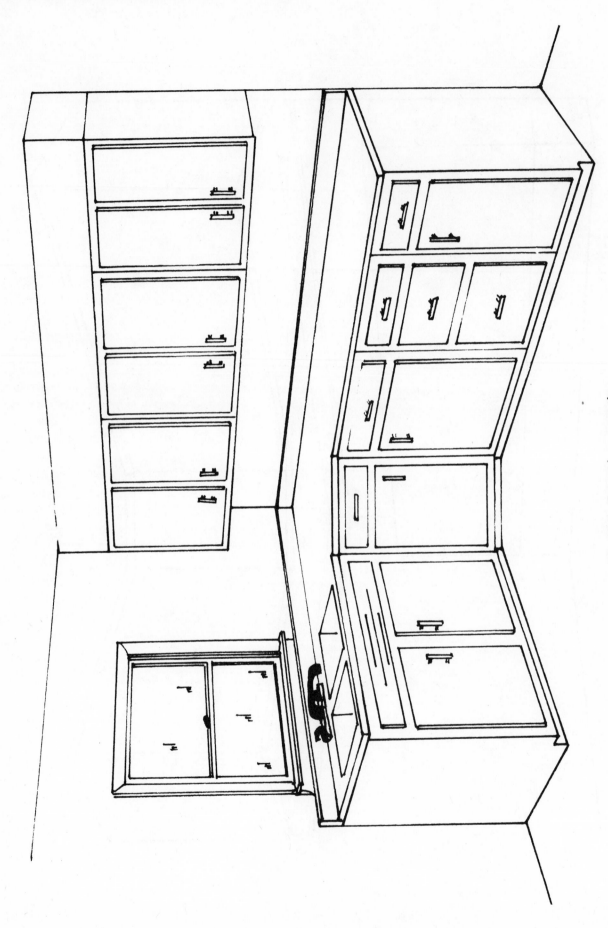

Fig. 37-6. Two-point perspective.

38
GENERAL REVIEW—
ESTIMATING A RANCH HOUSE

The following pages include specifications and working drawings for a single-story residence. This will provide you an opportunity to review and apply the knowledge gained through the use of this book.

Review the specifications carefully and compare then with the appropriate drawings as you prepare the material list. In cases of conflict between the specifications and the working drawings, it should be remembered that the specifications take precedence over the drawings, and that detail drawings take precedence over large scale drawings.

It should be noted that construction specifications may be written in a variety of forms and styles. The specifications for this review use the Description of Materials, Form 2005, from the United States Federal Housing Administration. These specifications describe the material to be used, but do not provide additional information about the construction of the house. When a question arises about construction techniques, you should rely on the knowledge of accepted practices, which we have used throughout this book.

A partially completed list of materials is included. It is only necessary that you determine the amount required. This house is very similar to the one you just drew. The floor plan is the same, except that this plan uses a chimney and no fireplace and the windows are casements and double hung; that the back porch comes even with the main house foundation, where the one we drew was two feet wider and the porch extended beyond the foundation wall; and that the house we drew had wood shingles on the roof and the rafters were 2 x 6s placed 16 inches o.c. and this house uses trusses.

The specifications, the working drawings, and the material list follow.

FHA Form 2005
VA Form 26-1852
Rev. 2/75

U. S. DEPARTMENT OF HOUSING AND URBAN DEVELOPMENT
FEDERAL HOUSING ADMINISTRATION

For accurate register of carbon copies, form
may be separated along above fold. Staple
completed sheets together in original order.

Form Approved
OMB No. 63–R0055

DESCRIPTION OF MATERIALS

No. _____
(To be inserted by FHA or VA)

☐ Proposed Construction

☐ Under Construction

Property address _____ REVIEW HOUSE _____ City _____ Hometown _____ State _____ USA

Mortgagor or Sponsor _____ First Main Bank _____ 101 Main Street
(Name)   (Address)

Contractor or Builder _____ John Doe _____ 225 Main Street
(Name)   (Address)

INSTRUCTIONS

1. For additional information on how this form is to be submitted, number of copies, etc., see the instructions applicable to the FHA Application for Mortgage Insurance or VA Request for Determination of Reasonable Value, as the case may be.
2. Describe all materials and equipment to be used, whether or not shown on the drawings, by marking an X in each appropriate check-box and entering the information called for in each space. If space is inadequate, enter "See misc." and describe under item 27 or on an attached sheet. THE USE OF PAINT CONTAINING MORE THAN ONE HALF OF ONE PERCENT LEAD BY WEIGHT IS PROHIBITED.
3. Work not specifically described or shown will not be considered unless

required, then the minimum acceptable will be assumed. Work exceeding minimum requirements cannot be considered unless specifically described.
4. Include no alternates, "or equal" phrases, or contradictory items. (Consideration of a request for acceptance of substitute materials or equipment is not thereby precluded.)
5. Include signatures required at the end of this form.
6. The construction shall be completed in compliance with the related drawings and specifications, as amended during processing. The specifications include this Description of Materials and the applicable Minimum Property Standards.

---

1. EXCAVATION:

Bearing soil, type _____ Sand and Loam

2. FOUNDATIONS:

Footings: concrete mix _____ ; strength psi _____ 3000 _____ Reinforcing _____ none

Foundation wall: material _____ 8x8x16 & 8x12x16 Conc. Block _____ Reinforcing _____ every 3 courses

Interior foundation wall: material _____ none _____ Party foundation wall _____ none

Columns: material and sizes _____ 3" steel w/ base & cap _____ Piers: material and reinforcing _____ none

Girders: material and sizes _____ 6x8 wood, built-up _____ Sills: material _____ 2x6 fir

Basement entrance areaway _____ none _____ Window areaways _____ corrugated steel

Waterproofing _____ 1/2" cement plaster & 2 coats asphalt _____ Footing drains _____ 4" plastic

Termite protection _____ none

Basementless space: ground cover _____ none _____ ; insulation _____ none _____ ; foundation vents _____ none

Special foundations _____

Additional information: _____ Sisalkraft vapor barrier under basement floor. 6x6-10/10 welded wire mesh on porches and garage

3. CHIMNEYS:

Material _____ brick & block _____ Prefabricated (make and size) _____

Flue lining: material _____ T.C. _____ Heater flue size _____ 8x8 _____ Fireplace flue size _____ none

Vents (material and size): gas or oil heater _____ ; water heater _____

Additional information: _____

4. FIREPLACES: none

Type: ☐ solid fuel; ☐ gas-burning; ☐ circulator (make and size) _____ Ash dump and clean-out _____

Fireplace: facing _____ ; lining _____ ; hearth _____ ; mantel _____

Additional information: _____

5. EXTERIOR WALLS:

Wood frame: wood grade, and species _____ Fir _____ ☐ Corner bracing. Building paper or felt _____ doublekraft

Sheathing _____ plywood _____ ; thickness _____ 1/2" _____ ; width _____ ; ☐ solid; ☐ spaced _____ " o. c.; ☐ diagonal; _____

Siding _____ ; grade _____ ; type _____ ; size _____ ; exposure _____ "; fastening _____

Shingles _____ 16" Red Cedar _____ ; grade _____ #1 _____ ; type _____ 5x _____ ; size _____ 16" _____ ; exposure _____ 7" _____ "; fastening _____

Stucco _____ ; thickness _____ "; Lath _____ ; weight _____ lb.

Masonry veneer _____ Brick _____ Sills _____ Brick _____ Lintels _____ none _____ Base flashing _____

Masonry: ☐ solid ☒ faced ☐ stuccoed; total wall thickness _____ "; facing thickness _____ 4" _____ "; facing material _____ Brick

Backup material _____ ; thickness _____ "; bonding _____

Door sills _____ Oak _____ Window sills _____ Pine _____ Lintels _____ Base flashing _____

Interior surfaces: dampproofing, _____ none _____ coats of _____ ; furring _____

Additional information: _____

Exterior painting: material _____ Nonchalking Shake & Shingle paint _____ ; number of coats _____ 2

Gable wall construction: ☒ same as main walls; ☐ other construction _____

270

6. FLOOR FRAMING:

Joists: wood, grade, and species _____2x8 fir_____ ; other _____ ; bridging ___metal___ ; anchors _____

Concrete slab: ☒ basement floor; ☐ first floor; ☐ ground supported; ☐ self-supporting; mix _____ ; thickness ___4"___ ";

reinforcing ___6x6–10/10 welded wire___ ; insulation _____ ; membrane ___Sisalkraft___

Fill under slab: material ___sand___ ; thickness ___6"___ ". Additional information: _____

7. SUBFLOORING: *(Describe underflooring for special floors under item 21.)*

Material: grade and species ___APA CD Int. w/ Ext glue___ ; size ___5/8"___ ; type ___plywood___

Laid: ☐ first floor; ☐ second floor; ☐ attic _____ sq. ft.; ☐ diagonal; ☐ right angles. Additional information: _____

8. FINISH FLOORING: *(Wood only. Describe other finish flooring under item 21.)*

Location	Rooms	Grade	Species	Thickness	Width	Bldg. Paper	Finish
First floor	Liv. - BR	Select	Oak	1"	3"	deadening felt	2 coats varnish
Second floor							
Attic floor ___ sq. ft.							

Additional information: _____

9. PARTITION FRAMING:

Studs: wood, grade, and species ___Fir___ size and spacing ___2x4–16" OC___ Other _____

Additional information: _____

10. CEILING FRAMING:

Joists: wood, grade, and species ___Trusses___ Other _____ Bridging _____

Additional information: _____

11. ROOF FRAMING:

Rafters: wood, grade, and species _____ Roof trusses (see detail): grade and species ___See details___

Additional information: _____

12. ROOFING:

Sheathing: wood, grade, and species ___1/2" APA CD Int w/ Ext glue plywood___ ; ☐ solid; ☐ spaced _____" o.c.

Roofing ___Asphalt strip___ ; grade _____ ; size _____ ; type ___240 lb.___

Underlay ___Asphalt felt___ ; weight or thickness ___15 lb___ ; size ___36"___ ; fastening _____

Built-up roofing _____ ; number of plies _____ ; surfacing material _____

Flashing: material ___Aluminum___ ; gage or weight ___28 gauge___ ; ☐ gravel stops; ☐ snow guards

Additional information: _____

13. GUTTERS AND DOWNSPOUTS: none

Gutters: material _____ ; gage or weight _____ ; size _____ ; shape _____

Downspouts: material _____ ; gage or weight _____ ; size _____ ; shape _____ ; number _____

Downspouts connected to: ☐ Storm sewer; ☐ sanitary sewer; ☐ dry-well. ☐ Splash blocks: material and size _____

Additional information: _____

14. LATH AND PLASTER

Lath ☐ walls, ☐ ceilings: material _____ ; weight or thickness _____ Plaster: coats _____ ; finish _____

Dry-wall ☒ walls, ☒ ceilings: material ___gypsum___ ; thickness ___1/2"___ ; finish ___Tape & Joint system___ ;

Joint treatment ___according to manufacturer's specifications___

15. DECORATING: *(Paint, wallpaper, etc.)*

Rooms	Wall Finish Material and Application	Ceiling Finish Material and Application
Kitchen	2 semi-gloss enamel	2 semi-gloss enamel
Bath	2 semi-gloss enamel	2 semi-gloss enamel
Other	2 flat latex	2 flat latex

Additional information: _____

16. INTERIOR DOORS AND TRIM:

Doors: type ___Flush___ ; material ___Birch veneer___ ; thickness ___1 3/8"___

Door trim: type ___Colonial___ ; material ___pine___ Base: type ___colonial___ ; material ___pine___ ; size ___3"___

Finish: doors ___Varnish___ ; trim ___primed & painted___

Other trim *(item, type and location)* _____

Additional information: _____

17. WINDOWS:

Windows: type ___D.H. & Csmnt___ ; make ___Wilson___ ; material ___pine___ ; sash thickness ___1 3/8"___

Glass: grade ___insul.___ ; ☐ sash weights; ☐ balances, type _____ ; head flashing _____

Trim: type ___Colonial___ ; material ___pine___ Paint ___2 coats___ ; number coats _____

Weatherstripping: type ___factory-installed___ ; material ___vinyl___ Storm sash, number _____

Screens: ☒ full; ☐ half; type _____ ; number ___10___ ; screen cloth material ___aluminum___

Basement windows: type ___2-light___ ; material ___steel___ ; screens, number _____ ; Storm sash, number _____

Special windows _____

Additional information: _____

18. ENTRANCES AND EXTERIOR DETAIL:

Main entrance door: material ___pine___ ; width ___3'-0"___ ; thickness ___1 3/4"___ Frame: material ___pine___ ; thickness ___5/4"___

Other entrance doors: material ___pine___ ; width ___2'-8"___ ; thickness ___1 3/4"___ Frame: material ___pine___ ; thickness ___5/4"___

Head flashing _____ Weatherstripping: type ___interlocking___ ; saddles ___oak___

Screen doors: thickness ___none___ ; number _____ ; screen cloth material _____ Storm doors: thickness _____" ; number _____

Combination storm and screen doors: thickness ___1 1/8"___ ; number ___2___ ; screen cloth material ___aluminum___

Shutters: ☐ hinged; ☐ fixed. Railings _____ , Attic louvers _____

Exterior millwork: grade and species ___#1 pine___ Paint ___primed & painted___ ; number coats ___2___

Additional information: _____

19. CABINETS AND INTERIOR DETAIL:

Kitchen cabinets, wall units: material ___Kingswood Oakmont – Sutherland___ ; lineal feet of shelves _____ ; shelf width _____

Base units: material ___Kingswood Oakmont___ ; counter top ___Plastic laminate___ ; edging ___Plastic Laminate___

Back and end splash ___4"___ Finish of cabinets ___Factory finished___ ; number coats _____

Medicine cabinets: make ___none___ ; model _____

Other cabinets and built-in furniture ___Bathroom vanity RV-48___

Additional information: _____

20. STAIRS:

STAIR	TREADS		RISERS		STRINGS		HANDRAIL		BALUSTERS	
	Material	Thickness	Material	Thickness	Material	Size	Material	Size	Material	Size
Basement ___	fir	5/4"	pine	3/4"	fir	3/4"	pine	1 1/2"	none	
Main ___										
Attic ___										

Disappearing: make and model number _____

Additional information: _____

21. SPECIAL FLOORS AND WAINSCOT:

	LOCATION	MATERIAL, COLOR, BORDER, SIZES, GAGE, ETC.	THRESHOLD MATERIAL	WALL BASE MATERIAL	UNDERFLOOR MATERIAL
FLOORS	Kitchen ___				
	Bath ___				

	LOCATION	MATERIAL, COLOR, BORDER, CAP. SIZES, GAGE, ETC.	HEIGHT	HEIGHT OVER TUB	HEIGHT IN SHOWERS (FROM FLOOR)
WAINSCOT	Bath ___				

Bathroom accessories: ☐ Recessed; material _____ ; number _____ ; ☐ Attached; material _____ ; number _____

Additional information: _____

22. PLUMBING:

FIXTURE	NUMBER	LOCATION	MAKE	MFR'S FIXTURE IDENTIFICATION NO.	SIZE	COLOR
Sink ___	1	Kitchen	Kessler	SA-410 Titan	21x32	ss
Lavatory ___	1	Bath	Kessler	K-2208 Continental	16" round	blue
Water closet ___	1	Bath	Kessler	K-3512—PBA		blue
Bathtub ___	1	Bath	Kessler	K-786-SA Royal	5'-0"	blue
Shower over tub △	1	Bath	Nisson	7004-T		chrome
Stall shower △ ___						
Laundry trays ___						

△ ☒ Curtain rod   △ ☐ Door   ☐ Shower pan: material _____

Water supply: ☒ public; ☐ community system; ☐ individual (private) system. ★

272

Sewage disposal: ☒ public; ☐ community system; ☐ individual (private) system.★

★*Show and describe individual system in complete detail in separate drawings and specifications according to requirements.*

House drain (inside): ☐ cast iron; ☐ tile; ☐ other ___copper___ House sewer (outside): ☒ cast iron; ☐ tile; ☐ other _____

Water piping: ☐ galvanized steel; ☒ copper tubing; ☐ other _____ Sill cocks, number __2__

Domestic water heater: type ___electric___ ; make and model ___Williams WG-40___ ; heating capacity ___4000 Watts___

_____ gph. 100° rise. Storage tank: material _____ ; capacity __40__ gallons.

Gas service: ☐ utility company; ☐ liq. pet. gas; ☐ other _____ Gas piping: ☐ cooking; ☐ house heating.

Footing drains connected to: ☐ storm sewer; ☐ sanitary sewer; ☐ dry well. Sump pump; make and model _____ ; capacity _____ ; discharges into _____

23. HEATING:

☐ Hot water. ☐ Steam. ☐ Vapor. ☐ One-pipe system. ☐ Two-pipe system.

☐ Radiators. ☐ Convectors. ☐ Baseboard radiation. Make and model _____

Radiant panel: ☐ floor; ☐ wall; ☐ ceiling. Panel coil: material _____

☐ Circulator. ☐ Return pump. Make and model _____ ; capacity _____ gpm.

Boiler: make and model _____ Output _____ Btuh.; net rating _____ Btuh.

Additional information: _____

Warm air: ☐ Gravity. ☒ Forced. Type of system ___Mitchell, upflow oil fired___

Duct material: supply ___galv.___ ; return ___galv.___ Insulation _____ , thickness _____ ☐ Outside air intake.

Furnace: make and model ___Mitchell___ Input __125,000__ Btuh.; output __100,000__ Btuh.

Additional information: ___1500 cfm___

☐ Space heater; ☐ floor furnace; ☐ wall heater. Input _____ Btuh.; output _____ Btuh.; number units _____

Make, model _____ Additional information: _____

Controls: make and types ___Sherwood T861A clock thermostat___

Additional information: _____

Fuel: ☐ Coal; ☐ oil; ☐ gas; ☐ liq. pet. gas; ☐ electric; ☐ other _____ ; storage capacity __1,000 gal.__

Additional information: _____

Firing equipment furnished separately: ☐ Gas burner, conversion type. ☐ Stoker: hopper feed ☐; bin feed ☐

Oil burner: ☐ pressure atomizing; ☐ vaporizing _____

Make and model _____ Control _____

Additional information: _____

Electric heating system: type _____ Input _____ watts; @ _____ volts; output _____ Btuh.

Additional information: _____

Ventilating equipment: attic fan, make and model _____ ; capacity _____ cfm.

kitchen exhaust fan, make and model _____

Other heating, ventilating. or cooling equipment _____

24. ELECTRIC WIRING:

Service: ☐ overhead; ☒ underground. Panel: ☐ fuse box; ☒ circuit-breaker; make ___Cunningham___ AMP's __200__ No. circuits __30__

Wiring: ☐ conduit; ☐ armored cable; ☒ nonmetallic cable; ☐ knob and tube; ☐ other _____

Special outlets: ☒ range; ☒ water heater; ☐ other _____

☐ Doorbell. ☒ Chimes. Push-button locations ___Front door___ Additional information: _____

25. LIGHTING FIXTURES:

Total number of fixtures __11__ Total allowance for fixtures, typical installation, $ __400.00__

Nontypical installation _____

Additional information: _____

26. INSULATION:

LOCATION	THICKNESS	MATERIAL, TYPE, AND METHOD OF INSTALLATION	VAPOR BARRIER
Roof			
Ceiling	R-21	Fiberglass stapled to trusses	Kraft
Wall	R-11	Fiberglass stapled to studs	Kraft
Floor			

27. MISCELLANEOUS: (Describe any main dwelling materials, equipment, or construction items not shown elsewhere; or use to provide additional information where the space provided was inadequate. Always reference by item number to correspond to numbering used on this form.)

_____
_____
_____
_____
_____

273

_____
_____
_____
_____
_____
_____

HARDWARE: *(make, material, and finish.)* ____ Welch, brass finish. Allowance of $500.00 for purchase of finish hardware, including bathroom mirrors. _____

SPECIAL EQUIPMENT: *(State material or make, model and quantity.   Include only equipment and appliances which are accept-able by local law, custom and applicable FHA standards.   Do not include items which, by established custom, are supplied by occupant and removed when he vacates premises or chattles prohibited by law from becoming realty.)*_____
_____
_____
_____
_____
_____
_____
_____

PORCHES: Porches of same construction as house. _____
_____

TERRACES: _____
_____
_____

GARAGES: Attached — of same construction as house. _____
_____

WALKS AND DRIVEWAYS:
Driveway: width ___9'-0"___ ; base material __crushed stone__ ; thickness __3"__ "; surfacing material ___Asphalt___ ; thickness __2"__ "
Front walk: width __3'-6"__ ; material ___Conc.___ ; thickness __4"__ ". Service walk: width _____ ; material _____ ; thickness _____ "
Steps: material _____ ; treads _____"; risers _____". Cheek walls _____

OTHER ONSITE IMPROVEMENTS:
(Specify all exterior onsite improvements not described elsewhere, including items such as unusual grading, drainage structures, retaining walls, fence, railings, and accessory structures.)
_____
_____
_____

LANDSCAPING, PLANTING, AND FINISH GRADING:
Topsoil ___6___ " thick: ☒ front yard; ☒ side yards; ☒ rear yard to _____125_____ feet behind main building.
Lawns *(seeded, sodded, or sprigged)*: ☒ front yard __seeded__ ; ☒ side yards __seeded__ ; ☒ rear yard __seeded__
Planting: ☐ as specified and shown on drawings; ☐ as follows:

_____ Shade trees, deciduous, _____" caliper.	_____ Evergreen trees. _____' to _____', B & B.
_____ Low flowering trees, deciduous, _____' to _____'	_____ Evergreen shrubs. _____' to _____', B & B.
_____ High-growing shrubs, deciduous, _____' to _____'	_____ Vines, 2-year _____
_____ Medium-growing shrubs, deciduous, _____' to _____'	_____
_____ Low-growing shrubs, deciduous, _____' to _____'	_____

IDENTIFICATION.—This exhibit shall be identified by the signature of the builder, or sponsor, and/or the proposed mortgagor if the latter is known at the time of application.

Date_____   Signature _____

Signature _____

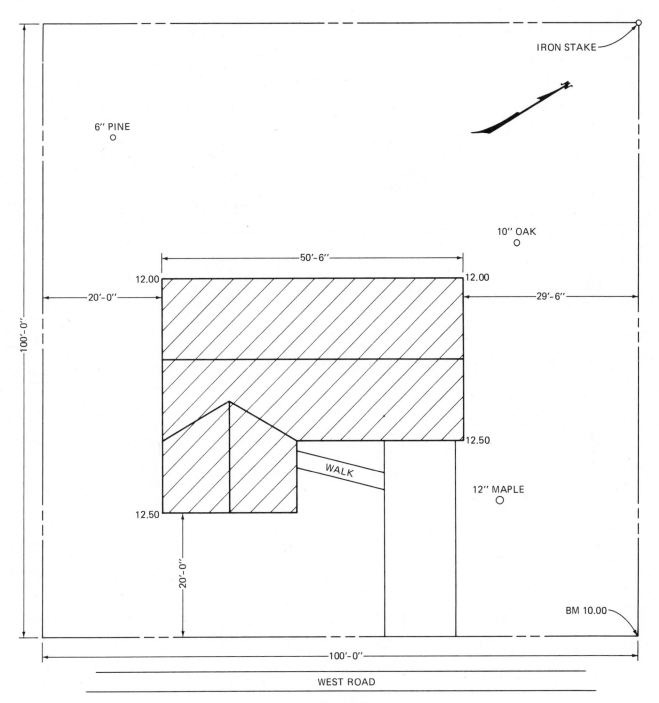

IRON STAKE

6" PINE
○

10" OAK
○

12.00   12.00

50'-6"

20'-0"   29'-6"

100'-0"

12.50

WALK

12" MAPLE
○

12.50

20'-0"

BM 10.00

100'-0"

WEST ROAD

PLOT PLAN
SCALE 1/16" = 1'-0"

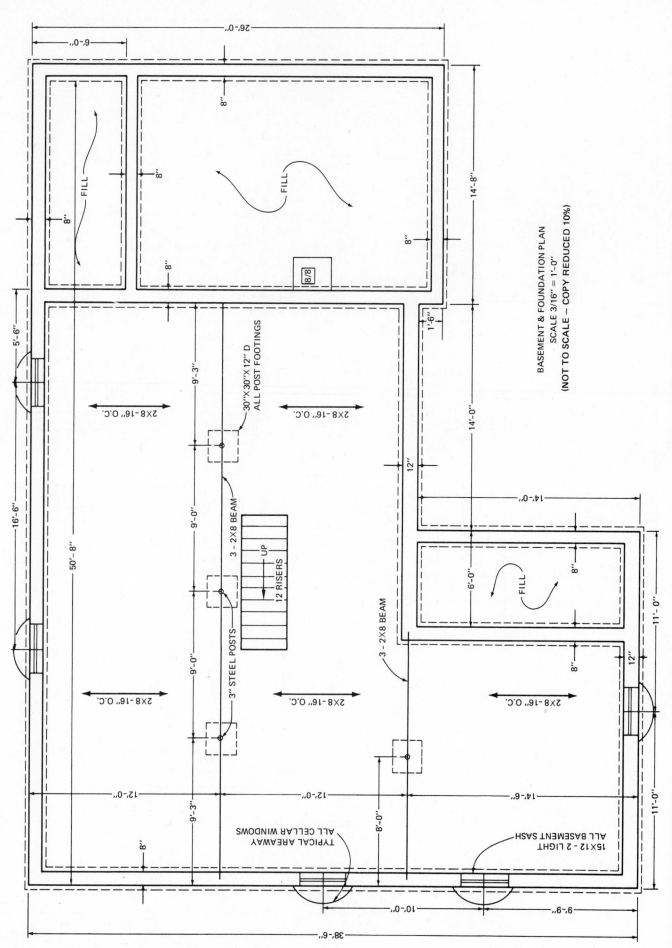

BASEMENT & FOUNDATION PLAN
SCALE 3/16" = 1'-0"
(NOT TO SCALE – COPY REDUCED 10%)

276

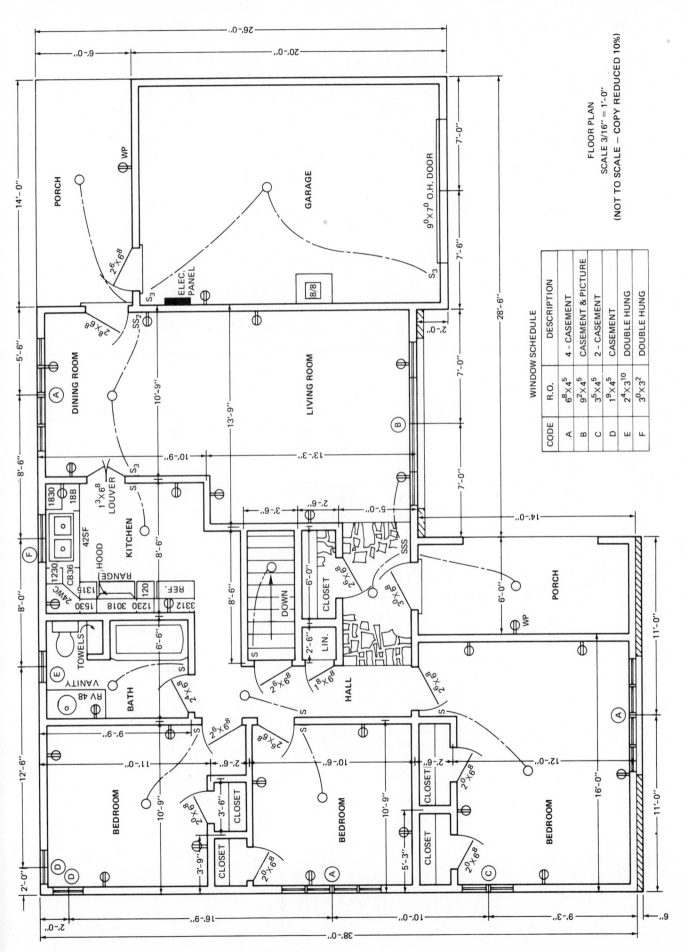

WINDOW SCHEDULE

CODE	R.O.	DESCRIPTION
A	$6^8 \times 4^5$	4 – CASEMENT
B	$9^2 \times 4^5$	CASEMENT & PICTURE
C	$3^5 \times 4^5$	2 – CASEMENT
D	$1^9 \times 4^5$	CASEMENT
E	$2^4 \times 3^{10}$	DOUBLE HUNG
F	$3^0 \times 3^2$	DOUBLE HUNG

FLOOR PLAN
SCALE 3/16" = 1'-0"
(NOT TO SCALE – COPY REDUCED 10%)

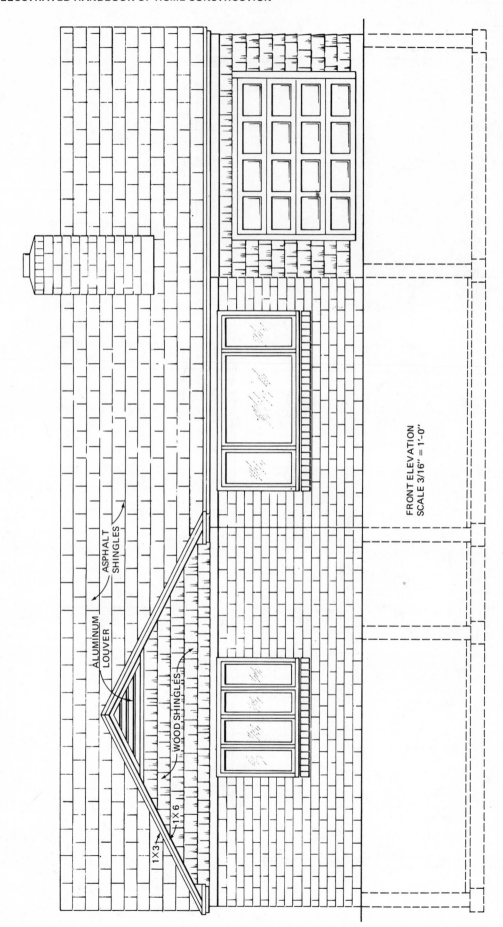

FRONT ELEVATION
SCALE 3/16″ = 1′-0″

ASPHALT SHINGLES

ALUMINUM LOUVER

WOOD SHINGLES

1 × 6

1 × 3

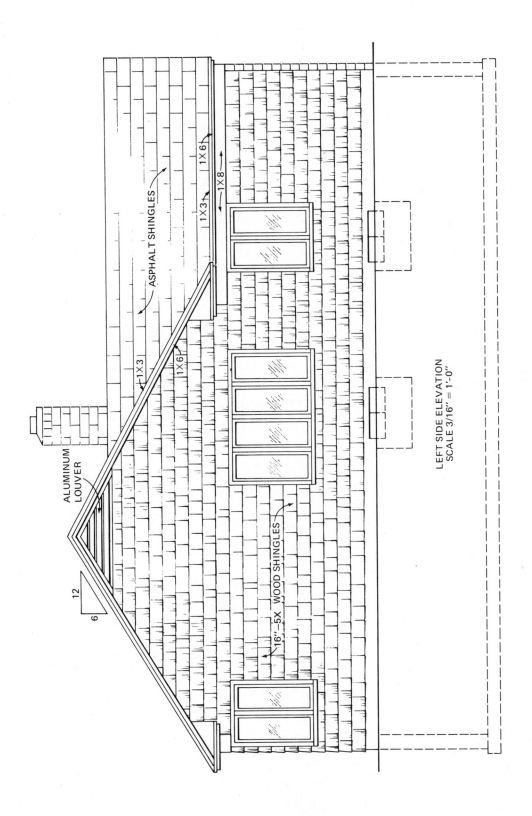

ASPHALT SHINGLES

ALUMINUM LOUVER

1X3

1X6

1X3

1X8

1X6

16''—5X WOOD SHINGLES

12

6

LEFT SIDE ELEVATION
SCALE 3/16'' = 1'-0''

ALUMINUM LOUVER

WOOD SHINGLES

RIGHT SIDE ELEVATION
SCALE 3/16" = 1'-0"

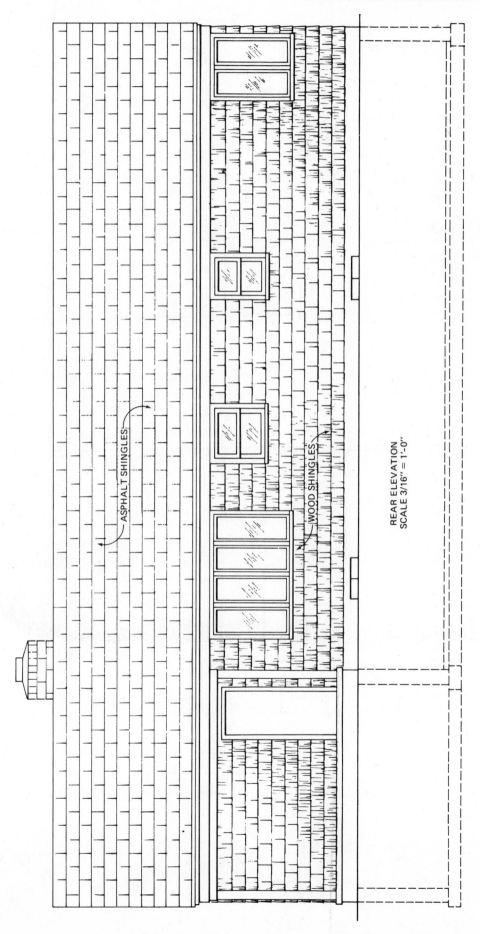

ASPHALT SHINGLES

WOOD SHINGLES

REAR ELEVATION
SCALE 3/16" = 1'-0"

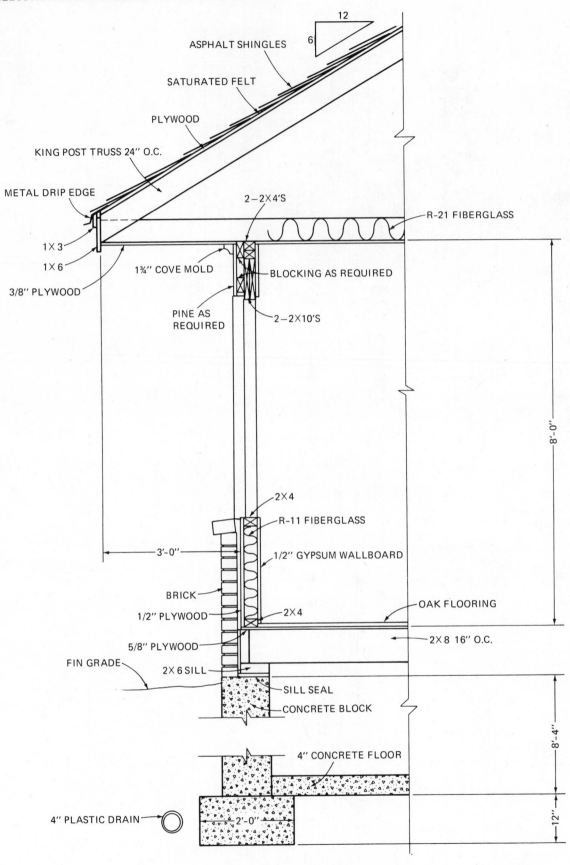

ASPHALT SHINGLES

SATURATED FELT

PLYWOOD

KING POST TRUSS 24″ O.C.

METAL DRIP EDGE

1 X 3

1 X 6

3/8″ PLYWOOD

1¾″ COVE MOLD

PINE AS REQUIRED

2 – 2X4'S

BLOCKING AS REQUIRED

2 – 2X10'S

R-21 FIBERGLASS

12
6

2X4

R-11 FIBERGLASS

3'-0″

1/2″ GYPSUM WALLBOARD

BRICK

1/2″ PLYWOOD

5/8″ PLYWOOD

FIN GRADE

2X6 SILL

2X4

OAK FLOORING

2X8 16″ O.C.

SILL SEAL

CONCRETE BLOCK

4″ CONCRETE FLOOR

4″ PLASTIC DRAIN

2'-0″

8'-0″

8'-4″

12″

SECTION THRU LIVINGROOM WINDOW
SCALE 1/2″ = 1'-0″

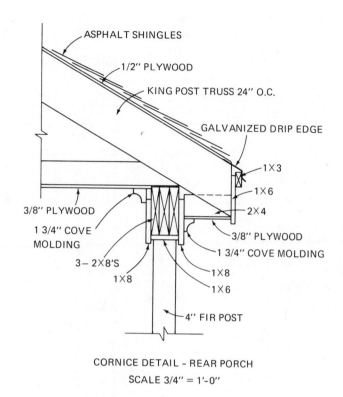

ASPHALT SHINGLES

1/2'' PLYWOOD

KING POST TRUSS 24'' O.C.

GALVANIZED DRIP EDGE

1 X 3

1 X 6

2 X 4

3/8'' PLYWOOD

1 3/4'' COVE MOLDING

3/8'' PLYWOOD

1 3/4'' COVE MOLDING

3 — 2X8'S

1 X 8

1 X 8

1 X 6

4'' FIR POST

CORNICE DETAIL - REAR PORCH
SCALE 3/4'' = 1'-0''

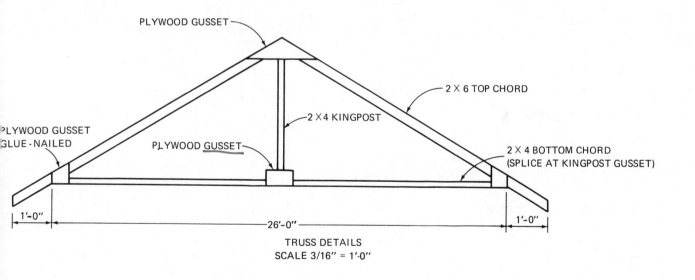

PLYWOOD GUSSET

PLYWOOD GUSSET
GLUE - NAILED

PLYWOOD GUSSET

2 X 6 TOP CHORD

2 X 4 KINGPOST

2 X 4 BOTTOM CHORD
(SPLICE AT KINGPOST GUSSET)

1'-0''

26'-0''

1'-0''

TRUSS DETAILS
SCALE 3/16'' = 1'-0''

Screens indicate where information is required.

Building ______________________

Location ______________________

Architect ______________________

Estimator ______________________

Sheet ______ of ______

Date ______________

Description		Unit	Quantity	Unit Price	Total Material Cost	Labor	Total
FTG. FORM LUMBER	1x8						
WOOD STAKES - 12/BUNDLE							
TRANSIT MIX CONCRETE - FTG.							
TRANSIT MIX CONCRETE - FLOORS							
PORCHES & WALK							
CONCRETE BLOCKS - FOUNDATION	8x8x16						
CONCRETE BLOCKS - FOUNDATION	8x12x16						
SOLID BLOCKS - FOUNDATION	4x8x16						
REINFORCEMENT - MASONRY WALLS							
HALF SASH BLOCKS							
FULL SASH BLOCKS							
STEEL CELLAR SASH	15x12						
GALVANIZED AREAWAYS							
MASONRY CEMENT - BLOCKS		BAGS					
PLASTIC DRAIN PIPE	4"	FT.					
#2 CRUSHED STONE							
PORTLAND CEMENT - PARGING		BAGS					
MASON'S LIME - PARGING		BAGS					
ASPHALT FOUNDATION COATING		GAL.					
MASONRY SAND							
SISALKRAFT RAPOR BARRIER							
WIRE MESH - CONC. FLOORS							
LALLY COLUMNS							
FLUE LINERS							
CHIMNEY BLOCKS							
COMMON BRICKS - CHIMNEY							
COMMON BRICKS - VENEER							

Building _____

Location _____

Architect _____

Estimator _____

Sheet _____ of _____

Date _____

Description	Unit		Quantity	Unit Price	Total Material Cost	Labor	Total
MASONRY CEMENT-							
BRICK MORTAR	BAGS						
GALVANIZED WALL TIES							
FIR- FRONT BEAM							
FIR- MAIN BEAM							
FIR- STAIR STRINGERS							
FIR- STAIR TREADS							
#1 COM. PINE - STAIR							
RISERS	1 x 8 x 12						
PINE HANDRAIL MOLDING							
SILL SEAL							
FIR - SILL							
FIR - BOX SILL							
FIR - JOISTS							
FIR - JOISTS							
FIR - STAIR HEADERS							
GALVANIZED BRIDGING	PCS.						
CD PLYWOOD w/EXTERIOR							
GLUE - SUBFLOOR	5/8						
FIR - PLATES							
FIR - STUDS HOUSE							
FIR-STUDS GARAGE & PORCH							
FIR-GARAGE DOOR HEADER							
FIR-WINDOW & DOOR HEADERS	2 x 10	LIN. FT					
FIR-BEAM FRONT PORCH	2 x 10 x 10						
FIR-BEAM REAR PORCH							
FIR-BEAM REAR PORCH							
FIR-POST REAR PORCH							
$\frac{6}{12}$ x 26' TRUSS- GARAGE & L.R.							
$\frac{6}{12}$ x 26' TRUSS- (no tail							
ONE END)							

Building ——————————————————

Location ——————————————————

Architect ——————————————————

Estimator ——————————————————

Sheet ——— of ———

Date ——————————

Description	Unit	Quantity	Unit Price	Total Material Cost	Labor	Total
$\frac{6}{12}$ × 22' TRUSS - FRONT WING						
FIR - JACK RAFTERS 2×6	LIN. FT.					
FIR RIDGE AT JACK RAFTERS 2×8×10						
$\frac{9}{12}$ × 26' GABLE TRUSS						
$\frac{9}{12}$ × 22' GABLE TRUSS						
ALUMINUM LOUVERS $\frac{6}{12}$ PITCH						
CD EXTERIOR GLUE PLYWOOD - ROOF SHEATHING ½"						
15 LB. SATURATED FELT 500'	RLLS.					
GALVANIZED DRIP EDGE						
240 LB. ASPHALT ROOF SHINGLES	SQ.					
28 GA. ALUMINUM FLASHING 20"×50'	RLL.					
PINE CORNICE 1×3						
PINE FASCIA 1×6						
AC EXT. PLYWOOD - SOFFIT ⅜"						
PINE FRIEZE 1×6						
COVE MOLDING - FRONT CORNICE 1¾"						
CD EXT. GLUE PLYWOOD - WALL SHEATHING ½"						
BUILDING PAPER - WALLS	RLL.					
5× CEDAR SHINGLES 16"						
PINE - REAR PORCH BEAM						
PINE - REAR PORCH BEAM						
PINE - REAR PORCH BEAM						
PINE - REAR PORCH BEAM						
PINE - FRONT PORCH TRIM 1×6	LIN. FT.					
AC EXT. PLYWOOD - PORCH CEILINGS ⅜						

Building _____

Location _____

Architect _____

Estimator _____

Sheet _____ of _____

Date _____

Description		Unit	Quantity	Unit Price	Total Material Cost	Labor	Total
COVE MOLDING - PORCHES	$1\frac{3}{4}$						
4-CASEMENT WINDOW	$6^8 \times 4^5$						
CASEMENT-PICTURE WINDOW	$1^9 \times 4^5 / 5^8 \times 4^5$						
2 CASEMENT WINDOWS	$3^5 \times 4^5$						
CASEMENT WINDOWS	$1^9 \times 4^5$						
D. H. WINDOW	$2^4 \times 3^{10}$						
D. H. WINDOW	$3^0 \times 3^2$						
ALL WINDOWS w/ INSULATED							
GLAZING & SCREENS							
PRE - HUNG EXT. DOOR	$3^0 \times 6^8 \times 1\frac{3}{4}$						
PINE, LEFT- HAND SWING							
PRE - HUNG EXT. DOOR	$2^8 \times 6^8 \times 1\frac{3}{4}$						
PINE, RIGHT- HAND SWING							
PRE-HUNG EXT. DOOR,	$2^6 \times 6^8 \times 1\frac{3}{4}$						
PINE, RIGHT- HAND SWING							
ALUMINUM COMBINATION -	$3^0 \times 6^8 \times 1\frac{1}{8}$						
STORM & SCREEN DOOR							
ALUMINUM COMBINATION -	$2^8 \times 6^8 \times 1\frac{1}{8}$						
STORM & SCREEN DOOR							
PRE-HUNG INTERIOR FLUSH DOOR	$2^6 \times 6^8 \times 1\frac{3}{8}$						
BIRCH, LEFT-HAND SWING							
COLONIAL CASING							
PRE-HUNG INTERIOR FLUSH DOOR							
BIRCH, RIGHT-HAND SWING	$2^6 \times 6^8 \times 1\frac{3}{8}$						
COLONIAL CASING							
PRE-HUNG INTERIOR FLUSH DOOR	$2^0 \times 6^8 \times 1\frac{3}{8}$						
BIRCH, RIGHT HAND SWING							
COLONIAL CASING							
PRE-HUNG INTERIOR FLUSH DOOR	$1^9 \times 6^8 \times 1\frac{3}{8}$						
BIRCH, LEFT HAND SWING							
COLONIAL CASING							

Building _____

Location _____

Architect _____

Estimator _____

Sheet _____ of _____

Date _____

Description	Unit	Quantity	Unit Price	Total Material Cost	Labor	Total
PRE-HUNG INTERIOR FLUSH DOOR $2^4 \times 6^8 \times 1^3/8$						
BIRCH, LEFT HAND SWING,						
COLONIAL CASING						
INTERIOR DOOR JAMBS $2^6 \times 6^8$	SCT.					
LOUVERED DOORS $1^3 \times 6^8 \times 1^1/8$						
DOUBLE-SWING HINGES	PR.					
OVERHEAD GARAGE DOOR						
w/ HARDWARE $9^0 \times 7^0$						
OVERHEAD DOOR FRAME $9^0 \times 7^0$						
R-21 FIBERGLASS INSULATION						
R-11 FIBERGLASS INSULATION						
GYPSUM WALLBOARD $1/2"$						
WALLBOARD COMPOUND 5 GAL.	CAN					
PERFORATED TAPE 250'	ROLL					
DEADENING FELT-FLOORS 500'	ROLL					
OAK STRIP FLOORING 1×3						
UNDERLAYMENT PLYWOOD $1/2"$	SHEET					
VINYL TILE - KITCHEN $12" \times 12"$	SQ. FT.					
ADHESIVE - VINYL TILE	GAL.					
CERAMIC FLOOR TILE $1" \times 1"$	SQ. FT.					
SLATE FLOORING	SQ. FT.					
ADHESIVE FOR TILE & SLATE	GAL.					
GROUT FOR TILE & SLATE 5 LB.	PKG.					
AC EXT. PLYWOOD -						
TUB AREA $1/2"$						
CERAMIC WALL TILE $4^1/4 \times 4^1/4$	SQ. FT.					
CERAMIC CAP TILES 2×6	PC.					
MARBLE THRESHOLD - BATH $2'-4"$						
COLONIAL BASE MOLDING $3"$						
STOOL - WINDOW TRIM						
APRON - WINDOW TRIM						

Building _____

Location _____

Architect _____

Estimator _____

Sheet _____ of _____

Date _____

Description	Unit	Quantity	Unit Price	Total Material Cost	Labor	Total
COLONIAL CASING - WINDOWS						
CHROME ADJUSTABLE						
CLOSET RODS						
#2 PINE SHELVING	1×12	UN. PT.				
AD INT. PLYWOOD - LIN.						
CLOSET & BATH						
NAIL & FASTENER ALLOWANCE						
1830 WALL CABINET						
1230 WALL CABINET						
24 WC CORNER WALL						
CABINET						
1530 WALL CABINET						
3018 WALL CABINET						
3312 WALL CABINET						
D12 BASE CABINET W/						
DRAWERS						
B15 BASE CABINET						
CB 36 BASE CABINET						
SF 42 SINK FRONT						
B18 BASE CABINET						
LAMINATED PLASTIC						
COUNTER TOP						
VANITY W/ LAMINATED						
PLASTIC TOP	48"					
FINISH HARDWARE						
ALLOWANCE						

APPENDIX

The following labor estimating tables provide a quick method for determining the amount of time required to complete various types of construction.

Time requirements shown are based on average production per man hour, for quality workmanship, under average conditions. Time requirements do not include supervision.

How to use labor estimating tables

Locate in the tables the type of work to be done. Tabulate skilled and unskilled hours. Add these figures to get the number of hours required per estimating unit, (100 sq. ft., 100 lin. ft., etc.) Multiply this total by the wage rate per hour prevailing in your area. This is the labor cost per estimating unit. Multiply the unit cost by the number of units involved. Repeat this process for all phases of the job to get total labor cost.

If you desire you can also develop a "price per sq. ft." for labor on each type of construction. Or, by combining the various totals of a given job you may develop a "per cu. ft." price for the labor required. Once these figures are prepared they may be used for future projects of the same kind of construction.

EXAMPLE: Extra room, size 10 ft. x 16 ft., 160 square feet area.

Construction, Wood floor with 2 in x 8 in.. joist 12 in. on center with braces. Plywood subfloor with paper, 3/8 in. or 1/2 in. prefinished oak floor.

Labor rates, skilled $13.00, unskilled $9.75 per hr.

From the tables use the following figures:

	Skilled	Unskilled
Joist labor	5 hrs. (one unit)	2 hrs. (one unit)
Subfloor labor	1 hr. "	1/2 hr. "
Finished floor labor	4 hrs. "	1/2 hr. "
(160 sq. ft. =	10 hrs.	3 hrs.
1.6 estimating units) x 1.60 S.F.		x 1.60 S.F
	16 hrs.	4.8 hrs.
16 hrs. skilled @	13.00 208.00	
4.8 hrs. unskilled @	9.75 46.80	
	Total $254.80 labor cost for	
	160 sq. ft.	

Reduced to a square ft. price, the labor cost for this type floor construction is $1.593.

EXCAVATIONS, BACKFILLS, CONCRETE WALLS AND SLABS

HOURS PER UNIT

	Machine	Unskilled
Excavation, 100 cubic yards.		
Rates based on average type dry, solid soil.		
Handwork		130
Machine work (3/4 yard dipper continuous operation)		
Power shovel	1	
Backhoe	1 1/2	
Bulldozer (medium size tractor and blade)	3	
Backfilling, 100 cubic yards.		
Loose soil		
Machine work (see above)		
Bulldozer or backhoe	1	
Power tamping	2	

HOURS PER UNIT

	Machine	Unskilled
Handwork		65
Hand tamping		80
Ditching, 100 lineal feet,		
Based on trench size of 12" x 24" or 2 cubic feet per lineal foot.		
Machine work (trencher)	2/3	
Handwork		10
Sewers and Drains, 100 lineal feet.		
Laying in 2' ditch and covered		
3' & 4' vitrified, 2' lengths		6
6" vitrified, 2' lengths		7
8" vitrified, 2' lengths		8
10" vitrified, 2' lengths		9
12" vitrified, 2' lengths		10
(Deduct 25% if 4 ft. lengths. Add 10% if asphalt, rubber or cement joints.)		
3-4-6" plastic, fibre 10' joints		4 3/4
3 & 4" drain tile, 1' lengths		5

HOURS PER UNIT

	Machine	Unskilled
6″ drain tile, 1′ lengths		5 1/2
8″ drain tile, 1′ lengths		6

Footers (Excavating), 100 lineal feet.
Based on 8″ x 16″ footer.

	Machine	Unskilled
Machine work (trencher)	1/3	
Handwork		5

(Add or deduct for other dimensions of ditches or footers.)

Footers (Placing), 100 lineal feet.
Based on 8″ x 16″ footer.

		Machine	Unskilled
Setting forms to level grade.	Wood	2	2
	Steel	1 1/2	1 1/2
Placing reinforcing rods			1/4
Placing key forms		1	
Placing ready mixed concrete		1/2	4

(Average conditions and wheeling distance to forms)

	Machine	Unskilled
If vibrated concrete		1/3
Laying drain tile		1
Placing 12″ porous fill over tile		

Foundations (Concrete), 100 square feet.
Average type construction to 8′ heights. 8″ to 12″ thick walls. Normal openings included.

		Machine	Unskilled
Setting 2′ x 8′ sectional forms.	Wood	1 1/4	3/4
	Steel	1	3/4
Building forms			
(plywood construction)		2	3/4
(1″ x 8″ sheathing and 2″ x 4″ or 2″ x 6″)		4	1 1/2
Placing reinforcing steel			
(1/2″ rods spaced 12″ x 12″ or wire mesh reinforcing)			1 1/2
Corbeling, chamfering or setbacks (up to 4″ x 6″)		1/2	
Placing concrete			
(Ready mixed concrete under average conditions)			
8″ walls		1	3
If vibrated concrete			1/2
12″ walls		1	4
If vibrated concrete			3/4
Removing forms, ties, etc.			
Sectional forms		1/2	1/2
Built in place forms		1	1/2
Hand rubbing walls, (minor blemishes)			1/2
Cleaning, oiling sectional forms			1

(Adjust rates for extra wheeling distance and handling or foundation heights over 8 Rate may change for unusual type forms or form hardware and vibrators attached to forms.)

Foundations (Masonry), 100 square feet.
(Average conditions, struck joints, common bond, openings included)

HOURS PER UNIT

	Machine	Unskilled

Foundations (Masonry) — Cont.

	Machine	Unskilled
8 x 8 x 16 concrete masonry units	6	6
10 x 8 x 16 concrete masonry units	6 1/2	6 1/2
12 x 8 x 16 concrete masonry units	7	7

(Rate based on heavy units. Deduct 10% for medium weight and 20% for lightweight units)

	Machine	Unskilled
Placing masonry reinforcing		1/2
8″ solid brick walls	11	11
12″ solid brick walls	16	16
Loadbearing structural tile		
5 x 8 x 12 laid flat	5 1/2	5 1/2
8 x 12 x 12	7	7
10 x 12 x 12	7 1/2	7 1/2
12 x 12 x 12	8	8
Waterproofing		
Cement plaster, 1 coat	2	
Membrane (felt, polyethylene cemented to wall)		3
Tar or asphalt, 1 coat Brush coat		1 1/4
Trowel coat		2 1/2

Concrete (Walls), 1 cubic yard.
Based on sections 12″ to 36″ thick. 20 to 100 cubic yard projects. Average type construction. Ready mixed concrete dumped in forms to 8′ heights.

		Machine	Unskilled
Setting 2′ x 8′ sectional forms	wood	3/4	1/4
	steel	1/2	1/4
Building forms			
Plywood construction		1	1/2
1″ x 8″ sheathing and 2″ x 4″ or 2″ x 6″		2	1
Placing reinforcing steel			
12″ x 12″ spacing 3/4″ rods wired			1/2
Placing concrete		1/4	1 1/2
If vibrated concrete			1/4

(Adjust vibration rate for unusual reinforcing problems and vibrators placed on forms.)

Concrete Slab Construction, 100 sq. ft.
Base Preparation

	Machine	Unskilled
Handwork after machine grading		1/2
(Not required if hand excavated)		
Placing, grading, tamping base material. Stone, Slag, Sand, Gravel, Cinders, etc.	1 1/4	
Placing reinforcing wire mesh or rods		1/2
Vapor Barrier—Polyethylene, sisal reinforced paper felt or other non-rigid material		1/2
Perimeter Insulation		
1″ x 12″ x 2″ x 24″ rigid in 8′ strips or sheets		1/2

	Skilled	Unskilled
Concrete (Slab Construction) — Cont.		
Slab Insulation		
Up to 2″ rigid or semi-rigid insulation laid on base under entire floor area	1/2	1/4
Insulating Concrete, 100 square feet		
Lightweight vermiculite, perlite, pumice (Not finish troweled)		
4″ thick	3/4	3/4
6″ thick	1	1
8″ thick	1 1/2	1 1/2
(Finish troweled floor use figures below for concrete slab installation)		
Concrete Slab Pouring, 100 square feet.		
3″ thick	1 1/2	1 1/2
4″ thick	1 1/2	1 1/2
If vibrated concrete		1/4
5″ thick	1 1/2	1 1/2
6″ thick	2	2
If vibrated concrete		1/3
8″ thick	2 1/2	2 1/2
If vibrated concrete		1/2

(Time based on the use of ready mixed concretes delivered to the site, no forming, finished or troweled surface on prepared base. Includes placing of 1/4″ or 1/2″ asphalt, rubber or fibre expansion joint and normal blocking of 4′ squares. Add for wheeling concrete if required. Adjust for machine finishing.)

CONCRETE FLATWORK, STEPS

Walks, Driveways, Patios, 100 square feet.
Based on hand work under average conditions. Placed on prepared grade.

	Skilled	Unskilled
Walks (4″ concrete) Average 4′ width		
Grading, leveling 4″ to 6″ base materials		1 1/4
Forming: metal or wood and placing joints	3/4	1/2
Pouring and finishing ready mixed concrete	1 1/2	1 1/2
If vibrated concrete		1/4
Removing forms		1/4
For other walk surfaces see brick and patio block under heading of Floors.		
Driveways 6″ concrete, to 16′ widths on prepared grade		
Grading, leveling 4″ to 8″ base materials		1 1/2
Forming: metal or wood & placing asphalt, rubber or fibre expansion joints.	1	1/2
Placing reinforcing mesh	1/2	
Pouring and finishing using ready mixed concrete	2	2

	Skilled	Unskilled
Walks, Driveways, Patios — Cont.		
If vibrated concrete		1/3
Removing forms		1/2
Adjust for machine work & special curing methods, weather protection, etc.		
Porches and Patios		
Concrete slab construction on prepared base, shored wood formed or metal deck.		
Placing reinforcing mesh		1/2
Placing reinforcing rods		
8″ x 8″ centers 3/8″ rods		1
Pouring ready mixed concrete troweled finish		
4″ thick	1 1/2	1 1/2
5″ thick	1 3/4	1 3/4
6″ thick	2	2
If vibrated concrete		1/4
Placing corrugated metal deck over steel or concrete joists	1/2	1/2
Complete forming and shoring to 8′ heights	8	4
Removing forms	1/2	1
Steps, 10 sq. ft. tread area.		
(On roughed-in concrete base)		
12″ treads, 8″ risers.		
Brick treads and risers, tooled joints		
Treads and risers	5	2 1/2
Treads only	3	1 1/2
Concrete masonry treads and risers (4 x 8 x 12 solids or equivalent)	2	1
Stone or precast concrete treads on concrete base.		
one piece to 4″ thick cut to size	1	1
Rough slate or stone treads to 4″ thick	2	2
Concrete step treads to 12″ wide, 4″ to 6″ thick		
Forming	1	
Pouring ready mixed concrete	1/2	1/2
If vibrated concrete		1/4
Removing forms and finishing	1/2	1/2
Concrete steps, risers & treads (see labor required for cheek walls under Foundations, concrete or masonry)		
Forming steps 8″ rise 12″ treads	2	1

	HOURS PER UNIT	
	Skilled	Unskilled

	HOURS PER UNIT	
	Skilled	Unskilled

	Skilled	Unskilled
Steps — Continued		
Pouring steps and risers	1	1/2
If vibrated concrete		1/4
Removing forms and finishing	1	1/2
Reinforced concrete steps 8″		
rise to 12″ treads to 20″ length,		
4 ft. wide to 8 ft. height		
Complete forming and placing steel	6	1 1/2
Pouring ready mixed concrete	2	1
If vibrated concrete		1/3
Removing forms and finishing	1	1 1/2
Wood steps and stairs		
(see frame construction)		
Masonry, 100 square feet.		
(All rates under masonry cover standard		
or modular units		
Face brick — based on 3/8″ flush		
joints. Normal openings, sills,		
headers. To 16″ heights.		
4″ brick veneer		
Stretcher bond	12	8
Stacked bond	15	10
Soldier course	15	10
Header course	18	12
Roman size (12″) Stretcher bond	12	18
Stacked bond	15	10
Norman size (12″) Stretcher bond	10	7
Stacked bond	13	9
SCR (1/2″ joints) Stretcher bond	9	6
Double brick Stretcher bond	8	6
Adjust for special bonds, tooled		
raked, struck or special joints,		
special color patterns, curved or fancy		
walls or brick sizes. Add 10% for		
glazed brick.		
Glass block — 1/4″ joints, 3-7/8″		
thick block		
5 3/4″ x 5 3/4″	20	10
7 3/4″ x 7 3/4″	15	8
11 3/4″ x 11 3/4″	11	6
4 3/4″ x 11 3/4″	20	
Cleaning masonry — brick, tile,		
concrete, stone etc. using muriatic		
acid and water, (no scaffolding		
included).		
Rough surfaced walls	1 1/2	3/4
Smooth surfaced walls	1	1/2
Concrete masonry (other than		
foundation work) — average weight		
regular or patterned units, stretcher		
bond hand tooled joints. Rein-		
forcing mesh. To 16′ heights. Normal		
openings, sills, headers, control joints		
included.		

	Skilled	Unskilled
Masonry — Continued		
4 x 8 x 12	5	5
2 x 8 x 16	3	3
3 x 8 x 16	3 1/2	3 1/2
4 x 8 x 16	4	4
6 x 8 x 16	4 1/2	4 1/2
8 x 8 x 16	5	5
Add 25% for stacked, ashler or		
other bonds.		
Masonry fill — Vermiculite insulation		
8″ walls		1/2
12″ Walls		3/4
Figure cavity walls based on		
space to be filled at the rate of		
5 minutes per 4 cubic feet for		
openings over 2″ wide.		
Stone work—coursed ashler		
4″ — 5″ widths. Random lengths.		
End cuts made on job.		
2 1/4″ thick	16	16
5″ thick	12	12
Random widths 2 1/4″ to 10 1/2″		
thick	14	14
Random widths 5″ to 10 1/2″ thick	10	10
Random ashler cut to size		
2 1/4″ to 5″ thick	10	10
5″ to 10 1/2″	7	7
Rustic rubble stone 6″		
to 18″ size cut and fit on		
job	16	16
Back plastering masonry work		
(all types) 1 coat work	1 1/2	1
Backcoating masonry work (all types)		
1 coat work asphalt type trowel		2 1/2
brush		1 1/4
Liquid (transparent water-		
proofing) — brick, concrete		
masonry, stone		
1 coat application brush coat		1
spray coat		1/2
Painting concrete masonry,		
brick, tile, stone, concrete.		
Cement base paints 1 coat brush work		1 1/2
spray work		3/4
roller work 9″		1
Back-up walls — normal		
openings.		
4″ Common brick	8	6
8″ Common brick	10	8
Concrete masonry		
Average weight units		
Reinforcing mesh included		
4 x 8 x 12 ″ ″ ″	5	5

	HOURS PER UNIT	
	Skilled	Unskilled

Masonry — Cont.

Reinforcing mesh included

	Skilled	Unskilled
4 x 8 x 16 ” ” ”	4	4
6 x 8 ” ” ”	4 1/2	4 1/2
8 x 8 ” ” ”	5	5
10 x 8 ” ” ”	6	6
12 x 8 ” ” ”	7	7

Structural tile

	Skilled	Unskilled
4 x 5 x 12	5 1/2	5 1/2
5 x 8 x 12	4 1/2	4 1/2
4 x 12 x 12	4 1/2	4 1/2
6 x 12 x 12	5	5
8 x 12 x 12	6	6
10 x 12 x 12	6 1/2	6 1/2
12 x 12 x 12	7	7

Add 10% to all types for cavity walls, adjust for special sizes, shapes, unusual wall designs.

Gypsum partition tile (partition walls) 3/8" joints 12 x 30" units, unusual openings

	Skilled	Unskilled
2" thickness	3	2 1/2
3" thickness	3 1/2	3
4" thickness	4	3 1/2
6" thickness	5	4

Screen or decorative walls, flush joints.

	Skilled	Unskilled
8 x 8 to 8 x 12 Face 4" thick concrete	5	5
8" thick concrete	6	6
8 x 8 to 16 x 16 Face 4" thick concrete	6	6
8" thick concrete	7	7
20 x 20 to 24 x 24 Face 8" thick concrete	6	6
8 x 8 to 8 x 12 Face 4" thick clay	5	5

Adjust for unusual wall designs, joints, patterns or unit types.

Glazed Masonry Work, 100 pieces.

Interior or exterior. Tooled joints. Up to 8' heights.

5-1/16" x 7-3/4" Face

	Skilled	Unskilled
1-3/4" Soap Stretcher	4 1/2	4 1/2
3-3/4" Stretcher	5	5
5-3/4" Stretcher	5 1/2	5 1/2
7-3/4 " Stretcher	6	6

5-1/16" x 7-3/4" Face

	Skilled	Unskilled
1-3/4" Soap Stretcher	6 1/2	6 1/2
3-3/4" Stretcher	7	7
5-3/4" Stretcher	7 1/2	7 1/2
7-3/4" Stretcher	8	8

Glazed Masonry - Cont.

7-3/4" x 15-3/4" Face

	Skilled	Unskilled
1-3/4" Soap Stretcher	10	10
3-3/4" Stretcher	11	11

Glazed concrete masonry

	Skilled	Unskilled
2 x 8 x 16	10 1/2	10 1/2
4 x 8 x 16	11 1/2	11 1/2
6 x 8 x 16	12 1/2	12 1/2
8 x 8 x 16	14	14
12 x 8 x 16	17	17

Both types add 25% if glazed two sides. Double rates for all shapes.

Cutting masonry units (Average power saw) Clay tile, brick, concrete masonry.

	Skilled	Unskilled
To 32" perimeter units	1	
32" to 48" perimeter units	1 1/2	

COPING, LINTELS, BEAMS, COLUMNS

Wall Coping, 100 lineal feet.

Vitrified

	Skilled	Unskilled
9" – 13"	6	6
18"	7	7

Includes placing corners and ends.

Precast concrete or cut stone 4" thick to 16" widths;

	Skilled	Unskilled
4–5' lengths	6	5

Sills—Lintels, 100 lineal feet.

Door—window. Precast concrete or stone cut to size to 5' lengths.

	Skilled	Unskilled
To 4" x 8"	5	2
6" x 8" – 8" x 8"	5 1/2	2
8" x 10" – 8" x 12"	6	3

Concrete Beams, Lintels, 100 lineal feet

Poured in place to 8' heights

	Skilled	Unskilled
8" x 8" setting forms and placing steel rods	7 1/2	3 1/2
8" x 12" setting forms and placing steel rods	8	4

Pouring ready mixed concrete

	Skilled	Unskilled
8" x 8"	1	3
8" x 12"	1	4
Vibration		1/4
Removing forms	1	1/2
Finishing concrete		2

Concrete Masonry Pilasters, 100 lineal feet.

Laying units. Placing reinforcing rods. Filling with ready mixed concrete to 16' heights. Hand work.

	Skilled	Unskilled
8" x 8" Core types	9 1/2	17

	HOURS PER UNIT	
	Skilled	Unskilled

Concrete Masonry Pilasters, Cont.

	Skilled	Unskilled
12" x 12" Core types	11 1/2	23 1/2

Adjust for additional heights and Mechanical filling methods and vibration if required.

Concrete Columns, 100 lineal feet.

Poured in place to 8" heights, setting forms and placing steel.

	Skilled	Unskilled
8" x 8" – 8" x 12" – 12" x 12"	6	3
Pouring ready mixed concrete		
8" x 8"	1	3
8" x 12"	1	4
12" x 12"	1	5
Vibration		1/3
Removing forms	3/4	1/2
Finishing concrete		2

Adjust for additional heights or vibrators attached to forming.

Foundation Block,
(For slab construction) 100 lineal feet.

	Skilled	Unskilled
6" x 12" x 16" U and J types or 8" x 8" x 16" header block	2	3

Concrete Masonry Beams and Lintels, 100 lineal feet.

Laying units. Placing reinforcing rods. Filling with ready mixed concrete. Hand work to 8' heights.

	Skilled	Unskilled
8" x 8" x 16" units continuous beams	5	7
8" x 8" x 16" units over openings. Shoring included.	6	8
8" x 16" x 12" units continuous beams	12	16
8" x 16" x 12" units over openings. Shoring included.	13	16 1/2
Vibration		1/3

Adjust for additional heights and mechanical filling methods or vibrators attached to forms.

CHIMNEYS, FIREPLACES

Chimneys, per foot of height
with flue liner, to 30' heights.
Inside chimneys with normal face brick topping. No scaffolding provided. Footer or Base (use rates under Footers).

Brick	Skilled	Unskilled
1 8" x 8" flue	3/4	3/4
2 8" x 8" flues	1 1/4	1 1/4
1 8" x 12" flue	3/4	3/4
2 8" x 12" flues	1 1/2	1 1/2
1 12" x 12" flue	1	1
2 12" x 12" flues	1 3/4	1 3/4

	HOURS PER UNIT	
	Skilled	Unskilled

Chimneys, Fireplaces Cont.

Adjust for other sizes and combinations. Add 25% for average face brick outside chimneys. Flue liner sizes may vary. Use nearest size.

Concrete masonry chimney units	Skilled	Unskilled
8" x 8" flue size	1/4	1/4
8" x 12" flue size	1/3	1/3

Deduct 10% for units without flue liner or for unlined round types. Add 10% for solid masonry or 2 unit types.

Fireplaces, per fireplace.
Based on average design and type 6' wide and 5' high, 36" x 30" opening.

Hearth construction	Skilled	Unskilled
Concrete base	1/2	1
Brick hearth on concrete base	2	2
Tile hearth on concrete base 2" x 2" to 6" x 6"	2	2
Brick work	10	10
Firebrick lining and damper	4	2
Setting steel circulator	1	1

Double above time for double faced or 3 way installation. Use 1 1/2 times above for corner projecting fireplace. Adjust for unusual brick patterns or other facing materials.

	Skilled	Unskilled
Tile or glass facing 4" x 4" to 8" x 8" sizes	4	4
Setting factory mantel (to prepared wall)	1	1/2

FRAMING, SHEATHING, DECKING

Floor Joists, Wood, 100 square feet.

Joist Size	Spacing	Approx. Span*	Skilled	Unskilled
2" x 6"	12"	8'	4 1/2	1 1/2
	16"	7'	4	1 1/2
	18"	6'–6"	4	1 1/4
	24"	6'	3 1/2	1
2" x 8"	12"	11'	6	2
	16"	10'	4 1/2	1 3/4
	18"	9'	4	1 1/2
	24"	8'	3 1/2	1
2" x 10"	12"	14'	6 1/2	2 1/2
	16"	12'	6	2 1/4
	18"	11'	5 1/2	2
	24"	10'	5	1 3/4
2" x 12"	12"	17'	7 1/2	2 3/4
	16"	15'	7	2 1/2
	18"	14'	6 1/2	2 1/4
	24"	12'	6	2

	HOURS PER UNIT	
	Skilled	Unskilled

Framing Cont.

Labor rates cover both platform or balloon type construction, also sill, plate, edge joist and bridging. Add 25% for spans over 12', wood girder if required also if above 1st floor level.

*Spanning based on 100 pounds per square foot total load using Douglas Fir or equal.

Steel Bar Joist, 100 square feet.

Based on 36" spacing to 20' spans. Types 2 & 3 lightweight. To 2nd floor levels.

	Skilled	Unskilled
8"	1	1
10"	1 1/4	1
12"	1 1/2	1 1/4

Steel Beams, 100 square feet.

Light weight type to 20' spans; to 2nd floor levels.

	Skilled	Unskilled
6" – 7" widths	1	1
8" – 10" widths	1 1/4	1
12"	1 1/2	1 1/4

Concrete Joists, 100 square feet.

Based on 24" spacing to 12' lengths, at ground level.

	Skilled	Unskilled
3" x 8"	1	2
3" x 10"	1 1/4	2 1/4
4" x 12"	1 1/2	3

Add 25% for 14 to 20' lengths.
Add 50% for 2nd floor construction.

Precast Concrete Floor Slabs, 100 square feet.

On concrete joist or steel beams.

	Skilled	Unskilled
1" to 2" thick 24" x 30"	1 1/4	1 1/4
1" to 2" thick 24" x 60"	3/4	1

Adjust for various sizes and types.

Precast and Prestressed Concrete Floor Beams, 100 square feet.

Over steel or concrete beams. To 20' lengths. Erected to second floor level. Includes top grouting. Hollow beams.

	Skilled	Unskilled
6" x 12" Beams	1/2	1
6"–8" x 16" Beams	1/3	2/3
6" –8" x 48" Beams	1/4	1/2

Adjust for other sizes. Add 25% for spans 20' to 30'. Add machine or crane time.

Commercial Store Fronts and Window Construction.

Based on total front or window area. Average type construction. Excluding glass.

	HOURS PER UNIT	
	Skilled	Unskilled

Commercial Store Fronts Cont.

	Skilled	Unskilled
Rough framing wood	8	2
Metal work over rough framing using rolled light metal stock	20	5
Metal work over rough framing or free standing using medium or heavy extruded stock	30	10

Adjust for unusual types, patterns and designs or extra metal cutting and fitting.

Studding, 100 square feet.

Includes normal openings, average type outside walls, frame or veneer construction. Plates, headers, fillers, bracing, firestops, girts included.

	Skilled	Unskilled
2" x 4" 12" centers 8' to 12' heights	3	1/2
12" centers 12' to 20' heights	4	3/4
16" centers 8' to 12' heights	2 1/2	1/2
16" centers 12' to 20' heights	3 1/2	3/4
24" centers 8' to 12' heights	2	1/2
24" centers 12' to 20' heights	3	3/4
2" x 6" 12" centers 8' to 12' heights	4 1/2	3/4
12" centers 12' to 20' heights	5 1/2	1
16" centers 8' to 12' heights	3 1/2	3/4
16" centers 12' to 20' heights	4 1/2	1
24" centers 8' to 12' heights	3 1/2	1

Add 25% for irregular or cut up walls and shed or gable dormers.

Deduct 25% for interior stud partition walls.

SCR construction (Furring)

	Skilled	Unskilled
2" x 2" 12" centers 8' to 12' heights	2 1/2	1/2
2" x 2" 16" centers 8' to 12' heights	2	1/2

Ceiling Joists, 100 square feet.

Normal type frame construction. 1st and 2nd floor levels. Includes bridging, trimmers to 16' spans. Normal openings.

	Skilled	Unskilled
2" x 6" 12" centers	5 1/2	2

		HOURS PER UNIT	
		Skilled	Unskilled
Ceiling Joists Cont.			
2″ x 6″	16″ centers	5	2
	18″ centers	4 1/2	1 1/2
	24″ centers	3	1
2″ x 8″	12″ centers	7	2
	16″ centers	6	2
	18″ centers	5 1/2	1 1/2
	24″ centers	4	1
2″ x 10″	12″ centers	8	2 1/2
	16″ centers	7	2 1/2
	18″ centers	6 1/2	2
	24″ centers	5	1 1/2
2″ x 12″	12″ centers	9	3
	16″ centers	8	3
	18″ centers	7 1/2	2 1/2
	24″ centers	6	2

For open beam ceilings

		Skilled	Unskilled
3″ x 6″-8″	16″ centers	7	2 1/2
	18″ centers	6 1/2	2
	24″ centers	5	1 1/2
4″ x 6″-8″	16″ centers	7 1/2	2 1/2
	18″ centers	7	2
	24″centers	5	1 1/2

All of above add 25% for spans over 16′and heights above normal 2nd floor ceiling level.

Rafters, 100 square feet.

Average type construction, normal pitch and flat, gable ends, to 22′ lengths.

		Skilled	Unskilled
2″ x 4″	12″ centers	3	3/4
	16″ centers	2 1/2	3/4
2″ x 6″	12″ centers	3 1/2	1
	16″ centers	3	1
	18″ centers	2 1/2	1
	24″ centers	2	3/4
2″ x 8″	12″ centers	4	1
	16″ centers	3 1/2	1
	18″ centers	3	1
	24″ centers	2 1/2	3/4

Open beam rafters

		Skilled	Unskilled
3″ x 6″ - 8″ 16″ centers Flat construction		7	3 1/2
24″ centers Flat construction		6	3
4″ x 6″ - 8″ 16″ centers Flat construction		8	4
24″ centers Flat construction		7	3 1/2

(Add 25% for pitched roofs over 2″ pitch)

All of above add 50% for cut up and hip roof types.

Rafters   Cont.

Add 25% for rafters on shed, gable or hip type roofs on dormers and for rafters over 22′ long. Deduct 10% for rafters with short or plain overhang.

Sheathing, 100 square feet.

Average type construction, frame or veneer, normal openings.

		Skilled	Unskilled
1″ x 6″ — 8″ Wood			
sheathing	flat roofs	1	1/2
	pitched roofs	1 1/2	3/4
	steep, cut-up or hip roofs	3	1 1/2
	sidewalls	1 1/2	3/4

(Add 25% for diagonal sheathing on sidewalls)

		Skilled	Unskilled
2″ x 6″ — 8″ / 3″ x 6″			
	flat roofs	3	1 1/2
	pitched roofs	4	2 1/2

(Add 25% if exposed underside on beamed ceiling construction)

Plywood, rigid insulating sheathing to 25/32″ thick

		Skilled	Unskilled
4′ x 8′ sheets	flat roofs	1/2	1/4
	pitched roofs	3/4	1/2
	steep, cut-up or hip roofs	1 1/4	1
	sidewalls	1	1/2

Gypsum sheathing

		Skilled	Unskilled
2′x 8′ panels	sidewalls	1 1/4	1/2

Strip sheathing

		Skilled	Unskilled
1″ x 2″ — 3″ — 4″pitched roofs		1	1/2

(Deduct 33-1/3% for use of automatic or power nailers or staplers.)

Insulating sheathing

		Skilled	Unskilled
1″ to 3″ thick	flat roofs	3/4	1/2
	pitched roofs	1 1/4	1

(Add 25% if finished open beam ceiling construction)

Asbestos covered insulating sheathing 19/16″ to 2″ thick, nailed or screwed

		Skilled	Unskilled
4 x 8 panels. Frame construction	flat roofs	1 1/2	1
	pitched roofs	2	1 1/2

(Adjust for various types of fasteners)

Bar Rib Lath, 100 square feet

For poured concrete or gypsum over concrete, bar joist or steel beams.

HOURS PER UNIT

	HOURS PER UNIT	
	Skilled	Unskilled
Bar Rib Lath Cont.		
3/4" rib lath.*	1/2	1/2
Steel Decking, 100 square feet.		
Sheets 2' x 8' for floor or roof construction.		
18 or 20 gauge.*	1/2	3/4
Paper Backed Wire Mesh, 100 square feet.		
For floor construction.		
3" x 4" mesh, 12 gauge, rolls 4' x 125'	1/4	1/4
*(Add 25% for 2nd or 3rd floor construction)		
Stairways (Frame), per stairway		
Average type straight flight to 4' wide and 12' long.		
Rough cutting, framing, placing	5	2
(Add 50% for 2 flight type)		
Roof Trusses, (wood) per truss.		
Average type 1 floor construction, using precut lumber.		
24" spacing 14' to 20' length	1/2	1/2
22' to 32' length	3/4	3/4
SUBFLOORS, FINISHED FLOORS		
Subfloors on Wood Joists, 100 sq. ft.		
1" x 8" Lumber	1 1/2	1/2
Plywood 3/8" to 3/4"	1	1/2
(Deduct 50% for use of automatic or power nailers or staplers)		
Insulation, 100 sq. ft.		
Blanket type placed between joists	3/4	
Underlayment, 100 sq. ft.		
Board-type (Hardboard, Fiber Board, Plywood, etc.) 1/4" to 1/2" thick	1	
Wood Finish Flooring, 100 sq. ft.		
Softwood strip		
2 1/4" face	3	1/2
3 1/2" face	2	1/2
Hardwood strip 7/8"		
1 1/4" face	4 1/2	3/4
2 1/2" face	4	1
Hardwood strip 1/2" - 3/8"		
1 1/2" face	4	1/2
2 1/4" face	3 1/2	1/2
(Deduct 25% for use of automatic nailers)		
(Includes placing paper or felt under floor.)		
Sanding, machine	1	
Finishing		
Three liquid applications to all unfinished wood floors	1 1/2	
Prefinished hardwood		

	HOURS PER UNIT	
	Skilled	Unskilled
Wood Finish Flooring, Cont.		
(Strip, plank, block)		
1/2" − 3/8" x 1 1/2" face	4	1/2
Plank to 6" wide	2 1/2	3/4
Block 9" x 9"	2	3/4
(Either nailed or cemented on prepared subfloor)		
Resilient Type Flooring, 100 Sq. ft.		
Roll linoleum (plain)	3	1/2
Patterns, borders	4	1
Tile, (asphalt, vinyl, rubber, linoleum, cork, etc.)		
4" x 4" size	5	1
6" x 6"	4	1
9" x 9"	3	1
6" x 12"	3	1
12" x 12"	2 1/2	1
9" x 18"	2 1/2	1
Strip (to 8")	3	1
(Includes laying felt or underlayment)		
Ceramic Tile on Concrete, 100 sq. ft.		
1/2" to 2", paper backing	4	4
2" x 2"	5	5
4" x 4"	4	4
6" x 6"	3	3
(Includes laying the necessary underlayment on any base.)		
Slate on Concrete, 100 sq. ft.		
Up to 1" thick.		
Random cut sizes to 12" x 12"	2	2
Rough uncut slabs	1 1/2	1 1/2
Marble on Concrete, 100 sq. ft.		
4 x 6 to 12 x 12 tile		
Cement bed preparation	2	4
Setting tile	8	8
Machine finishing	8	
Brick Floor on Concrete, 100 sq. ft.		
(Standard size on edge)		
Laid in mortar, mortared joints, basket weave or common bond.	12	6
Herringbone or fancy patterns	16	8
(Standard size laid flat)		
Basket weave or common bond	8	4
Herringbone or fancy patterns	12	6
(Use 1/3 of above rates if laid in sand without mortar.)		
Adjust rates for other brick sizes.		
Patio Blocks on Concrete, 100 sq. ft.		
Laid in mortar, mortared joints up to 4" thick. Square, rectangular, octagonal or odd shapes.		

	HOURS PER UNIT	
	Skilled	Unskilled
Patio Blocks Cont.		
6 x 12 to 8 x 16	4	2 1/2
12 x 12 to 12 x 18	3 1/2	2
12 x 24	2 1/2	2
24 x 24	2	2

Use 1/3 of above rates if laid in sand base without mortar.

ROOFING

Asphalt Roofing, 100 sq. ft. (Square)

	Skilled	Unskilled
Roll, Plain Surface Pitched roofs	3/4	1/4
Mineral Surface Pitched roofs	1	1/4
Felt Underlayment		1/4
Shingles, Individual type	3	1
Individual type Dutch lap	1 3/4	1
Interlocking	2	1
Strip 10" x 36"	2	1
Strip 12" x 36"	2 1/4	

(Deduct 33-1/3% for use of automatic or power nailers or staplers)

Asbestos Shingles, 100 sq. ft. (Square)

	Skilled	Unskilled
American method 8" x 16"	4 1/2	2 1/2
Dutch lap hexagonal 16" x 16"	2 1/2	1 1/2
Colonial method 10" x 24" to 12" x 30"	2 1/4	1

Wood Shingles, 100 sq. ft. (Square)

	Skilled	Unskilled
16" Approx. 5" exposure	4	1 1/2
18" Approx. 6" exposure	3 1/2	1 1/4
24" Approx. 8" exposure	3	1

(Add 35% for staggered or thatched butts or double coursing.)

Slate, 100 sq. ft. (Square)

	Skilled	Unskilled
16" x 8"	5	2 1/2
18" x 9"	4	2
20" x 10"	3 1/2	2
22" x 12"	3	1 1/2
Random widths to 3/8" thick	7	3 1/2
Graduated slate to 3/4" thick	8	4

Clay Tile, 100 sq. ft. (Square)

	Skilled	Unskilled
Spanish, Mission or Shingle type	5	2 1/2
Interlocking tile type	6	3

(Add for stripping and underlayment)

Metal Sheets, 100 sq. ft. (Square)

Pitched roofs.

	Skilled	Unskilled
Corrugated aluminum or galvanized steel in sheets. Over wood framing.	1 1/4	3/4
Crimped types, aluminum or steel in sheets over wood framing	1 1/2	3/4
Rolls over flat areas. Copper or tin 14" widths. Flat seams	2 1/2	2 1/2

Corrugated Asbestos Sheets, 100 sq. ft. (Square) (Screws or Fasteners)

	Skilled	Unskilled
1/4 x 42—8' length over open rafters or steel	2	2

	HOURS PER UNIT	
	Skilled	Unskilled
Built-Up, 100 sq. ft. (Square)		
(Includes topping of slag, stone, gravel or chips)		
3 Ply over wood or insulating deck	1 1/2	3/4
3 Ply over gypsum or concrete deck	1 3/4	1
4 Ply over wood or insulating deck	1 3/4	1
4 Ply over gypsum or concrete deck	2	1 1/4

Concrete Slabs, Lightweight, 100 sq. ft. (Square)

On steel beams or bar joist. 2nd floor levels. Add for mechanical hoisting equipment.

	Skilled	Unskilled
Flat type 1" to 2" thick 24" x 5'	1	2
Channel type.		
2 3/4" thick 24" x 8'	1 1/4	2 1/2
3 3/4" thick 18" x 8'	1 1/2	3
3 3/4" thick 24" x 8'	1 1/4	2 1/2

NOTE: Add 25% to all roofing estimates for steep pitched or cut-up roofs and unusual conditions. Above rates include time for average type ridges, hips and valleys. Underlayment or scaffolding not included. Add for preparation of surface, eaves, etc. when estimating reroofing jobs.

Metal Work 100 lineal feet

	Skilled	Unskilled
Eaves and gutters, metal, standard sizes.	5	5
(Deduct 1/3 for interlocking prefit types)		
Downspouts 2" to 6" round or square	4	2
Valleys 20" width metal	4	
Flashing Parapet walls, chimneys and dormer sides, etc.	4	
Raggles for flashing (cutting only in set up masonry walls.)	1	4

Wood Gutters, 100 lineal feet

	Skilled	Unskilled
3" x 5" and 4" x 6"	5 1/2	4
5" x 7"	6 1/2	5

Cant Strips, 100 lineal feet

	Skilled	Unskilled
3" x 6"	1 1/2	1/4

SIDING

Wood Siding (horizontal), 100 sq. ft.

Shiplap, patterns, rustic types

	Skilled	Unskilled
1" x 3 1/2 – 4"	2 1/2	3/4
1" x 4 1/2 – 5"	2 1/4	3/4
1" x 6 – 8"	2	1

Lap, bevel or bungalow types

	Skilled	Unskilled
1/2" x 8"	2 1/2	3/4
3/4" x 10"	2 1/4	3/4
3/4" x 12"	2	3/4

Add 25% for cuts to fit ends and mitered corners. Includes metal corners if required.

	HOURS PER UNIT	
	Skilled	Unskilled

Siding Cont.

Vertical patterned types

1″ x 6-8″	3	1 1/2
1″ x 10-12″	2 3/4	1 1/4

Board and batten

1″ x 6-8″	3 1/2	1 1/2
1″ x 10-12″	3	1 1/4

(Includes horizontal stripping over studs)

Wood shingles

16″ approximately 5″ exposure	5	1 1/2
18″ approximately 6″ exposure	4 1/2	1 1/2
24″ approximately 8″ exposure	3 1/2	1 1/2

Add 25% for special patterns or double coursing. Adjust for fasteners or special nailers.

Asphalt Siding*, 100 sq. ft.

Brick, Stone, Shingle patterns.

Roll types (15″ widths)	2	1/2

Panel types (insulating)

10-7/8″ x 43″	3	1
14″ x 43″	2 1/2	1

(Deduct 25% for use of automatic or power nailers or staplers)

Composition Siding*, 100 sq. ft.

Hardboard, wood fibre etc., 1/4″-5/16″, plain or factory painted. 10″ - 12″ widths. Horizontal.

Self venting and spacing metal furring.	1 3/4	1/2
10″- 12″ widths. Horizontal.	1 1/4	1/2
16″ widths. Horizontal	1	1/2
10″ - 12″ widths. Horizontal Shadow edge furring	2	1/2
16″ widths. Horizontal. Shadow edge furring	1 3/4	1/2
10″ - 12″ widths. Vertical Board & Batten	2	1
16″ widths. Vertical Board & Batten	1 3/4	1

(Deduct 25% for use of automatic or power nailers or staplers)

Asbestos Siding*, 100 sq. ft.

8″ x 32″ to 12″ x 24″	2	1

Plain Sheet Types*

3/16″ and 1/4″ thick, 48″ wide 8′ long	1 1/2	1 1/2

Corrugated Sheets*

1/4″ x 42″ - 8′ lengths	1 1/4	1 1/4

* Includes corners, trim, etc.

	HOURS PER UNIT	
	Skilled	Unskilled

Shingle Backer, 100 sq. ft.

8 3/4″ to 11 3/4″ 48″ lengths	1 1/4	1/2
13 1/2″ to 15 1/2″ 48″ lengths	1	1/2

(Deduct 25% for use of automatic or power nailers or staplers)

Lattice strip furring	1/4	1
3/4″ x 2″ strip furring	1/2	1/4

Paper, Felt, Etc., 100 sq. ft. | **1/4** | **1/4** |

Aluminum Siding, 100 sq. ft.

Clapboard types 8″

Horizontal plain	1 1/2	1 1/2
insulated	2	2
Vertical plain	2	2
insulated	2 1/2	2 1/2

(Includes felt, paper, foil, moisture barrier, corners, fittings and trim)

Corrugated sheets, 8′ lengths

32″ coverage sheets, over wood

frame	1	1
over steel frame	2	1 1/2

(Deduct 10% for 45″ coverage sheets)

Galvanized Steel Siding

Corrugated Sheets, 100 sq. ft.

8′ sheets, 26″ coverage, over

wood frame	1 1/4	1
over steel frame	2	2

Adjust for unusual cutting and fitting and heights over 20′.

To all siding jobs add 25% for unusual conditions, cut up walls, bays and gables.

Adjust for wall preparation on all types of remodeling projects.

Stucco, 100 sq. yards.

Cement stucco 3 coats. Float finish over tile, concrete, brick, metal lath or stucco base. Average openings included to 16′ heights. Scaffolding provided.

Grey, white portland cement or colored prepared stucco	25	18

Add 10% skilled time for textured finish

Add 15% skilled time for troweled finish, unusual finished surfaces or patterns.

Based on average conditions, normal openings. New frame construction.

	HOURS PER UNIT	
	Skilled	Unskilled
Rigid Board Types, 100 sq. ft.		
4' x 8' to 1" thick		
Flat roofs	1/2	1/4
Pitched roofs	3/4	1/2
Steep pitched or cut-up roofs	1	3/4
(Add 10% for 2" & 3" thickness)		
Sidewalls	1	1/2
(Deduct 33 1/3% for automatic or power nailers or fasteners)		
Non-Rigid Types, 100 sq. ft.		
Batt type 2 to 4" thick between studs		
15-19-23" x 24" batts	3/4	1/2
15-19-23" x 48" batts	1/2	1/2
Strip types 2 to 4" thick, between studs		
15" widths	1/2	1/2
Blanket types 1 to 4" thick, between studs		
16-20" widths, 4 to 8' lengths	3/4	1/2
24-33" widths, 4 to 8' lengths	1/2	1/2
Wide stock over studs or sheathing	1/3	1/3
(Add 25% for ceiling installations)		
Reflective Types, 100 sq. ft.		
1 sheet stripped in place between studs	1/2	1/4
Multiple, accordian sheets	3/4	1/4
Pouring Types, 100 sq. ft.		
Between 4" studs when accessible	1	1
Over ceilings 4 to 6" thick	1	1/2
Poured in cavities concrete masonry or cavity type walls to 4" space	1	1/2
Semi-Rigid Types, 100 sq. ft.		
Applied over frame or masonry walls using adhesives or asphaltic cements	2	2
(Add 25% for each additional 2" in thickness)		
(Add 25% for ceiling or unusual wall installations all thicknesses)		
Vapor Seals, Moisture Barriers, 100 sq. ft.		
Roll or sheet types, aluminum, polyethylene, papers		
Tacked in place	1/2	1/4
Cemented on frame or masonry surfaces	3/4	1/2

Adjust all rates for extra time and conditions on unusual projects and remodeling jobs.

WALLS AND CEILINGS

Average type construction, normal openings. Walls and ceilings combined. Necessary working level planking or scaffolding included.

	HOURS PER UNIT	
	Skilled	Unskilled
Plaster Bases, 100 square yards		
Gypsum lath 3/8" - 1/2" 16" x 48"	6	1 1/2
3/8" - 1/2" 24" to 12'	4	1 1/2
Insulating lath 1/2" 18" x 48"	7	2
Metal lath, average size sheets, over studs	4	1 1/2
(Adjust for staples or mechanical fasteners)		
Metal lath, average size sheets, over masonry	5	1 1/2
Paper backed wire or expanded mesh	3 1/2	1 1/2
Foamed plastic, 12" x 9' planks, 1-2" thick cemented	8	1 1/2
Plaster, average 1/2" - 5/8" grounds, 100 square yards.		
Scratch coat over gypsum or insulating lath	2	2
Scratch coat over masonry or gypsum tile	2 1/2	2
Scratch coat over metal lath or mesh	3	2
Brown coat over average scratch coats	3	2
Sand finish coat*	4	2
White coat lime finish*	6	2
* Add 10% if color mixed on job.		
Prepared color finish	5	2
Keenes cement finish, smooth	4 1/2	2
Keenes cement finish, 4" x 4" tile pattern	5 1/2	3
Acoustical Plaster		
(1 coat over suitable base— handwork)	4	2
2 coat work over gypsum or insulating lath	10	5
3 coat work over gypsum or insulating lath	13	6
2 coat work over metal lath, mesh	11	6
3 coat work over metal lath, mesh	14	6
2 coat work over masonry, gypsum tile concrete	10 1/2	5
3 coat work over masonry, gypsum tile concrete	13 1/2	6
2 coat work over foamed plastic base	10	5
3 coat work over foamed plastic base	13	6

Deduct 25% unskilled time for factory mixed plasters. Add 10% for heavily wood fibred plasters or heavy plasters. Adjust for special finishes of all types, machine application and for work over radiant heating cables or tubing, etc.

	Skilled	Unskilled		Skilled	Unskilled
Metal Lath and Plaster Partitions, 100 square yards.			**Plywood Panels, Cont.**		
Hollow 4″ thick installed in masonry or concrete work			Plank type paneling tongue and grooved 6″ to 16″ widths	3	1
2 coat work plastered 1 side only	41	10	**Patterned Paneling, 100 square feet**		
2 coat work plastered both sides	52	16	3/4″ x 6″ to 12″ widths, horiz. or vertical	3 1/2	1
3 coat work plastered 1 side only	44	10	**Beaded Wood, 100 square feet.**		
2 coat work plastered both sides	55	18	1/2″ x 3 1/2″	3	1
(Add for various partition widths)			**Insulating Board, 100 square feet.**		
2 coat work both sides	46	12	Rigid panels 4′ x 8′ to 12′	1	1
3 coat work both sides	52	18	Plank type	1 1/2	1
Gypsum Lath and Plaster Partitions, 100 square yards.			**Insulating or Acoustical Tile 100 sq. ft.**		
Solid 2″ thick, includes floor and ceiling runners, concrete or masonry construction			Plain or decorated (Nailed, stapled, clipped to metal strips)		
2 coat work, 2 sides	50	16	12″ x 12″	3	1
3 coat work 2 sides	53	18	12″ x 24″	2	1
Suspended Ceilings, 100 square yards			24″ x 24″	1 1/2	1
Hung from concrete or steel, 1 1/2″ main runners with 3/4″ channels on 12″ centers, metal lath wired or clipped on.			16″ x 32″	1	1
2 coat work complete ceiling	45	15	(Furring not included)		
3 coat work complete ceiling	48	17	Add 25% for patterns and adhesive application.		
Less 10% if hung from wood joist or rafters.			**Wall Tile, 100 square feet.**		
Plaster Bonding, 100 square yards.			Applied with adhesives to prepared walls, plain patterns. Includes base and cap. Metal trim.		
Over concrete and masonry surfaces			Plastic 4″ x 4″	5	1
Asphaltic type brushed on		5	9″ x 9″	3	1
Cement base types brushed on		6	Metal 4″ x 4″	6	1
Cement base types troweled on 1/8″ to 1/2″	4	4	Ceramic 4″ x 4″	8	2
Adjust and add for chipping, roughing up or other wall preparations if required.			(Add 50% if applied in mortar bed)		
Gypsum Board, Walls, 100 square feet.			Adjust all work for unusual conditions, materials, designs, etc.		
3/8″ - 1/2″ 5/8″ thick			**Hardboard, 3/16″ - 1/4″ - 5/16″, 100 sq. ft.**		
4′ x 8′ to 10′ panels	1	1	Plain 4′ x 8′, 8′ to 12′ lengths.		
Finishing joints	1		Nailed	1 1/4	1
Finishing joints 2 ply application			4′ x 8′, 8′ to 12′ lengths.		
2nd layer cemented in place finished	3 1/2	1 1/2	Nailed with metal fittings	2 1/2	1
Gypsum Board, Ceilings, 100 square feet.			**Metal Sidewalls and Ceilings, 100 square feet.**		
Plain 4′ x 6′ to 12′ lengths	1 1/2	1 1/2	Patterned types	5	2
Wood veneer or other finishes	2	1 1/2	(Add for furring and intricate designs)		
Plank type 16″ widths	2 1/2	1 1/2	**Curtain Wall Panels, 100 square feet.**		
(Adjust for screws or unusual fastening methods)			1-9/16″ and 2″ thickness		
Plywood Panels to 4′ x 12′ panels, 100 square feet.			4′ x 8′ - 9′ - 10′ - 12′ panels	4 1/2	4
			(Includes 2″ x 4″ studs and plates floor and ceiling) (Screws or nails)		
Plain joints or tongue and grooved	1	1	Adjust for special fastening methods.		
Covered joints with moulding	3	1	**Furring, Plaster Grounds, Up to 3/4″ x 4″ wood strips, 100 square feet.**		
Fitted joints cemented	4	1			

	HOURS PER UNIT	
	Skilled	Unskilled

	HOURS PER UNIT	
	Skilled	Unskilled

Furring, Plaster, Cont.

	Skilled	Unskilled
Frame construction 12" centers	1	
16" centers	3/4	
24" centers	1/2	
Nailable masonry construction		
12" centers	1 1/4	
16" centers	1	
24" centers	3/4	

Adjust for installation on concrete walls
and for use of power activated powder
or explosive type automatic fasteners. Add
25% all types of work if figuring ceilings only.

**Moldings, Trim, and Accessories,
100 lineal feet.**

	Skilled	Unskilled
Plastering and corner beads, corner lath, metal furring, screeds, picture molding, bullnose beads, etc.	2/3	1/3
Door and window casing beads (metal) metal trim, corners, mitered and fitted before plastering or used with dry wall construction	2 1/2	1

WINDOWS AND DOORS

Window Sizes:

A To 3' x 3'. Single unit

B Over 3' to 3' x 5'-6" Single unit

C Over 3' to 6' x 5'-6" Double unit

**D Over 3' to 6' to 8' x 6' Triple unit and
picture windows**

Wood Windows, per each unit

	Skilled	Unskilled
Window frames, assembled from stock sections		
Size A	1/2	1/4
B	3/4	1/2
C	1 1/4	1/2
D	1 1/2	1/2
Setting window frames, includes handling and bracing when required.		
Size A	1/2	1/4
B	3/4	1/2
C	1	1/2
D	1 1/2	1/2
Fitting and hanging double hung sash,		
Size A	1/2	1/4
B	3/4	1/2
C	1 1/2	1/2
D	2 1/2	3/4
Fitting and hanging casement sash, Per opening.		
Size A 1 sash	1/2	1/4
B 2 sash	1	1/4
C 3 sash	1 1/2	1/2
D 4 sash	2	1/2

Setting complete window units.
All types glazed. To prepared
openings.

	Skilled	Unskilled
Size A	3/4	1/4
B	1	1/2
C	1 1/2	1/2
D	1 3/4	3/4

Metal Windows, per each unit.

All residential types, fins or
wood surround attached. Setting
sash units. To prepared openings.

	Skilled	Unskilled
Size A	3/4	1/2
B	1 1/4	1/2
C	1 3/4	3/4
D	2	1

(Adjust for unusual installations
and glazed units.)

**Basement Sash, wood or metal, per
each unit**

	Skilled	Unskilled
		1/4
Setting poured in place basement frames & sash		1/2
Setting utility metal sash		
4 light open		1/2
6 light open		3/4

Commercial Projected.

Pivoted or security type metal
sash. Bracing and handling to
second floor heights. Fixed
or vented.

	Skilled	Unskilled
Small units to 2' x 4'	3/4	1/4
2' x 4' to 4' x 8'	1 1/2	1/2
4' x 8' to 6' x 8'	2	3/4
6' x 8' to 8' x 8'	2 1/2	1

(Add 25% for architectural
projected or heavy awning
types)

Add 25% to all work if second
floor installation only.

Adjust for plate glass or twin
glass glazing where work is
figured glazed.

Trim, Interior, per each unit.

Single member type, casings,
stool, apron, stops. Soft wood.

	Skilled	Unskilled
Size A	1	
B	1 1/4	
C	1 1/2	
D	2	

(Add 25% for wood
jambs over plaster
returns)

Plain wood Stool and Apron.

	Skilled	Unskilled
All sizes		1/2

	HOURS PER UNIT	
	Skilled	Unskilled

Trim, Interior, Cont.

	Skilled	Unskilled
Plain Wood Stool only. All sizes	1/4	
Adjust for various types trim.		
Metal Stools to 4'	3/4	
Metal Stools 4' to 8'	1	
Glass, Marble, 1 piece precut stools to 4'	1	
Ceramic tile stools to 4' plain		1 1/2

Trim, Exterior, per each window

	Skilled	Unskilled
Size A	1/2	
B	3/4	1/4
C	1	1/4
D	1 1/2	1/4
(Includes flashing at head if required)		

Shutters, per pair

Wood, Metal

	Skilled	Unskilled
Small	1/2	
Medium	3/4	
Large	1	
(Add 50% if over masonry construction)		

Storm Sash & Screens, (Prefit wood comb.) per unit

	Skilled	Unskilled
Size A	1/4	1/4
B	1/2	1/4
C	3/4	1/4
D	1	1/4
(Add 25% for metal or plastic)		

Doors (Residence types)

	Skilled	Unskilled
Setting exterior door frames. Wood residential types. Standard sizes	1	1/4
Setting front entrance door frames with patterned side panels to 5' widths	2	1/4
Setting front entrance door frames with side light panels to 5' widths	2 1/2	1/4
Cutting and setting 2" x 6" - 8" frames from stock material 3' x 7'	1	1/4
Same as above 8' x 8' to 10' x 8'	1 1/2	1/2
10' x 10' to 16' x 10'	2	3/4
Interior door jambs and heads. Assembling from stock sections.		
Standard sizes	3/4	1/4
Setting interior door frames. Standard sizes	1	1/4
Same as above small closet	1/2	
large closet	3/4	
Setting sliding door pockets in stud walls. Standard sized doors	3/4	1/4
Building sliding door pockets & setting track		
1 pocket, single door	2 1/2	
2 pockets, double door	4 1/2	

Glass Sliding Doors, setting complete units.

	Skilled	Unskilled
To 4' widths	2	
4' to 6' widths	2 1/2	
6' to 8' widths	3	

Metal Door Frames, setting each.

	Skilled	Unskilled
Standard size	3/4	1/4
Double size	1 1/2	1/2

Outside Door Trim, 1 piece setting each.

	Skilled	Unskilled
Standard size	3/4	

Inside Door Trim, 1 piece, setting each.

	Skilled	Unskilled
Standard size doors	1 1/4	
Small closet doors	1/2	
Large closet doors	3/4	

Thresholds, setting each.

	Skilled	Unskilled
Wood	1/4	
Metal	1/4	
Ceramic, Marble	3/4	

Hanging Doors, per each.

Exterior, wood, standard sizes. All types. 3 butts.

	Skilled	Unskilled
1 3/4"	1	1/2
metal, prefit	3/4	1/4
Interior metal, prefit 1-3/8"	1/2	1/4
wood 1-3/8"	3/4	1/2
wood sliding 1-3/8"	1/2	1/2
metal sliding 1-3/8"	1/2	1/4
metal bifolding, pair	3/4	1/4
wood bifolding, pair	3/4	1/4
wood french, pair	1	1/4
closet small, 1-1/8"-1-3/8" thick	1/2	
closet large, 1-1/8"-1-3/8" thick	3/4	
folding fabric or slatted types	3/4	
Exterior combination storm/ screen wood complete	1	1/4
Complete prefit metal or plastic	3/4	1/4
(Adjust all door rates for special designs and types)		

Gargage and heavy doors.

	Skilled	Unskilled
1 3/4" hinged 4' x 8'	2	1/2
sliding 4' x 8'	2 1/2	1/2
sliding 8' x 8'	3	1/2

Setting overhead doors, wood 1 3/8", metal or glass fiber. (complete)

	Skilled	Unskilled
8' x 6'-6" to 8' x 8'	3	1
to 10' x 7' & 8'	3 1/2	1
12' x 7' & 8'	4	1
12' x 10' & 12'	4 1/2	1 1/2
16' x 6'-6" to 8'	5 1/2	2

(Adjust for metal doors, add 25% for 1 3/4" doors)

	HOURS PER UNIT	
	Skilled	Unskilled

Weatherstripping, per opening.
Standard type metal for average
size door. — 3/4

Louvers, Vents, per each.
Metal, screened or regular.

	Skilled	Unskilled
Small medium sizes	1/4	
Large sizes	1/2	
Half circle types	1	

Access Doors, prefit metal.

	Skilled	Unskilled
To 24" x 30"	1/2	
Over to 36" x 48"	3/4	

Metal-Clad Doors, complete masonry or steel construction. (Underwriters)

		Skilled	Unskilled
Swinging	Single	4	2
	Double	6	2
Sliding	Single	6	2
	Double	8	2

Adjust all window and door rates
for transoms if required. Add 25%
hardwood trim. Adjust for masonry
construction unless indicated and for
unusual conditions. Rates do not
include hardware unless shown as
complete.

MILLWORK AND TRIM

Normal openings. Frame construction.
Soft wood. First class work. Rooms
10' x 10' or over. Add 25% for hardwood.
Adjust if drilling or screws are required.

Wood Interior Trim, 100 lineal feet.

	Skilled	Unskilled
Base — 1 piece 2 1/2	1	
Shoe mold	1 1/2	
Chair rail	2 1/2	1
Picture mold	3	1
Cornices, crown, bed, cornice mold		
single member to 3 1/2" widths	6	2
single member over 3 1/2" widths	7	2
2 to 4 member	12	2
Ceiling beams, build in place		
3" x 6" to 6" x 12"	18	2
Closet shelving 3/4" x 12" stock		
includes cleats	3	1

Metal Interior Trim, 100 lineal feet.
Applied either before or after plastering.
Applied with screws or clipped-on.

	Skilled	Unskilled
1 piece base	3 1/2	1
2 piece base	4 1/2	1
Cove mold	2	1
Chair rail	2 1/2	1
Picture mold	3	1
Cornice to 8" width with preformed		
mitered corners	6	2
Closet shelving. Adjustable lengths		
12" and 16" widths	2	1

	HOURS PER UNIT	
	Skilled	Unskilled

Metal Interior Trim Cont.
Adjust time for all trim if applied
over masonry walls or over plaster
over masonry.

Wood Exterior Trim, 100 lineal feet.

	Skilled	Unskilled
Corner board, verge boards,		
fascia, frieze to 3/4" x 8"	3	2
Shingle mold, bed mold	1 1/2	1
Soffits, 3/4" x 6"-8" to 24" widths	7	1
Plywood, hardboard, asbestos		
pegboard 1/4" x 12" x 24"	4	1
(Add for screens if used)		
Metal with screens or vents	3	1
Cornices, 2 members to 12" widths	6	1

Porch Rail, 100 lineal feet.

	Skilled	Unskilled
Top, bottom, balusters	3	1 1/2

Stairs, Exterior Wood, per flight.
Plain open stairs. 4' widths to
12' lengths,

	Skilled	Unskilled
single flight	4	2
double flights	6	2

Porch Columns to 10' heights with caps and bases where required.

	Skilled	Unskilled
4' x 4"-6" x 6" solid	1/2	1/2
built up square to 8" x 8"	1 1/2	1
round hollow to 12"	1	1
round turned solid	3/4	1/2

Mantels, per unit.

	Skilled	Unskilled
Setting average type factory built		
mantel units. To prepared walls	2 1/2	1 1/2

Cabinets, Cupboards, 100 sq. ft. face area
Average type work. Includes base and mold
as required.

	Skilled	Unskilled
Setting factory built base cabinets,		
cases, range, oven sections. Also		
broom and utility cabinets, vanities	4	2
Hanging top cabinets	3	2

Counter Tops, 10 sq. ft. surface area
Placing factory built tops over
cabinet bases without sinks

	Skilled	Unskilled
3/4" Plywood, lumber	1/6	1/6
1" to 2" Maple	1/2	1/4
Stainless Steel	1/5	1/5

Covering sink, base, vanity tops,
plain type using metal trim.
Cutting, fitting, cementing

	Skilled	Unskilled
Laminated plastic	3/4	
Linoleum	1/2	
Ceramic tile 4" x 4"	1 1/2	
(Add 50% if sinks or lavatories		
included)		

**Misc. Factory-Built Cabinets
per each unit.**

	HOURS PER UNIT	
	Skilled	Unskilled

Factory Built Cabinets, Cont.

Placing or setting to rough openings

	Skilled	Unskilled
Package receivers, through-wall types. Average size.	1	
Medicine and wall cabinets —		
Small	1/2	
Large	3/4	
Ironing board, broom, shoe cabinets, etc.	1/2	
Clothes chutes, complete, (except basement bin)	1 1/2	
Bath and Kitchen Accessories. (Soap dish, towel racks, paper holder, hand bars, etc.)	1/4	
Setting prefit lightweight shower doors — plastic glazing	1	
heavy weight plate glass	2	

GLAZING

Average type work, to second floor heights. Single strength, double strength.

Glazing Wood Sash and Doors.

Using putty, per 10 lights

Approximate glass size	Skilled	
8″ x 10″ to 12″ x 14″	1	
16″ x 20″ to 20″ x 28″	2	
30″ x 36″ to 36″ x 40″	2 1/2	
40″ x 48″ to 48″ x 60″	4 1/2	

Deduct 1/3 if glazed with wood stops.

Glazing Metal Sash.

Using putty or plastic glazing compound, per 10 lights

Approximate glass size. Square inches per light	Skilled	
Up to 300 square inches	1 1/4	
300 to 600 square inches	2 1/2	
600 to 900 square inches	3 1/4	
900 to 1200 square inches	4 1/2	
1200 to 1800 square inches	5 1/2	
1800 to 2400 square inches	6 1/2	

Deduct 1/3 if glazed with metal or plastic stops or strips.

Adjust all rates for twin-pane lights, plate, wired, ribbed or special types of glass and other unusual conditions.

Store or Commercial Building Fronts, 100 square feet.

	Skilled	
Setting plate glass over 4′ x 5′ to 8′ x 10′	10	
Over 8′ x 10′ to 10′ x 15′	15	

PAINTING — **Skilled**

Based on 1st coat over new work unless indicated otherwise. Surface areas.

Outside Walls, 100 square feet.

Oil, stain or rubberized types.

	HOURS PER UNIT	
	Skilled	Unskilled

	Skilled	Unskilled
Wood sidings	1	
Wood shingles (stain)	1 1/4	
Burned smooth brick	1 1/4	
Burned rough brick	1 3/4	
Concrete masonry, concrete, stucco	1	

(Add 10% if masonry paints)

(Reduce 50% if spray work-large projects)

(Reduce 25% if roller work)

Inside Walls and ceilings, 100 square feet.

Flat, casein, or rubberized types.

	Skilled	Unskilled
Sizings	1/4	
Plaster, white coat, dry wall	1/2	
Insulating plank, panels, tile, rough raw finish	1	
Plywood, lumber, plasterboard,	3/4	
composition board, smooth finishes	3/4	

(Deduct 10% if calcimine or similar types. Add 25% all work if ceilings only)

(Reduce 50% if spray work, large projects)

(Reduce 25% if roller work)

Cabinets, Cupboards.

Vanities, cabinet or closet doors, bookcases. 1 side

	Skilled	Unskilled
Vanities, cabinet or closet doors, bookcases. 1 side	1	1/2

Wallpaper, 100 square feet.

Ordinary type walls.

	Skilled	Unskilled
Butt joint work, Average type paper	2	
Special papers and coated fabrics, etc.	3	

Floors, 100 square feet.

	Skilled	Unskilled
Filling, wiping	2/3	
Shellac, varnish, stain	1/2	
Painting — Wood types	1/2	
Concrete	1/3	

(Reduce 25% if roller work)

Roofs, 100 square feet.

Average conditions, smooth surfaces.

	Skilled	Unskilled
Asphalt, aluminum and other free flowing types	1/2	
Fibred semi-plastic types	1	
Wood shingles — stained	3/4	
painted	1	

(Adjust for unusual conditions, spray work etc.)

Trim, 1 coat per opening.

Inside and Outside

	Skilled	Unskilled
Windows and doors, average size	1/4	
Picture mold, chair rail, base, cornice, etc. Figure each item for each average size room as 1 opening	1/4	

HOURS PER UNIT

	Skilled	Unskilled
Doors, 1 coat per opening.		
All types, inside, outside 2 sides average size	1/2	
Combination storm doors	3/4	
Inside, medium size closet & cupboard	1/3	
Inside, small size closet & cupboard	1/4	
Windows — Based on wood double hung, casements, sliding, projected, awning, window-wall types with large lights. 1 side.		
Small sizes	1/4	
Medium sizes	1/3	
Large sizes	1/2	
Metal, Residential types average sized lights. 1 side. All types.		
Small size units	1/5	
Medium size units	1/4	
Large size units	1/3	
Adjust both wood and metal residential rates for small lights of glass where found.		
Metal, Commercial types, average sized lights. 1 side.		
Small openings	1/4	
Medium openings	1/3	
Large openings	1/2	
All window units figure each unit and combine time when two or more units are combined by mullions.		
Shutters. All types, per 2 sides	1/4	
Stairways. Complete open types, per stairway	4	
Wood Mantels, per unit	1/2	
Caulking, 100 lineal feet.		
Average type work using gun, around windows, door trim, etc.	1 1/4	
Adjust all painting time for enamels, special type paints, finishes and unusual conditions, old work, and spray painting. Deduct 10% for successive coats on large areas of inside walls, ceilings and outside walls.		
HARDWARE		
Lock Sets, per each unit.		
Outside doors, plain type sets	1/2	
Outside doors, front, fancy type sets	3/4	
Outside storm doors	1/2	
Outside door closers	1/2	
Inside doors softwood	1/5	

HOURS PER UNIT

	Skilled	Unskilled
Lock Sets, Cont.		
Inside doors hardwood	1/4	
Inside closet, cupboard, cabinet, etc.	1/10	
Inside sliding doors, swinging, single	1/5	
Inside sliding doors, swinging, double	1/4	
Door Accessories, per 10 units.		
Door bumpers, stops, surface bolts, night latches, closet hooks, handles, pulls, catches, and misc. small items. Simple installation	1/2	
Sash Hardware, per each unit.		
Double hung, locks and lifts per window	1/10	
Casement sash, locks and operators per sash	1/4	
Closet Rods, per rod	1/6	
Garage or heavy swinging doors, latches and lock sets, holders and top or bottom bolts, per door	1/2	

WORK DESCRIPTION	LABOR	MATERIAL	TOTAL
Site/Subgrade Prep.			
Construction utilities			
Heat			
Water			
Electricity			
Engineering Fees			
Temporary construction			
Clearing site			
Grading			
Fill material			
Foundation			
Layout			
Stakes			
Batter boards			
Line			
Nails			
Excavation			
Pier footings			
Continuous foundation			
Dimension			
Boards			
Stakes			
Nails			
Ties			
Wire			
Reinforcing steel			
Column reinforcement			
Concrete			
Curing			
Foundation drainage			
Drain tile			

WORK DESCRIPTION	LABOR	MATERIAL	TOTAL
Porous fill			
Waterproofing			
Concrete Floor Const.			
Backfill			
Top soil			
Aggregate			
Fill and tamp			
Aggregate			
Dirt			
Vapor Barrier			
Concrete forming			
Dimension			
Board			

WORK DESCRIPTION	LABOR	MATERIAL	TOTAL
Nails			
Reinforcing steel			
Control joint			
Expansion joint			
Screeds			
Concrete			
Curing			
Exterior/Interior Walls			
Wall panels			
Dimension			
Nails			
Reinforcing steel			
Chairs			

GLOSSARY

Aggregate — Materials such as sand, rock and vermiculite used to make concrete.

Ampacity — The current-carrying capacity of an electrical device or conductor.

Apron — Trim placed against the wall directly below the window stool.

Architect's scale — A flat or triangular scale used to make and read architectural drawings.

Areaway — Metal, concrete, or masonry construction to hold back earth and provide an opening around basement windows and doors.

Ash dump — A hatch placed in the bottom of a fireplace which disposes ashes into the ash pit.

Awning window — A window that is hinged near the top so that the bottom opens outward.

Backfill — Earth placed against a building wall after the foundations are in place.

Backsplash — The raised lip on a countertop to help prevent water from running down the back of the cabinet.

Balloon framing — Type of construction in which the studs are continuous from the sill to the top plate, under the roof rafters. The second-floor joists are supported by a strip or ribbon, which is cut into the studs.

Balusters — Vertical pieces which support a handrail.

Balustrade — An assembly of balusters (vertical supports) and a handrail; commonly found on open stairways.

Barefoot — Method of construction in which the rafters rest on top of the ceiling joists.

Base cabinets — Any kitchen cabinet that rests on the floor.

Bid bond — Guarantees that the contractor will not withdraw the bid for a specified period of time.

Board foot — (board measure) One hundred forty-four cubic inches of wood, or the quantity contained in a piece which measures 12" x 12" x 1" thick.

Box sill — The header joist placed at right angles to the ends of the floor joists.

Boxed rake — A rake cornice that extends beyond the end of the building and is enclosed with a fascia and soffit.

Branch — (electricity) Circuits that carry the current to various parts of the building.

Branch — (plumbing) The piping from a fixture to the point in the plumbing system where it joins piping from other fixtures.

Bridging — The lateral bracing of the floor joists.

Btu — British thermal unit is the amount of heat required to raise the temperature of one pound of water one degree Fahrenheit.

Btuh — British thermal unit per hour.

Butt hinges — The hinges on which the doors swing. (sometimes referred to as *butts*).

Casement window — A window that is hinged at one side so that the opposite side opens outward.

Centerline — An actual or imaginary line that equally divides the surface or sides of something in half.

Chimney cap — Special masonry construction at the top of a chimney to protect the masonry from the elements.

Collar beams — Horizontal members which tie opposing rafters together, usually installed approximately half way up the rafter.

Common brick — Clay brick which is formed and then baked to a hard material. Common bricks generally used as structural material where appearance is not important.

Concrete — Building material consisting of fine and coarse aggregates bonded together by cement.

Conductor — Electrical wire; a cable may contain several conductors.

Consolidate — Refers to concrete and means to work to remove voids or air pockets.

Contour lines — Lines on a map or plot plan to identify the elevation. These lines describe the slope of the ground.

Contract — An agreement in which one party agrees to perform certain work and the other party agrees to pay for the work and services.

Cornice — The assembly of boards and moldings used in combination with each other to provide a finish to the ends of the rafters which extend beyond the face of the outside walls.

Cornice return — Where the level cornice at the eaves turns the corner at the end of the house.

Course — A single row of building materials such as concrete blocks or roofing shingles.

Damper – A door installed in the flue of a fireplace or furnace to regulate the draft.

Dampproofing – Construction technique which helps prevent moisture from seeping through to the inside of a foundation wall.

Datum – The permanent points in a city or county establishing elevations above sea level.

Double-hung window – A window consisting of two sash which slide up and down past one another.

Drywall – A type of interior wall covering using gypsum wallboard and special compound to conceal the joints.

Face brick – Shale brick which is formed and then baked to a hard material. They are dense and will not absorb water. Face brick is usually used for exterior facing.

Fascia – The part of a cornice that covers the ends of the rafters. Usually a 1 x 8 board.

Finish hardware – All of the hardware which is exposed to view in the completed house.

Float – To smooth concrete with a tool called a float. Floating is done before the concrete cures and does not produce as smooth a surface as trowelling.

Floor plan – An architectural drawing showing the layout of rooms, location of windows and doors, and the location of special equipment.

Flue – A duct for carrying off smoke and gases of combustion. Chimneys normally have a fire clay flue lining.

Flush doors – A door of any size having both surfaces smooth and flat.

Footing – Concrete construction which may be found under foundation walls, columns, or chimneys and which distribute the weight of these structures to prevent settling.

Frieze – A horizontal board directly below the cornice and above the siding. A frieze is not used on all construction.

Gable – (gable end) The triangular area between the rafters and top wallplate at the ends of a building with a gable roof.

Gable studs – Vertical framing members placed between the end rafters on a gable roof and the two plates of the end wall.

Grout – Special cement for tile joints.

Gypsum wallboard – Drywall material composed of reconstituted gypsum encased in paper.

Headers – Used to support the ends of framing members which have been cut to make an opening for stairways, chimneys, fireplaces, doors, and windows.

Hip rafter – The rafter running from the corner of a building to the ridge of a hip roof.

Insulated glazing – Two pieces of glass with air space between them.

Jack rafter – Spans the space between the wall plates and the hip rafters, or the ridge and valley rafter.

Jambs – Side members of a window or door frame.

Joists – Horizontal members that support the floor.

K factor – The rate at which a material conducts heat.

Lally column – Steel pipe used as a support post.

Lintel – Structural member that spans a clear opening to support the structure above the opening. Can be found above doors, windows, and fireplace openings.

Lock set – An assembly consisting of a latching mechanism.

Masonry cement – Cement which is specially prepared to make mortar.

Miter joint – A joint with pieces cut at 45° angle to form a right angle corner.

Mortar – Material made from portland cement and additives to bond masonry units together.

Mullion – The vertical piece between two windows when they are mounted side-by-side. Windows with mullions are called mullion windows.

Nominal size – The size by which materials are specified. Actual size is usually slightly smaller.

Oakum – A fibrous rope-like material used for caulking joints in cast iron soil pipe.

Orthographic-projection – A method of drawing used by draftsmen to show three-dimensional objects on two-dimensional paper. (also referred to as third-angle projection.)

Panel doors – Any door having decorative panels on either surface.

Parging – Thin coat of cement plaster used to smooth rough masonry walls.

Payment bond – Guarantees that all the contractor's debts relating to the job will be paid so there will be no liens against the property.

Penny size – (abbreviated d) Penny refers to the length of a nail.

Performance bond – Guarantee issued by a bonding company stating that the work will be done according to the plans and specifications.

Perimeter drain – An underground drain pipe around the footings.

Plate – A horizontal piece, the same width as a stud placed at the bottom and top of a stud wall.

Platform framing – A method of framing in which each floor is framed separately, with the subfloor in place before the wall and partition studs of that floor are erected.

Plot plan – A drawing that provides all of the information necessary to clear the lot, the size of the lot, and the location of the building.

Plywood — A wood material made up of an odd number of layers laminated into a sandwich-like panel.

Polyurethane varnish — A clear plastic coating material which is extremely durable.

Portland cement — Finely powdered limestone material which is mixed with water to bond aggregates in concrete. Also can be used to mix mortar.

Primer — Type of paint used for the first coat on surfaces that do not hold paint well.

Quantity takeoff — List of materials required to complete the construction of a building.

R value — Thermal resistance: ability to resist the flow of heat.

Rafter — The framing members of a roof. The rafters extend from the wall to the ridge of the roof.

Rake — The inclined portion of a cornice at the ends of a roof.

Resilient flooring — vinyl, vinyl-asbestos, linoleum, or other synthetic floor covering which provide a smooth surface.

Ridge board — The top member of the roof framing which runs the length of the roof.

Rough plumbing — The concealed portion of the plumbing system. This includes all piping and fitting up to the finished wall or floor surface.

Rowlock — Pattern of bricklaying in which bricks are laid on edge.

Run of Rafter — The horizontal distance covered by one rafter.

Run of stairs — The horizontal distance covered by a set of stairs.

Sash — Frame holding the glass of a window.

Saturated felt — Paper-like felt which has been treated with asphalt to make it waterproof.

Screed — A straight board used to level concrete.

Sectional view — A drawing which shows what one would see after making a vertical cut through the building.

Sheathing — Boards applied horizontally or diagonally, plywood, or fiber board applied to the outside face of the studs.

Sill — A single piece of wood laid flat on the top of the foundation wall.

Sill cock — (sometimes called a hose bibb). A faucet mounted outside of the building to which a hose may be attached.

Sill seal — Man-made material which is laid on top of the building sill to prevent small gaps between the foundation and the sill.

Site-constructed — Built on the job.

Sliding window — A window consisting of two sash which slide from side to side past one another.

Soffit — In cornice construction, the surface of the bottom or underside is called the soffit.

Soil pipe — Cast iron pipe used for waste plumbing.

Soldier — Pattern of bricklaying in which bricks are stood on end.

Span of Roof — The horizontal distance covered by a roof; the width of the building.

Specifications — A document which conveys important information to the builder that cannot be shown on the working drawings. Specifications explain quality, color, and finishes of materials to be used, as well as the workmanship to be expected.

Square — 100 square feet.

Stack — The large vertical pipe in the main house drain into which the branches run.

Stair carriage — The supporting frame member under the treads and risers.

Stair risers — The vertical pieces between the stair treads.

Stair stringers — The inclined pieces at the ends of the treads and risers which support the stairs.

Stair tread — The top surface of a step in stair construction.

Stair winders — Tapered treads used to make a turn without a landing.

Standard glazing — Single piece of glass in a window.

Stool — Trim which forms the finished window sill.

Stop molding — Small molding that stops the door from swinging through the opening as it is closed. Also used to hold window sash in place.

Stud — Vertical framing member that supports the weight of the ceiling upper floors, and the roof.

Subfloor — The layer of the floor which is applied to the joists. The subfloor is covered with finished flooring or underlayment.

Sweat — Method of soldering used in plumbing. The fittings are called sweat fittings.

Termite shield — Metal shield installed at the top of the foundation to prevent termites from entering the wood framing.

Thermostat — A switch which is activated by changes in temperature. All heating systems use thermostats.

Trap — U-shaped fitting which prevents sewer gas from entering the house.

Trimmers — The joists at the ends of the headers.

Truss — A factory-build assembly used to support a load over a wide span. Trusses are often used in place of rafters.

U factor — The combination of all of the K factors in any given type of construction.

Underlayment — Any material installed on the subfloor that will give a smooth surface to receive the finish floor covering.

Valley rafter — Carries the ends of the jack rafters where two roof surfaces meet to form a valley.

Vanity — A bathroom vanity is a cabinet in which a lavatory is installed.

Vapor barrier — Material used to help prevent the movement of water vapor through building surfaces.

Varnish — Combination of resin, oil, thinner, and dryer which makes a transparent coating. Most varnish does not contain pigment.

Vent pipe — Allows atmospheric pressure into the plumbing system and prevents vacuum from building up as the waste water is discharged.

Vertical contour interval — The vertical distance between the contour lines which show the change in elevation.

Wall cabinets — Any cabinet that is mounted on a wall above a countertop or base cabinet.

Waste plumbing — Includes the pipes and fittings that carry the waste water from the fixtures to the house septic system or municipal sewer and the pipes and fittings that make up the vent system.

Wood stain — Similar to paint in the fact that it contains a vehicle and pigment. It is not opaque, however.

Working drawings — Drawings containing all dimensions and information necessary to carry a job through to a successful completion.

Building _____

Location _____

Architect _____

Estimator _____

Sheet __1__ of __6__

Date _____

Description		Unit	Quantity	Unit Price	Total Material Cost	Labor	Total
FTG. FORM LUMBER	1 × 8	lin. ft.	521				
WOOD STAKES - 12/BUNDLE			11				
TRANSIT MIX CONCRETE - FTG.		Cu. Yd.	8½				
TRANSIT MIX CONCRETE - FLOORS		Cu. Yd.	21				
PORCHES & WALK							
CONCRETE BLOCKS - FOUNDATION	8 × 8 × 16		1965				
CONCRETE BLOCKS - FOUNDATION	8 × 12 × 16		420				
SOLID BLOCKS - FOUNDATION	4 × 8 × 16		198				
REINFORCEMENT - MASONRY WALLS		Cu. Yd.	964				
HALF SASH BLOCKS			10				
FULL SASH BLOCKS			10				
STEEL CELLAR SASH	15 × 12		5				
GALVANIZED AREAWAYS			5				
MASONRY CEMENT - BLOCKS		BAGS	102				
PLASTIC DRAIN PIPE	4"	FT.	182				
#2 CRUSHED STONE		TON	10				
PORTLAND CEMENT - PARGING		BAGS	24				
MASON'S LIME - PARGING		BAGS	24				
ASPHALT FOUNDATION COATING		GAL.	72				
MASONRY SAND		Cu. Yd.	10				
SISALKRAFT VAPOR BARRIER		Sq. Ft.	1551				
WIRE MESH - CONC. FLOORS		Sq. Ft.	1551				
LALLY COLUMNS	3" × 8'-4"		4				
FLUE LINERS	8" × 8"		13				
CHIMNEY BLOCKS			15				
COMMON BRICKS - CHIMNEY			162				
COMMON BRICKS - VENEER			1884				

Building _____

Location _____

Architect _____

Date _____

Estimator _____

Description	Unit	Quantity	Unit Price	Total Material Cost	Labor	Total
MASONRY CEMENT -						
BRICK MORTAR	BAGS	16				
GALVANIZED WALL TIES		178				
FIR - FRONT BEAM 2×8×16		3				
FIR - MAIN BEAM 2×8×10		12				
FIR - STAIR STRINGERS 2×10×10		2				
FIR - STAIR TREADS 5/8×10×3		11				
#1 COM. PINE - STAIR						
RISERS 1×8×12		3				
PINE HANDRAIL MOLDING	lin. ft.	12				
SILL SEAL	lin. ft.	149				
FIR - SILL 2×6	lin. ft.	149				
FIR - BOX SILL 2×8	lin. ft	149				
FIR - JOISTS 2×8×16		13				
FIR - JOISTS 2×8×14		56				
FIR - STAIR HEADERS 2×8×10		4				
GALVANIZED BRIDGING	PCS.	138				
CD PLYWOOD W/EXTERIOR						
GLUE - SUBFLOOR 5/8		34				
FIR - PLATES 2×4		1206				
FIR - STUDS HOUSE 2×4×8		295				
FIR - STUDS GARAGE & PORCH 2×4×10		71				
FIR - GARAGE DOOR HEADER 2×10×10		2				
FIR - WINDOW & DOOR HEADERS 2×10	LIN. FT	163				
FIR - BEAM FRONT PORCH 2×10×10		2				
FIR - BEAM REAR PORCH 2×8×8		3				
FIR - BEAM REAR PORCH 2×8×16		3				
FIR - POST REAR PORCH 2×4×9		3				
6/12 × 26' TRUSS - GARAGE & L.R.		14				
6/12 × 26' TRUSS - (NO TAIL						
ONE END)		12				

Building _____

Location _____

Architect _____

Date _____

Estimator _____

Description					Unit	Quantity	Unit Price	Total Material Cost	Labor	Total
6/12 x 22' TRUSS - FRONT WING						7				
FIR - JACK RAFTERS	2 x 6				LIN. FT.	44				
FIR RIDGE AT JACK RAFTERS	2 x 8 x 10					1				
9/12 x 26' GABLE TRUSS						2				
9/12 x 22' GABLE TRUSS						1				
ALUMINUM LOUVERS 6/12 PITCH						3				
CD EXTERIOR GLUE PLYWOOD ROOF SHEATHING	1/2"					66				
15 LB. SATURATED FELT	500'				RLLS.	5				
GALVANIZED DRIP EDGE					LIN. FT.	197 OR 20 PCS.				
240 LB. ASPHALT ROOF SHINGLES					SQ.	24				
28 GA. ALUMINUM FLASHING	20" x 50'				RLL.	1				
PINE CORNICE	1 x 3				LIN. FT.	217				
PINE FASCIA	1 x 6				LIN. FT.	217				
AC EXT. PLYWOOD - SOFFIT	3/8"				SHt.	4				
PINE FRIEZE	1 x 6				LIN. FT.	58				
COVE MOLDING - FRONT CORNICE	1 3/4"				LIN. FT.	31				
CD EXT. GLUE PLYWOOD - WALL SHEATHING	1/2"				SHt.	60				
BUILDING PAPER - WALLS					RLL.	4				
5x CEDAR SHINGLES	16"				SQ.	16.7				
PINE - REAR PORCH BEAM	1 x 8 x 16					2				
PINE - REAR PORCH BEAM	1 x 8 x 12					1				
PINE - REAR PORCH BEAM	1 x 6 x 16					1				
PINE - REAR PORCH BEAM	1 x 6 x 8					1				
PINE - FRONT PORCH TRIM	1 x 6				LIN. FT.	20				
AC EXT. PLYWOOD - PORCH CEILINGS	3/8					6				

Building _____

Location _____

Architect _____

Estimator _____

Description			Unit	Quantity	Unit Price	Total Material Cost	Labor	Total
COVE MOLDING - PORCHES	$1^{3}\!/\!_{4}$		LIN FT	88				
4 - CASEMENT WINDOW	6^{8} x 4^{5}			3				
CASEMENT - PICTURE WINDOW	1^{9}x4^{5}/5^{8}x4^{5}			1				
2 CASEMENT WINDOWS	3^{5}x4^{5}			1				
CASEMENT WINDOWS	1^{9}x4^{5}			2				
D. H. WINDOW	2^{4} x 3^{10}			1				
D. H. WINDOW	3^{0}x 3^{2}			1				
ALL WINDOWS w/ INSULATED GLAZING & SCREENS								
PRE - HUNG EXT. DOOR PINE, LEFT- HAND SWING	3^{0}x6^{8}x$1^{3}\!/\!_{4}$			1				
PRE - HUNG EXT. DOOR PINE, RIGHT- HAND SWING	2^{8}x6^{8}x$1^{3}\!/\!_{4}$			1				
PRE - HUNG EXT. DOOR, PINE, RIGHT- HAND SWING	2^{6}x6^{8}x$1^{3}\!/\!_{4}$			1				
ALUMINUM COMBINATION - STORM & SCREEN DOOR	3^{0}x6^{8}x$1^{1}\!/\!_{8}$			1				
ALUMINUM COMBINATION - STORM & SCREEN DOOR	2^{8}x6^{8}x$1^{1}\!/\!_{8}$			1				
PRE-HUNG INTERIOR FLUSH DOOR BIRCH, LEFT-HAND SWING COLONIAL CASING	2^{6}x6^{8}x$1^{3}\!/\!_{8}$			1				
PRE-HUNG INTERIOR FLUSH DOOR BIRCH, RIGHT-HAND SWING COLONIAL CASING	2^{6}x6^{8}x$1^{3}\!/\!_{8}$			1				
PRE-HUNG INTERIOR FLUSH DOOR BIRCH, RIGHT HAND SWING COLONIAL CASING	2^{0}x6^{8}x$1^{3}\!/\!_{8}$			4				
PRE-HUNG INTERIOR FLUSH DOOR BIRCH, LEFT HAND SWING COLONIAL CASING	1^{9}x6^{8}x$1^{3}\!/\!_{8}$			1				

Building _____

Location _____

Architect _____

Estimator _____

Date _____

Description		Unit	Quantity	Unit Price	Total Material Cost	Labor	Total
PRE-HUNG INTERIOR FLUSH DOOR	$2^4 \times 6^8 \times 1^{3/8}$		1				
BIRCH, LEFT HAND SWING,							
COLONIAL CASING							
INTERIOR DOOR JAMBS	$2^6 \times 6^8$	SCT.	1				
LOUVERED DOORS	$1^3 \times 6^8 \times 1^{1/8}$		2				
DOUBLE-SWING HINGES		PR.					
OVERHEAD GARAGE DOOR							
W/ HARDWARE	$9^0 \times 7^0$		1				
OVERHEAD DOOR FRAME	$9^0 \times 7^0$		1				
R-21 FIBERGLASS INSULATION	50 SQ.FT.	BAGS	22				
R-11 FIBERGLASS INSULATION	70 SQ.FT.	ROLLS	16				
GYPSUM WALLBOARD	1/2"	SQ.FT.	4576				
WALLBOARD COMPOUND	5 GAL.	CAN	6				
PERFORATED TAPE	250'	ROLL	5				
DEADENING FELT-FLOORS	500'	ROLL	2				
OAK STRIP FLOORING	1x3	BD.FT.	1100				
UNDERLAYMENT PLYWOOD	1/2"	SHEET	8				
VINYL TILE-KITCHEN	12"x12"	SQ.FT.	104				
ADHESIVE-VINYL TILE		GAL.	1				
CERAMIC FLOOR TILE	1"x1"	SQ.FT.	50				
SLATE FLOORING		SQ.FT.	64				
ADHESIVE FOR TILE & SLATE		GAL.	2				
GROUT FOR TILE & SLATE	5 LB.	PKG.	2				
AC EXT. PLYWOOD-							
TUB AREA	1/2"		3				
CERAMIC WALL TILE	$4^{1/4} \times 4^{1/4}$	SQ.FT.	60				
CERAMIC CAP TILES	2x6	PC.	44				
MARBLE THRESHOLD-BATH	2'-4"		1				
COLONIAL BASE MOLDING	3"	LIN.FT.	414				
STOOL-WINDOW TRIM		LIN.FT.	50				
APRON-WINDOW TRIM		LIN.FT.	50				

Building _____

Location _____

Architect _____

Date _____

Estimator _____

Description	Unit		Quantity	Unit Price	Total Material Cost	Labor	Total
COLONIAL CASING - WINDOWS		LIN. FT.	124				
CHROME ADJUSTABLE							
CLOSET RODS			5				
#2 PINE SHELVING	1×12	UN. FT.	50				
AD INT. PLYWOOD - LIN.							
CLOSET & BATH			1				
NAIL & FASTENER ALLOWANCE							
1830 WALL CABINET			1				
1230 WALL CABINET			2				
24 WC CORNER WALL			1				
CABINET							
1530 WALL CABINET			1				
3018 WALL CABINET			1				
3312 WALL CABINET			1				
D12 BASE CABINET W/							
DRAWERS			1				
B15 BASE CABINET			1				
CB 30 BASE CABINET			1				
SF 42 SINK FRONT			1				
B18 BASE CABINET			1				
LAMINATED PLASTIC							
COUNTER TOP		LIN FT.	18'-3"				
VANITY W/ LAMINATED							
PLASTIC TOP	48"		1				
FINISH HARDWARE							
ALLOWANCE							

REFERENCES

There are numerous sources available to the estimating instructor. A few are listed with their addresses.

ASSOCIATIONS

American Association of Cost Engineers
308 Monongahela Building
Morgantown, WV 26505

American Institute of Architects
1735 New York Avenue, N.W.
Washington, DC 20006

International Conference of Building Officials
5360 S. Workman Mill Road
Whittier, CA 90601

National Association of Home Builders
1625 L Street, N.W.
Washington, DC 20036

PUBLICATIONS

Books

ARCHITECTURAL GRAPHIC STANDARDS,
  Ramsey and Sleeper
John Wiley & Sons, Inc.
605 Third Avenue
New York, NY 10016

BLUEPRINT READING AND SKETCHING FOR
  CARPENTERS – RESIDENTIAL, McDonnell and
  Ball
Delmar Publishers
50 Wolf Road
Albany, NY 12205

BUILDING COST FILE
Construction Publishing Company
2 Park Avenue
New York, NY 10016

BUILDING ESTIMATOR'S REFERENCE BOOK
Frank R. Walker Company
Chicago, IL 20656

BUILDING TRADES BLUEPRINT READING
Delmar Publishers
50 Wolf Road
Albany, NY 12205

CONSTRUCTION DICTIONARY
National Association of Women in Construction
Greater Phoenix Chapter #98
Phoenix, AZ

CONSTRUCTION ESTIMATING, Jones
Delmar Publishers
50 Wolf Road
Albany, NY 12205

GUIDE TO INFORMATION SOURCES IN THE
  CONSTRUCTION INDUSTRY, Godel
Construction Publishing Company
2 Park Avenue
New York, NY 10016

SWEET'S CATALOG FILE
McGraw-Hill Information System, Inc.
1221 Avenue of the Americas
New York, NY 10020

Periodicals

Architectural Design, Cost, and Data
Allan Thompson (Publishers)
P.O. Box 796
Glendora, CA 91740

Home Improvements
Peacock Business Press
200 S. Prospect Avenue
Park Ridge, IL 60068

Homebuilding Business
Gralla Publications
1501 Broadway
New York, NY 10036

House & Home
McGraw-Hill Publications
1221 Avenue of the Americas
New York, NY 10020

INDEX

A

Abbreviations, lumber industry, 78
    working drawings, 8
Active solar heat, 157
Agreement between Owner and Contractor,
    Standard Form, 26–32
Aggregates, 38, 62, 63
Air conditioning, 154
    labor estimates, 154
Air-entraining portland cement, 62
Air handling, principles of, 148
Alphabet of lines, 5, 6
Altitude, triangle, 2
Aluminum flashing, 93
American Institute of Architects, Standard
    Form of Agreement between Owner
    and Architect, 26–32
American Wire Gage, 8
Angle iron, fireplace, 70
Apron, 127, 175
Architects, 29
    general conditions concerning, in agree-
        ment between owner and contractor,
        29
    specifications prepared by, 33
Architect's scale, 4–5
Architectural drawings, 13–24
    alphabet of lines, 5, 6
    symbols and conventions, 6–9
Area, mathematics, 3–4
Areaways, 66, 165, 209
Arithmetic review, 1–3
Ash pit, chimneys, 14, 24, 60, 207
Asphalt, foundation coating, 75, 168
Asphalt roofing, 96–171
ASTM standards, 62
Awning windows, 124

B

Backfill, 59
Backsplash, 137
Balloon framing, 79
Balusters, 130, 131
Barefoot, rafters, 229, 239
Base, triangle, 2
Base cabinets, 137, 138

Baseboard, 113, 114
Basement and foundation plan, 14,
    207–212
Basement beams, 78–80
Basement stairs, 128
Basement windows, 66, 165
Basic arithmetic, 1–4
Bathroom, ceramic tile, 118, 119
    floor, 119
Bathtubs, 143
Bay windows, 125
Beams, collar, 93
    girders, 80, 81
Bed molding, 101
Bench marks, 57
Beveled siding, 106
Board foot, 77
Bond patterns, brick, 72
Bonded roof, 98
Border line, 5
Box beams, 80
Box cornice, paint, 133
Box nails, 141, 142
Box sill, 81, 82, 168
Boxed rake cornice, 101
Bow windows, 125
Branch electrical circuits, 160–161
Branch piping, 144
Break line, 5
Brick, 71–73
Brick veneer, 17, 21, 167
Brick work, costs, 180
    specifications, 38
Bridging, 83
Btu, 151–152
Building paper, 106, 171, 172
Built-up roofing, 98
Built-up wood girder, 80–81
Butt hinges, 141

C

Cabinet work, costs, 184
Cabinets, 136, 175
    types of, 136–139
Carpentry, costs, 180–185
    specifications, 39–40

Carpet strip, 113–114
Carriage, stairs, 129
Casement windows, 125
Casing, 127
Casing nails, 141
Cast iron soil pipe, 144
Cedar shingles, 106, 171
Ceiling, joists, 70, 170
Cellar sash, 165
Cellar stairs, 128
Cellars, perimeter drain, 75
Cement, 62
Centerline, 5
Central air conditioning, 154
Ceramic tile, 118–119
Change order, 31
Chimney construction, 68
    estimating, 69–70
    framing for opening, 84
Circle, area, 3
Circuits, electrical, 160–161
Clay drain tiles, 74
Closed stairs, 81
Closet rods, 174
Coarse aggregates, 62–63
Collar beams, 93
Colonial door, 120
Common brick, 71
Common nails, 141–142
Common rafter, 92
Comprehensive review, 271–291
Concrete block chimney, 69
    fireplace, 64–70
Concrete blocks, 64–68
    costs, 179
    kinds, 64
    mortar for, 166
Concrete floors and slabs, 62
    cost, 179
Concrete footings, 60–62
    cost, 179
    estimating, 61
Concrete walls, 66
Condensation, 110
Conductors, electrical wiring, 159
Conservation, energy, 155–158

Contour lines, plot plans, 57–58
Contract, defined, 25
    general conditions, 25–32
    general requirements, 36–37
    Standard Form of Agreement between
      Owner and Contractor, 26–32
Contractors, 25, 34
    specifications prepared by, 33
    Standard Form of Agreement with
      Owner, 26–32
Convector, 149
Conventions, working drawings, 7
Copper fittings, 144
Corner boards, 17–20
Corners, framing, 87
Cornice detail, 21
Cornice return, 21, 101
Cornices, 100–102
    estimating paint, 133
    porches, 109
Costs, labor, 178–185
    estimating tables, 293–311
Countertops, 137
Cove molding, 114
    stairs, 128
Crown molding, 114
Cutting plane line, 5

D

Dampers, fireplace, 70
Dampproofing foundation, 74–76
Datum, 57
Deadening felt, floors, 117–118
Decorating and painting, 132–135
Design temperature difference,
    151
Designing a house, 187–191
Dimension line, 6
Dimensions, 8–9
Doors, 120–123
    labor costs, 183–184
    moldings, 174
    paint, 134
    specifications, 41
    working drawings, 14, 15–16
Double-hung windows, 124
Draft, fireplace, 68–69
Drains, perimeter, 74–75
Drawings, architectural, 13–24
    architect's scale, 4–5
Drawings
    first floor plan, 192–200
    foundation plan, 207–212
    front elevation, 215–219
    half-section view, 213–214
    second floor plan, 201–206
    side and rear elevations, 200–
      223
    special details, 225–236
Drywall nails, 141
Duplex-headed nails, 142

E

Electric heating, 150
Electrical labor costs, 161
Electrical service, 159–162
Electrical symbols, 7, 161
Electrical wiring, 159–162

Elevation, contour lines, 58
    on drawings, 17–20, 219–221
    plans, 58
    symbols, 7
Energy conservation, 155–158
Engineer's scale, 58
Entrance floor, 15, 117
Equilateral triangle, 2
Equipment line, 6
Estimating, baseboards and carpet strip, 173
    basement windows, 165
    box sill, 168
    bricks, 167
    bridging, 169
    cabinets, 175
    ceiling joists, 170
    ceramic tile, 173
    collar beams, 170
    concrete, 164–165
    concrete blocks, 165
    cornices, 171
    cove molding, 173
    dampproofing, 168
    doors, 174
    electrical work, 159–162
    fireplace and chimney, 161–167
    floor joists, 169
    flooring and underlayment, 173
    footing, 164
    girders, 168
    gypsum wallboard, 172
    headers, 169
    heating loads, 152–154
    insulation, 172
    mortar, 166
    perimeter drain, 168
    plates, 169
    plumbing, 146–147
    porches, 109
    rafters, 172
    resilient flooring, 173
    roof sheathing, 171
    roof trusses, 170
    roofing, 171
    second floor materials, 169
    sheathing, 171
    siding, 171–172
    sill seal and sill, 168
    slate flooring, 173
    stairs, 175
    studs, 169
    subflooring, 169
    windows, 174–175
Estimator, legal aspects of building con-
    struction, 25
    site work, 163
Estimating check list, 163–171
Excavations, site work, 163
Exhaust hood, 137
Expenses, estimating labor costs, 178–186
Extension line, 5
Exterior doors, 120
Exterior grading, sitework, 58–59
Exterior lock sets, 140

F

Face brick, 71–72
Factory lumber, 77

FHA form, description of materials, 47–56
Fiberboard, 103
Fiberglas insulation, 110–111
Fine aggregates, 62–63
Finish hardware, 140
Finished plumbing, 145
Finishing, 41–42
    cabinets, 136–139
    ceramic tile, 118–119
    doors, 120–123
    flooring and underlayment, 115–118
    foundation dampproofing, 74–76
    hardware, 140
    insulation, 110
    moldings 113–115
    painting and decorating, 132–135
    roofing, 96–99
    slate and resilient floor covering, 115–
      118
    stairs, 128–131
    wallboard and plaster, 111–113
    windows, 124–127
Finishing nails, 141
Firebrick, 70
Fireplace construction, 68–69
    estimating, 69–70
    footings, 69
    quarry tile, 71
Flat roofs, 98
Floor framing, 82–83
Floor joists, 82
Floor plan, 15–16
Flooring, costs, 183
    slate and resilient, 183
    subflooring, 181
    varnish, 186
Flooring and underlayment, 183
Flooring nails, 141
Floors, concrete, 62, 164, 179
    costs, 179
Flue liners, 69
Flush doors, 120
Footings, concrete, 60–62
    cost, 179
    estimating, 61
Form work, cost, 179
Forms, *see* Printed forms
Foundations, concrete blocks, 64–66
    dampproofing, 74–76
    footings, 60–62
    plan, 21
Frame construction, 77–79
Framing, beginning, 80–82
    floors, 82–87
    porch, 109
    roof, 91–94
    wall, 86–90
Front elevation plan, 17, 219, 253
Furnace, chimney flue, 69
    estimating, 69

G

Gable roof, 93
Gable studs, 94
Gambrel roof, 95
General conditions, 29–32
    Agreement between Owner and Con-
      tractor, 26–28

General contractors, 26, 34
General requirements, specifications, 36, 37
Girders, 78–80
Glass, specifications, 41
Glossary, 312–315
Gothic roof, 95
Grading, sitework, 57
Ground-fault device, 162
Grout, 117–119
Gypsum lath, 112
Gypsum wallboard, 111–112

H

Half-open stairs, 130
Hardware, 140–143
Hardwood flooring, 117–118
    varnish, 133
Headers, floor, 84
    wall framing, 88
Hearth, fireplace, 15, 24
Heat pumps, 150
Heating, 148–154
    chimney, 69
Hinges, 141
Hip rafter, 92
Hip roof, 95
Horizontal reinforcement, 67
Hot air heating, 148–149
    thermostats, 150
Hot water heating, 149
    solar heating, 156
    thermostats, 150
House construction, 78–162
    general requirements, 36–37
    Standard Form of Agreement between
        Owner and Contractor, 26–28
Housed stringer, 124
Hypotenuse, 3

I

Index, specifications, 25
Infiltration heat losses, 152
Inside wall fireplace, 199
Inspectors, building, 29
Insulated glazing, 126
Insulation, 110–111
    labor costs, 182
    thermal resistance, 157
Insurance, agreement between owner and
    contractor, 31
Interior doors, 120
Interior grading, site work, 63
Interior lock sets, 140
Interior millwork, 120–131
Interior painting and decorating, 132–135
Interior trim, paint, 134
Interior water piping, 143–145
Invisible line, 5
Irregular polygons, area, 3, 4
Isosceles triangle, 2

J

Jack rafter, 92
Jack stud, 88
Job-hung doors, 121
Joint compound, 112
Joist hangers, 80

Joists, 82–84
    floor, 82
    porches, 109
    second floor, 88

K

K factor, 151
King-post truss, 91
Kitchen cabinets, 136–139
Kitchen perspective, 264–270
Kitchen floors, 116

L

Labor costs, 178–186
    estimating tables, 293–311
Landing nosing, 129
Lavatories, 145
Left side elevation plan, 18, 221
Legal aspects, 25
    Agreement between Owner and Con-
        tractor, 26–32
Lighting, electrical wiring, 159–162
Lightweight aggregates, 63
Line, mathematics, 1
Lines, alphabet of, 5–6
Lintel, 87
    wall framing, 88
Lock set, 140
Loose-pin hinges, 141
Louver doors, 121
Lumber, classifications, 77
    framing, 79
    measurement, 77

M

Mansard roof, 95
Marble thresholds, 116
Masonry cement, 62
    fireplace and chimney, 167
Masonry floors, 62–63
Masonry work, specifications, 37–38
Materials, complete list, 286–291
    description, form, 286
    pipes and fittings, 144
    symbols, 7
Measurement, 8
Mechanical, electrical wiring, 159–162
    heating, 148–154
    plumbing, 143–147
Mensuration, 1
Millwork, specifications, 39–41
Moisture control, 74–76
    specifications, 40
Moldings, 113
    common types, 114
Mortar, 166
    fireplace, 167
Mullion windows, 125

N

Nail popping, 141
Nails, 141–142
National Electrical Code, 159
Nominal size, lumber, 77

O

Oak flooring, grades, 117
    varnish, 133

Object line, 5
Oil-based paint, 132
Open divided scale, 4
Open stairs, 129
Orthographic-projection drawings, 10–12
Owner, Standard Form of Agreement with
    Contractor, 26–32

P

Painting, labor costs, 185–186
Painting and decorating, interior, 134–
    135
Panel doors, 120–121
Paneling, labor costs, 305
Parallelogram, area, 4
Parging, 75
Passive solar heat, 157
Penny nails, 141
Performance bond, 25
Perimeter, 2
Perimeter drain, 75
Perpendicular line, 1
Perspective drawings, 264–271
Pictorial drawings, 10–13
Picture windows, 125
Pigment, paint, 134–135
Pipes, perimeter drains, 74–75
    plumbing, 144–147
Piping layout, 146–147
Plan and section symbols, 7
Plan views, drawings, 11–13
Plane figures, 1
Plane surface, 1
Plaster, 112–113
    labor costs, 304
    parging, 75
Plastic pipe, 144
Plates, wall framing, 86
Platform framing, 79
Plot plan, 13, 277
Plumbing, 143–147
    labor costs, 178–186
Plumbing fixtures, 145
Plumbing symbols, 7
Plywood sheathing, 103
    grade-use chart, 104–105
Plywood subflooring, 84
Point, mathematics, 1
Polyurethane, 179
Porches, 108–109
    estimating paint, 133
Portland cement, 62
    mortar, 166
Posts, footings for, 22
Pre-hung doors, 121
Primer, 132

Q

Quantity takeoff, 241
    form, 286
Quarry tile, 71

R

R value, 151
Rafter shoe, 229
Rafters, porch, 229
    roof, 229

Raised hearth, 68
Rake cornice, 100
    paint, 133
Rear elevation plan, 19, 222, 256
Reinforcement, horizontal, 67
Residence, concrete blocks, 64–65
    specifications, 38
    Standard Form of Agreement between
        Owner and Contractor, 26–32
Resilient flooring, 116
Resistance, thermal, 151
Ridge board, 92
Right side elevation plan, 21
Right triangle, 3
Risers, stairs, 128
Roman brick, 72
Roof, porch, 109
Roof framing, 92
    labor costs, 181
Roof trusses, 91
    details, 91
Roof sheathing, 94–95
Roofing, 96–99
Roofing nails, 141–142
Rough lumber, 77
Rough plumbing, 143
Rowlock, brick position, 72

S

Sand, mortar, 166
Sanitary drainage piping, 144
Saturated felt, 96
Scale, architect's, 4
Scale drawings, 10
Scalene triangle, 2
Second floor framing, 88–90
Section symbols, 7
Sectional views, 12, 21
Shakes, siding, 106
Sheathing and siding, 103–106
Shed roof, 95
Shingles, roofing, 96–99
    siding, 106
Shiplap, 78
Shoe mold, 114
Shop lumber, 77
Shower bases, 143
Siding, 106
    estimating paint, 133
Sill, 81
Sill cocks, 145
Sill seal, 81
Site-constructed cabinet, 136
Site-constructed countertops, 137
Site work, 57–59
    costs, 178
    footings, 179
Slate flooring, 116–117
Sliding windows, 124
Soil pipe, 144
Solar heating 155–158
Soldier, brick position, 72
Sole plates, 86
Specifications, 35–46
    cabinets, 43
    carpentry and millwork, 39–40
    ceramic tile, 42

concrete footings, 37
doors, windows and glass, 41
electrical, 45
finish hardware, 43
general conditions and requirements,
    29–36
masonry work, 37–38
metals, 39
moisture protection, 40
painting, 41
plumbing, 43–44
site work, 36–37
Stack piping, 147
Stain, 133
Stair indicator line, 6
Stairs, 128–131
Stairwell, framing for, 84
Standard Form of Agreement between
    Owner and Contractor, 26–32
Steel support columns, 22
Stool, 127
Stop and waste valves, 144
Stop moldings, 114
Straight stairs, 128
Stringers, stairs, 129
Strip flooring, 117–118
Structural lumber, 77
Structural work, brick, 71–73
    chimneys, 68
    concrete block, 64
    concrete floors and slabs, 62
    cornices, 100–102
    fireplaces, 67–71
    floor framing, 82–85
    footings, 60–62
    girders, 80–82
    lumber and frame construction, 79
    metalwork, 39
    mortar, 38
    porches, 109
    roof framing, 91–93
    sheathing and siding, 103–107
    site work, 57–59
    wall framing, 86–90
Studs, gable, 94
    wall framing, 86
Subcontractors, 30
Subcontractors, in Agreement between
    Owner and Contractor, 34
Subflooring, 84–85
    second floor, 88–89
Sump pump, 74
Supply plumbing, 146
Support posts, 22
Surface, 1
Surfaced lumber, 77
Surveys required, 57
Sweat soldering, 145
Symbol section line, 5
Symbols and conventions, 7–8
    electrical, 7, 161

T

Tape, wallboard, 112
Temperature difference 151
Thermal conductivity and resistance,
    151

Thermal resistance, determining, 151
    values, 151
Thermostats, 150
Third angle projection, 10–11
Thresholds, 121
Tight cornice, 101
Tight rake cornice, 101
Tile, ceramic, 118–119
Tile work, costs, 183
Title page, specifications, 33
Top plates, 86
Traps, 144
Tread, stairs, 129
Trench work, 61
Triangles, 2
    area, 3
Triangular scales, 4–5
Trimmers, 83
Trusses, details, 91
    roof, 92

U

U Factor, 150–153
Underlayment, flooring, 115
Underlayment nails, 141

V

Valley rafter, 92
Vanities, 137
Vapor barrier, 179
Varnish, 135
Vehicle, paint, 152
Veneer, brick, 17, 21
Vent piping, 147
Ventilation, 108, 172
Vertex, triangle, 2
Vertical contour interval, 57

W

W truss, 91
Wall cabinets, 136–138
Wall framing, 86–90
Wall tile, 118–119
Wallboard, 111–112
    labor costs, 182
Walls, brick-veneer, 72
    concrete, 66
    concrete blocks, 64–65
    horizontal reinforcement, 66
    paint for, 133–134
Wash basins, 145
Waste layout, 147
Waste plumbing, 147
Water-based paint, 132
Water closets, 145
Water heater, 145
Water piping layout, 146
Water table, 74
Weatherhead, 159
Wide-plank flooring, 117
Windows, 124–127
    basement, 66
    labor costs, 180
    moldings, 127
    paint, 134
    sizes, 125

specifications, 41
 working drawings, 14-20
Wiring, electrical, 159-162
Wood girder, 80
Wood moldings, types, 114
Wood shingles, roofing, 97
Wood siding, 106
Wood stain, 133

Wood trim, costs, 184
Worked lumber, 77
Working drawings 14-20
 abbreviations used on, 8
 alphabet of lines, 5-6
 architect's scale, 4-5
 specifications, 25-46
 symbols and conventions, 7-9

Y

Yard lumber, 77

Z

Zoned heat, 150
Zoning laws, 57